An Introduction to Dynamic Meteorology

THIRD EDITION

This is Volume 48 in the
INTERNATIONAL GEOPHYSICS SERIES
A series of monographs and textbooks
Edited by RENATA DMOWSKA and JAMES R. HOLTON

AN INTRODUCTION TO DYNAMIC METEOROLOGY

Third Edition

JAMES R. HOLTON

Department of Atmospheric Sciences
University of Washington
Seattle, Washington

ACADEMIC PRESS
A Division of Harcourt Brace & Company
San Diego   New York   Boston
London   Sydney   Tokyo   Toronto

Copyright © 1992, 1979, 1972 by ACADEMIC PRESS, INC.

All Rights Reserved.

No part of this publication may be reproduced or transmitted in any form or by any means, electronic or mechanical, including photocopy, recording, or any information storage and retrieval system, without permission in writing from the publisher.

ACADEMIC PRESS, INC.
525 B Street, Suite 1900
San Diego, California 92101-4495

United Kingdom Edition published by
Academic Press Limited
24–28 Oval Road, London NW1 7DX

Library of Congress Cataloging-in-Publication Data

Holton , James R.
     An introduction to dynamic meteorology / James R. Holton. -- 3rd
ed.
          p. cm. -- (International geophysics series ; v. 48)
     Includes bibliographical references and index.
     ISBN   0-12-354355-X
     1. Dynamic meteorology.   I. Title.   II. Series.
     QC880.H65   1992
     551.5'153--dc20   91-40568
                                                                                           CIP

PRINTED IN THE UNITED STATES OF AMERICA
          94   95   96   97   BB   9   8   7   6   5   4   3

Contents

Chapter 8   Synoptic-Scale Motions II: Baroclinic Instability

Chapter 9   Mesoscale Circulations

Chapter 10   The General Circulation

Chapter 11   Tropical Dynamics

Chapter 12   **Middle Atmosphere Dynamics**

Chapter 13   **Numerical Modeling and Prediction**

Preface

During the past decade the science of dynamic meteorology has continued its rapid advance. Although there continue to be important new developments in the analysis and prediction of extratropical synoptic-scale systems, the scope of dynamic meteorology has broadened considerably. Important progress has been made in the understanding of mesoscale storms, in tropical dynamics, in the dynamics of climate, and in the dynamics of the middle atmosphere. Extratropical synoptic-scale forecasting, although it remains important, is no longer the major motivation for the study of atmospheric dynamics.

In this third edition of *An Introduction to Dynamic Meteorology* I have attempted to provide a text that reflects the full scope of modern dynamic meteorology, while providing a coherent presentation of the fundamentals. Approximately fifty percent of the text has been completely rewritten. As in the previous editions the emphasis is on physical principles rather than mathematical elegance. It is assumed that the reader has mastered the fundamentals of classical physics and has a thorough knowledge of elementary calculus. Some use is made of vector calculus. In most cases, however, the vector operations are elementary in nature so that the reader with little background in vector operations should not experience undue difficulties.

Much of the material included in this text is based on a two-term course sequence for seniors majoring in atmospheric sciences at the University of Washington. It would also be suitable for first-year graduate students with little previous background in meteorology.

The fundamentals of fluid dynamics necessary for understanding large-scale atmospheric motions are presented in Chapters 1–5. These have undergone only minor revisions, with the exception of Chapter 5, which has been rewritten to include discussion of the boundary layer turbulent kinetic energy budget and the mixed layer model of the boundary layer.

As in previous editions, Chapter 6 is devoted to quasi-geostrophic theory, which is still the fundamental unifying theory for large-scale extratropical motions. This chapter has been revised to provide increased emphasis on the role of potential vorticity conservation and to introduce the **Q**-vector approach for diagnosis of the ageostrophic circulation.

Chapter 7, which introduces the student to linear wave theory, has been expanded to include discussions of topographic Rossby waves, inertio-gravity oscillations, and the adjustment to geostrophic balance. In Chapter 8, linear wave theory is applied to the baroclinic instability problem. In addition to the two-level model of previous editions, the Eady instability model and the Rayleigh integral theorem are included.

Chapter 9, which discusses mesoscale circulations, is entirely new. A wide variety of mesoscale systems is analyzed, including fronts, convective storms, mountain waves, and hurricanes.

Chapters 10–12 are devoted primarily to applications of large-scale dynamics to understanding of the planetary-scale general circulation, the circulation of the tropics, and the circulation of the middle atmosphere, respectively. Chapter 10 has expanded discussions of longitudinally dependent time-averaged circulations and low-frequency variability. Chapter 11, in addition to the material on synoptic-scale tropical motions from earlier editions, includes introductions to equatorial wave theory and steady forced equatorial motions. Chapter 12 presents a completely revised summary of the large-scale dynamics of the stratosphere, including discussions of sudden stratospheric warmings and the equatorial quasi-biennial oscillation.

Finally, Chapter 13 is an introduction to modern numerical weather prediction techniques. It includes discussions of both finite difference and spectral modeling, as well as information on data assimilation and predictability.

I am indebted to a large number of colleagues and students for their continuing interest and suggestions. I am particularly grateful to the following colleagues for their helpful comments on drafts of various chapters of this edition: Dr. David Andrews, Dr. David Battisti, Dr. Dale Durran, Dr. Robert Houze, Dr. Åke Johansson, Dr. Erland Källén, Dr. Timothy Palmer, Dr. Richard Reed, Dr. Murry Salby, Dr. Adrian Simmons, Dr. Roland Stull, and Dr. John Wallace.

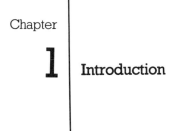

Chapter

1 | Introduction

1.1 The Atmospheric Continuum

Dynamic meteorology is the study of those motions of the atmosphere that are associated with weather and climate. For all such motions the discrete molecular nature of the atmosphere can be ignored, and the atmosphere can be regarded as a continuous fluid medium, or *continuum*. A "point" in the continuum is regarded as a volume element that is very small compared with the volume of atmosphere under consideration but still contains a large number of molecules. The expressions *air parcel* and *air particle* are both commonly used to refer to such a point. The various physical quantities that characterize the state of the atmosphere (e.g., pressure, density, temperature) are assumed to have unique values at each point in the atmospheric continuum. Moreover, these *field variables* and their derivatives are assumed to be continuous functions of space and time. The fundamental laws of fluid mechanics and thermodynamics, which govern the motions of the atmosphere, may then be expressed in terms of partial differential equations involving the field variables as dependent variables and space and time as independent variables.

The general set of partial differential equations governing the motions of the atmosphere is extremely complex; no general solutions are known to exist. To acquire an understanding of the physical role of atmospheric motions in determining the observed weather and climate, it is necessary to develop models based on systematic simplification of the fundamental governing equations. As we shall see in later chapters, the development of models appropriate to particular atmospheric motion systems requires careful consideration of the scales of motion involved.

1.2 Physical Dimensions and Units

The fundamental laws that govern the motions of the atmosphere satisfy the principle of *dimensional homogeneity*. That is, all terms in the equations expressing these laws must have the same physical dimensions. These dimensions can be expressed in terms of multiples and ratios of four dimensionally independent properties: length, time, mass, and thermodynamic temperature. To measure and compare the scales of terms in the laws of motion, a set of units of measure must be defined for these four fundamental properties.

In this text the international system of units (SI) will be used almost exclusively. The four fundamental properties are measured in terms of the SI *base units* shown in Table 1.1. All other properties are measured in terms of SI *derived units*, which are units formed from products or ratios of the base units. For example, velocity has the derived units of meter per second $(\mathrm{m\ s^{-1}})$. A number of important derived units have special names and symbols. Those that are commonly used in dynamic meteorology are indicated in Table 1.2. In addition, the supplementary unit designating a plane angle, the radian (rad), is required for expressing angular velocity $(\mathrm{rad\ s^{-1}})$ in the SI system.[1]

In order to keep numerical values within convenient limits it is conventional to use decimal multiples and submultiples of SI units. Prefixes used

Table 1.1 *SI Base Units*

Property	Name	Symbol
Length	Meter (metre)	m
Mass	Kilogram (kilogramme)	kg
Time	Second	s
Temperature	Kelvin	K

[1] Note that the *Hertz* measures frequency in *cycles* per second, not in radians per second.

Table 1.2 *SI Derived Units with Special Names*

Property	Name	Symbol
Frequency	Hertz	$Hz\ (s^{-1})$
Force	Newton	$N\ (kg\ m\ s^{-2})$
Pressure	Pascal	$Pa\ (N\ m^{-2})$
Energy	Joule	$J\ (N\ m)$
Power	Watt	$W\ (J\ s^{-1})$

to indicate such multiples and submultiples are given in Table 1.3. The prefixes of Table 1.3 may be affixed to any of the basic or derived SI units except the kilogram. Since the kilogram already is a prefixed unit, decimal multiples and submultiples of mass are formed by prefixing the gram (g), not the kilogram (kg).

Although the use of non-SI units will generally be avoided in this text, there are a few exceptions worth mentioning:

(1) In some contexts the time units minute (min), hour (h), and day (d) may be used in preference to the second in order to express quantities in convenient numerical values.

(2) The kilopascal (kPa) is the preferred SI unit for pressure. However, most meteorologists are accustomed to using the millibar (mb), which is equal to 100 Pa (0.1 kPa). Thus, for the reader's convenience pressures are here generally expressed in kilopascals followed by the equivalent in millibars [e.g., standard sea level pressure equals 101.325 kPa (1013.25 mb)]. However, to conform with conventional meteorological practice standard pressure surfaces will be referred to using the millibar (e.g., the 500-mb surface).

Table 1.3 *Prefixes for Decimal Multiples and Submultiples of SI Units*

Multiple	Prefix	Symbol
10^6	Mega	M
10^3	Kilo	k
10^2	Hecto	h
10^1	Deka	da
10^{-1}	Deci	d
10^{-2}	Centi	c
10^{-3}	Milli	m
10^{-6}	Micro	μ

(3)   Observed temperatures will generally be expressed using the Celsius temperature scale, which is related to the thermodynamic temperature scale as follows:

$$T_C = T - T_0$$

where T_C is expressed in degrees Celsius (°C), T is the thermodynamic temperature in kelvins (K), and $T_0 = 273.15$ K is the freezing point of water on the kelvin scale. From this relationship it is clear that one kelvin unit equals one degree Celsius.

1.3   Scale Analysis

Scale analysis, or scaling, is a convenient technique for estimating the magnitudes of various terms in the governing equations for a particular type of motion. In scaling, typical expected values of the following quantities are specified: (1) the magnitudes of the field variables; (2) the amplitudes of fluctuations in the field variables; and (3) the characteristic length, depth, and time scales on which these fluctuations occur. These typical values are then used to compare the magnitudes of various terms in the governing equations. For example, in a typical midlatitude synoptic[2] cyclone the surface pressure might fluctuate by 2 kPa (20 mb) over a horizontal distance of 2000 km. Designating the amplitude of the horizontal pressure fluctuation by δp, the horizontal coordinates by x and y, and the horizontal scale by L, the magnitude of the horizontal pressure gradient may be estimated by substituting $\delta p = 2$ kPa and $L = 2000$ km to get

$$\left(\frac{\partial p}{\partial x}, \frac{\partial p}{\partial y}\right) \sim \frac{\delta p}{L} = 1 \text{ kPa}/10^3 \text{ km} \quad (= 10 \text{ mb}/10^3 \text{ km})$$

Pressure fluctuations of similar magnitudes occur in other motion systems of vastly different scale such as tornadoes, squall lines, and hurricanes. Thus, the horizontal pressure gradient can range over several orders of magnitude for systems of meteorological interest. Similar considerations are also valid for derivative terms involving other field variables. Therefore, the nature of the dominant terms in the governing equations is crucially dependent on the horizontal scale of the motions. In particular, motions

---

[2] The term *synoptic* designates the branch of meteorology that deals with the analysis of observations taken over a wide area at or near the same time. This term is commonly used (as here) to designate the characteristic scale of the disturbances that are depicted on weather maps.

Table 1.4 *Scales of Atmospheric Motions*

Type of motion	Horizontal scale (m)
Molecular mean free path	10^{-7}
Minute turbulent eddies	10^{-2}–10^{-1}
Small eddies	10^{-1}–1
Dust devils	1–10
Gusts	10–10^2
Tornadoes	10^2
Cumulonimbus clouds	10^3
Fronts, squall lines	10^4–10^5
Hurricanes	10^5
Synoptic cyclones	10^6
Planetary waves	10^7

with horizontal scales of a few kilometers or less tend to have short time scales so that terms involving the rotation of the earth are negligible, while for larger scales they become very important.

Because the character of atmospheric motions depends so strongly on the horizontal scale, this scale provides a convenient method for classification of motion systems. In Table 1.4 examples of various types of motions are classified by horizontal scale for the spectral region from 10^{-7} to 10^7 m. In the following chapters scaling arguments will be used extensively in developing simplifications of the governing equations for use in modeling various types of motion systems.

1.4 The Fundamental Forces

The motions of the atmosphere are governed by the fundamental physical laws of conservation of mass, momentum, and energy. In Chapter 2 these principles are applied to a small volume element of the atmosphere in order to obtain the governing equations. However, before deriving the complete momentum equation it is useful to discuss the nature of the forces that influence atmospheric motions.

These forces can be classified as either *body forces* or *surface forces*. Body forces act on the center of mass of a fluid parcel; they have magnitudes proportional to the mass of the parcel. Gravity is an example of a body force. Surface forces act across the boundary surface separating a fluid parcel from its surroundings; their magnitudes are independent of the mass of the parcel. The pressure force is an example.

Newton's second law of motion states that the rate of change of momentum (i.e., the acceleration) of an object, as measured relative to coordinates fixed in space, equals the sum of all the forces acting. For atmospheric motions of meteorological interest, the forces that are of primary concern are the pressure gradient force, the gravitational force, and friction. These *fundamental* forces are the subject of the present section. If, as is the usual case, the motion is referred to a coordinate system rotating with the earth, Newton's second law may still be applied provided that certain *apparent* forces, the centrifugal force and the Coriolis force, are included among the forces acting. The nature of these apparent forces is discussed in Section 1.5.

1.4.1   The Pressure Gradient Force

We consider an infinitesimal volume element of air, $\delta V = \delta x\,\delta y\,\delta z$, centered at the point x_0, y_0, z_0 as illustrated in Fig. 1.1. Owing to random molecular motions, momentum is continually imparted to the walls of the volume element by the surrounding air. This momentum transfer per unit time per unit area is just the *pressure* exerted on the walls of the volume element by the surrounding air. If the pressure at the center of the volume element is designated by p_0, then the pressure on the wall labeled A in Fig. 1.1 can be expressed in a Taylor series expansion as

$$p_0 + \frac{\partial p}{\partial x}\frac{\delta x}{2} + \text{higher-order terms}$$

Neglecting the higher-order terms in this expansion, the pressure force acting on the volume element at wall A is

$$F_{Ax} = -\left(p_0 + \frac{\partial p}{\partial x}\frac{\delta x}{2}\right)\delta y\,\delta z$$

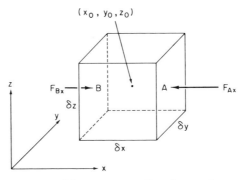

Fig. 1.1   The x component of the pressure gradient force acting on a fluid element.

where $\delta y \, \delta z$ is the area of wall A. Similarly, the force acting on the volume element at wall B is just

$$F_{Bx} = + \left(p_0 - \frac{\partial p}{\partial x} \frac{\delta x}{2} \right) \delta y \, \delta z$$

Therefore, the net x component of the pressure force acting on the volume is

$$F_x = F_{Ax} + F_{Bx} = - \frac{\partial p}{\partial x} \, \delta x \, \delta y \, \delta z$$

The mass m of the differential volume element is simply the density ρ times the volume: $m = \rho \, \delta x \, \delta y \, \delta z$. Thus, the x component of the pressure gradient force per unit mass is

$$\frac{F_x}{m} = - \frac{1}{\rho} \frac{\partial p}{\partial x}$$

Similarly, it can easily be shown that the y and z components of the pressure gradient force per unit mass are

$$\frac{F_y}{m} = - \frac{1}{\rho} \frac{\partial p}{\partial y} \quad \text{and} \quad \frac{F_z}{m} = - \frac{1}{\rho} \frac{\partial p}{\partial z}$$

so that the total pressure gradient force per unit mass is

$$\frac{\mathbf{F}}{m} = - \frac{1}{\rho} \nabla p \tag{1.1}$$

It is important to note that this force is proportional to the *gradient* of the pressure field, not to the pressure itself.

1.4.2   THE GRAVITATIONAL FORCE

Newton's law of universal gravitation states that any two elements of mass in the universe attract each other with a force proportional to their masses and inversely proportional to the square of the distance separating them. Thus, if two mass elements M and m are separated by a distance $r \equiv |\mathbf{r}|$ (with the vector $\mathbf{r}$ directed toward m as shown in Fig. 1.2), then the force exerted by mass M on mass m owing to gravitation is

$$\mathbf{F_g} = - \frac{GMm}{r^2} \left(\frac{\mathbf{r}}{r} \right) \tag{1.2}$$

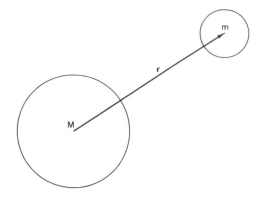

Fig. 1.2   Two spherical masses whose centers are separated by a distance r.

where G is a universal constant called the gravitational constant. The law of gravitation as expressed in (1.2) actually applies only to hypothetical "point" masses since for objects of finite extent, $\mathbf{r}$ will vary from one part of the object to another. However, for finite bodies (1.2) may still be applied if $|\mathbf{r}|$ is interpreted as the distance between the centers of mass of the bodies. Thus, if the earth is designated as mass M and m is a mass element of the atmosphere, then the force per unit mass exerted on the atmosphere by the gravitational attraction of the earth is

$$\frac{\mathbf{F}_g}{m} \equiv \mathbf{g}^* = -\frac{GM}{r^2}\left(\frac{\mathbf{r}}{r}\right) \tag{1.3}$$

In dynamic meteorology it is customary to use as a vertical coordinate the height above mean sea level. If the mean radius of the earth is designated by a and the distance above mean sea level is designated by z, then neglecting the small departure of the shape of the earth from sphericity, $r = a + z$. Therefore, (1.3) can be rewritten as

$$\mathbf{g}^* = \frac{\mathbf{g}_0^*}{(1 + z/a)^2} \tag{1.4}$$

where $\mathbf{g}_0^* = -(GM/a^2)(\mathbf{r}/r)$ is the value of the gravitational force at mean sea level. For meteorological applications $z \ll a$, so with negligible error we can let $\mathbf{g}^* = \mathbf{g}_0^*$ and simply treat the gravitational force as a constant.

1.4.3   THE VISCOUS FORCE

Any real fluid is subject to internal friction (viscosity), which causes it to resist the tendency to flow. Although a complete discussion of the resulting *viscous force* would be rather complicated, the basic physical concept can be illustrated by a simple experiment. A layer of incompressible fluid is confined between two horizontal plates separated by a distance l as shown in Fig. 1.3. The lower plate is fixed and the upper plate is placed into motion in the x direction at a speed u_0. Viscosity forces the fluid particles in the layer in contact with the plate to move at the velocity of the plate. Thus, at $z = l$ the fluid moves at speed $u(l) = u_0$, and at $z = 0$ the fluid is motionless. The force tangential to the upper plate required to keep it in uniform motion turns out to be proportional to the area of the plate, the velocity, and the inverse of the distance separating the plates. Thus, we may write $F = \mu A u_0 / l$ where μ is a constant of proportionality, the *dynamic viscosity coefficient*.

This force must just equal the force exerted by the upper plate on the fluid immediately below it. For a state of uniform motion, every horizontal layer of fluid of depth δz must exert the same force F on the fluid below. This may be expressed in the form $F = \mu A \, \delta u / \delta z$ where $\delta u = u_0 \, \delta z / l$ is the velocity shear across the layer δz. The viscous force per unit area, or *shearing stress*, can then be defined as

$$\tau_{zx} = \lim_{\delta z \to 0} \mu \frac{\delta u}{\delta z} = \mu \frac{\partial u}{\partial z}$$

where the subscripts indicate that τ_{zx} is the component of the shearing stress in the x direction owing to vertical shear of the x velocity component.

From the molecular viewpoint this shearing stress results from a net downward transport of momentum by the random motion of the molecules. Because the mean x momentum increases with height, the molecules passing downward through a horizontal plane at any instant carry more momentum than those passing upward through the plane. Thus, there is a net downward

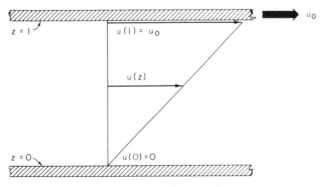

Fig. 1.3 One-dimensional steady-state viscous shear flow.

transport of x momentum. This downward momentum transport per unit time per unit area is simply the shearing stress.

In a similar fashion, random molecular motions will transport heat down a mean temperature gradient and trace constituents down mean mixing ratio gradients. In these cases the transport is referred to as *molecular diffusion*. Molecular diffusion always acts to reduce irregularities in the field being diffused.

In the simple two-dimensional steady-state motion example given above there is no net viscous force acting on the elements of fluid since the shearing stress acting across the top boundary of each fluid element is just equal and opposite to that acting across the lower boundary. For the more general case of nonsteady two-dimensional shear flow in an incompressible fluid, we may calculate the viscous force by again considering a differential volume element of fluid centered at (x, y, z) with sides $\delta x \, \delta y \, \delta z$ as shown in Fig. 1.4. If the shearing stress in the x direction acting through the center of the element is designated τ_{zx}, then the stress acting across the upper boundary on the fluid *below* may be written approximately as

$$\tau_{zx} + \frac{\partial \tau_{zx}}{\partial z} \frac{\delta z}{2}$$

while the stress acting across the lower boundary on the fluid *above* is

$$-\left[\tau_{zx} - \frac{\partial \tau_{zx}}{\partial z} \frac{\delta z}{2} \right]$$

(This is just equal and opposite to the stress acting across the lower boundary on the fluid *below*.) The net viscous force on the volume element acting in the x direction is then given by the sum of the stresses acting across the

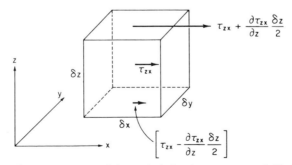

Fig. 1.4   The x component of the vertical shearing stress on a fluid element.

upper boundary on the fluid below and across the lower boundary on the fluid above:

$$\left(\tau_{zx} + \frac{\partial \tau_{zx}}{\partial z}\frac{\delta z}{2}\right)\delta y\,\delta x - \left(\tau_{zx} - \frac{\partial \tau_{zx}}{\partial z}\frac{\delta z}{2}\right)\delta y\,\delta x$$

Dividing this expression by the mass $\rho\,\delta x\,\delta y\,\delta z$ we find that the viscous force per unit mass owing to vertical shear of the component of motion in the x direction is

$$\frac{1}{\rho}\frac{\partial \tau_{zx}}{\partial z} = \frac{1}{\rho}\frac{\partial}{\partial z}\left(\mu\frac{\partial u}{\partial z}\right)$$

For constant μ, the right-hand side above may be simplified to $\nu\,\partial^2 u/\partial z^2$, where $\nu = \mu/\rho$ is the *kinematic viscosity coefficient*. For standard atmosphere conditions at sea level $\nu = 1.46 \times 10^{-5}$ m^2 s^{-1}. Derivations analogous to that shown in Fig. 1.4 can be carried out for viscous stresses acting in other directions. The resulting frictional force components per unit mass in the three Cartesian coordinate directions are

$$F_{rx} = \nu\left[\frac{\partial^2 u}{\partial x^2} + \frac{\partial^2 u}{\partial y^2} + \frac{\partial^2 u}{\partial z^2}\right]$$

$$F_{ry} = \nu\left[\frac{\partial^2 v}{\partial x^2} + \frac{\partial^2 v}{\partial y^2} + \frac{\partial^2 v}{\partial z^2}\right] \tag{1.5}$$

$$F_{rz} = \nu\left[\frac{\partial^2 w}{\partial x^2} + \frac{\partial^2 w}{\partial y^2} + \frac{\partial^2 w}{\partial z^2}\right]$$

For the atmosphere below 100 km ν is so small that molecular viscosity is negligible except in a thin layer within a few centimeters of the earth's surface where the vertical shear is very large. Away from this surface molecular boundary layer, momentum is transferred primarily by turbulent eddy motions. These will be discussed in Chapter 5.

1.5 Noninertial Reference Frames and "Apparent" Forces

In formulating the laws of atmospheric dynamics it is natural to use a *geocentric* reference frame, that is, a frame fixed with respect to the rotating earth. Newton's first law of motion states that a mass in uniform motion relative to a coordinate system fixed in space will remain in uniform motion in the absence of any forces. Such motion is referred to as *inertial motion*, and the fixed reference frame is an inertial, or absolute, frame of reference. It is clear however, that an object at rest or in uniform motion with respect to the rotating earth is not at rest or in uniform motion relative to a coordinate system fixed in space. Therefore, motion that appears to be inertial motion to an observer in the geocentric reference frame is really accelerated motion. Hence, the rotating frame is a *noninertial* reference frame. Newton's laws of motion can be applied in such a frame only if the acceleration of the coordinates is taken into account. The most satisfactory way of including the effects of coordinate acceleration is to introduce "apparent" forces in the statement of Newton's second law. These apparent forces are the inertial reaction terms that arise because of the coordinate acceleration. For a coordinate system in uniform rotation, two such apparent forces are required, the centrifugal force and the Coriolis force.

1.5.1 THE CENTRIFUGAL FORCE

A ball of mass m is attached to a string and whirled through a circle of radius r at a constant angular velocity ω. From the point of view of an observer in inertial space the speed of the ball is constant, but its direction of travel is continuously changing so that its velocity is not constant. To compute the acceleration we consider the change in velocity $\delta\mathbf{V}$ that occurs for a time increment δt during which the ball rotates through an angle $\delta\theta$ as shown in Fig. 1.5. Since $\delta\theta$ is also the angle between the vectors $\mathbf{V}$ and $\mathbf{V} + \delta\mathbf{V}$, the magnitude of $\delta\mathbf{V}$ is just $|\delta\mathbf{V}| = |\mathbf{V}|\delta\theta$. If we divide by δt and note that in the limit $\delta t \to 0$, $\delta\mathbf{V}$ is directed toward the axis of rotation, we obtain

$$\frac{d\mathbf{V}}{dt} = |\mathbf{V}|\frac{d\theta}{dt}\left(-\frac{\mathbf{r}}{r}\right)$$

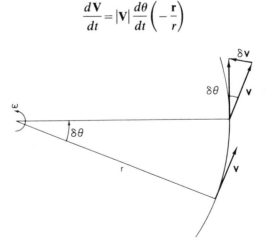

Fig. 1.5 Centripetal acceleration.

But $|\mathbf{V}| = \omega r$ and $d\theta/dt = \omega$, so that

$$\frac{d\mathbf{V}}{dt} = -\omega^2\mathbf{r} \qquad (1.6)$$

Therefore, viewed from fixed coordinates the motion is one of uniform acceleration directed toward the axis of rotation and equal to the square of the angular velocity times the distance from the axis of rotation. This acceleration is called the *centripetal acceleration*. It is caused by the force of the string pulling the ball.

Now suppose that we observe the motion in a coordinate system rotating with the ball. In this rotating system the ball is stationary, but there is still a force acting on the ball, namely the pull of the string. Therefore, in order to apply Newton's second law to describe the motion relative to this rotating coordinate system we must include an additional apparent force, the *centrifugal force*, which just balances the force of the string on the ball. Thus, the centrifugal force is equivalent to the inertial reaction of the ball on the string and just equal and opposite to the centripetal acceleration.

To summarize: Observed from a fixed system, the rotating ball undergoes a uniform centripetal acceleration in response to the force exerted by the string. Observed from a system rotating along with it, the ball is stationary and the force exerted by the string is balanced by a centrifugal force.

1.5.2 THE GRAVITY FORCE

A particle of unit mass at rest on the surface of the earth, observed in a reference frame rotating with the earth, is subject to a centrifugal force $\Omega^2\mathbf{R}$, where Ω is the angular speed of rotation of the earth[3] and $\mathbf{R}$ the position vector from the axis of rotation to the particle.

Thus, the weight of a particle of mass m at rest on the earth's surface, which is just the reaction force of the earth on the particle, will generally be less than the gravitational force $m\mathbf{g}^*$ because the centrifugal force partly balances the gravitational force. It is, therefore, convenient to combine the effects of the gravitational force and centrifugal force by defining a gravity force $\mathbf{g}$ (often called simply *gravity*) such that

$$\mathbf{g} \equiv \mathbf{g}^* + \Omega^2\mathbf{R} \qquad (1.7)$$

The gravitational force is directed toward the center of the earth, whereas the centrifugal force is directed away from the axis of rotation. Therefore,

---

[3] The earth revolves around its axis once every 23 h 56 min 4 s (86,164 s). Thus, $\Omega = 2\pi/(86,164\ s) = 7.292 \times 10^{-5}$ rad s^{-1}.

except at the poles and the equator, gravity is not directed toward the center of the earth. As indicated by Fig. 1.6, if the earth were a perfect sphere, gravity would have an equatorward component parallel to its surface. The earth has adjusted to compensate for this equatorward force component by assuming the approximate shape of a spheroid with an equatorial bulge, so that **g** is everywhere directed normal to the level surface. As a consequence, the equatorial radius of the earth is about 21 km larger than the polar radius. In addition, the local vertical, which is taken to be parallel to **g**, does not pass through the center of the earth except at the equator and poles.

Gravity can be represented in terms of the gradient of a potential function Φ, called the *geopotential*:

$$\nabla\Phi = -\mathbf{g}$$

But, since $\mathbf{g} = -g\mathbf{k}$ where $g \equiv |\mathbf{g}|$, it is clear that $\Phi = \Phi(z)$ and $d\Phi/dz = g$. If the value of the geopotential is set to zero at mean sea level, the geopotential $\Phi(z)$ at height z is just the work required to raise a unit mass to height z from mean sea level:

$$\Phi = \int_0^z g\, dz \tag{1.8}$$

1.5.3   THE CORIOLIS FORCE

Newton's second law of motion may be applied in rotating coordinates to describe an object at rest with respect to the rotating system provided that an apparent force, the centrifugal force, is included among the forces

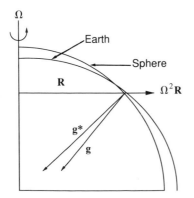

Fig. 1.6   The relationship between gravitation **g***, gravity **g**, and the shape of the earth.

acting on the object. If the object is moving with respect to the rotating system, an additional apparent force, the Coriolis force, is required in order that Newton's second law remain valid.

Suppose that an object is set in uniform motion with respect to an inertial coordinate system. If the object is observed from a rotating system with the axis of rotation perpendicular to the plane of motion, the path appears to be curved, as indicated in Fig. 1.7. Thus, as viewed in a rotating coordinate system there is an apparent force that deflects an object in inertial motion from a straight-line path. The resulting path is curved in a direction opposite to the direction of coordinate rotation. This deflection force is the Coriolis force. Viewed from the rotating system, the relative motion is an accelerated motion with the acceleration equal to the sum of the Coriolis force and the centrifugal force. The Coriolis force, which acts perpendicular to the velocity vector, can only change the direction of travel. However, the centrifugal force, which acts radially outward, has a component along the direction of motion, which increases the speed of the particle relative to the rotating coordinates as the particle spirals outward. Thus, in this example of inertial motion viewed from a rotating system the effects of both the Coriolis force and the centrifugal force are included.

Although the above example is simple enough to be readily comprehended in terms of everyday experience, it gives little insight into the meteorological role of the Coriolis force. The mathematical form for the Coriolis force due to motion relative to the rotating earth can be obtained by considering the motion of a hypothetical particle of unit mass that is free to move on a frictionless horizontal surface on the rotating earth. If the particle is initially at rest with respect to the earth, the only forces acting on it are the gravitational force and the apparent centrifugal force owing to the rotation of the earth. As we saw in Section 1.5.2, the sum of these two forces defines gravity, which is directed perpendicular to the local horizontal. Suppose

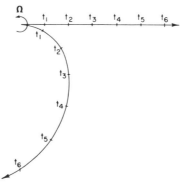

Fig. 1.7 Inertial motion as viewed from an inertial frame (straight line) and a rotating frame (curved line).

now that the particle is set in motion in the eastward direction by an impulsive force. Since the particle is now rotating faster than the earth, the centrifugal force on the particle will be increased. Letting Ω be the magnitude of the angular velocity of the earth, $\mathbf{R}$ the position vector from the axis of rotation to the particle, and u the eastward speed of the particle relative to the ground, the total centrifugal force is

$$\left(\Omega + \frac{u}{R}\right)^2 \mathbf{R} = \Omega^2 \mathbf{R} + \frac{2\Omega u \mathbf{R}}{R} + \frac{u^2 \mathbf{R}}{R^2} \tag{1.9}$$

The first term on the right is just the centrifugal force owing to the rotation of the earth. This is, of course, included in gravity. The other two terms represent *deflecting* forces, which act outward along the vector $\mathbf{R}$ (that is, perpendicular to the axis of rotation). For synoptic scale motions $|u| \ll \Omega R$ and the last term may be neglected in a first approximation. The remaining term in (1.9), $2\Omega u(\mathbf{R}/R)$, is the *Coriolis force* owing to relative motion parallel to a latitude circle. This Coriolis force can be divided into components in the vertical and meridional directions, respectively, as indicated in Fig. 1.8. Therefore, relative motion along the east–west coordinate produces an acceleration in the north–south direction given by

$$\left(\frac{dv}{dt}\right)_{\mathrm{Co}} = -2\Omega u \sin \phi \tag{1.10}$$

and an acceleration in the vertical given by

$$\left(\frac{dw}{dt}\right)_{\mathrm{Co}} = 2\Omega u \cos \phi \tag{1.11}$$

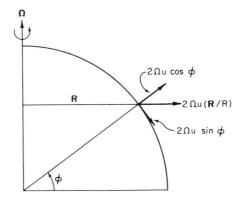

Fig. 1.8 Components of the Coriolis force owing to relative motion along a latitude circle.

where u, v, and w designate the eastward, northward, and upward velocity components, respectively, ϕ is latitude, and the subscript Co indicates that this is an acceleration owing only to the Coriolis force. A particle moving eastward in the horizontal plane in the Northern Hemisphere is deflected southward by the Coriolis force, whereas a westward-moving particle is deflected northward. In either case the deflection is to the right of the direction of motion. The vertical component of the Coriolis force (1.11) is ordinarily much smaller than the gravitational force, so its only effect is to cause a very minor change in the apparent weight of an object depending on whether the object is moving eastward or westward.

So far we have considered only the Coriolis force due to relative motion parallel to latitude circles. Suppose now that a particle initially at rest on the earth is set in motion equatorward by an impulsive force. As the particle moves equatorward it will conserve its angular momentum in the absence of torques in the east–west direction. Since the distance to the axis of rotation R increases for a particle moving equatorward, a relative westward velocity must develop if the particle is to conserve its absolute angular momentum. Thus, letting δR designate the change in the distance to the axis of rotation for a southward displacement from a latitude ϕ_0 to latitude $\phi_0 + \delta\phi$ (note that $\delta\phi < 0$ for an equatorward displacement), we obtain by conservation of angular momentum

$$\Omega R^2 = \left(\Omega + \frac{\delta u}{R + \delta R}\right)(R + \delta R)^2$$

where δu is the eastward relative velocity when the particle has reached latitude $\phi_0 + \delta\phi$. Expanding the right-hand side, neglecting second-order differentials, and solving for δu gives

$$\delta u = -2\Omega\, \delta R = +2\Omega a\, \delta\phi\, \sin \phi_0$$

where we have used $\delta R = -a\, \delta\phi\, \sin \phi_0$ (a is the radius of the earth). This relationship is illustrated in Fig. 1.9. Dividing through by the time increment δt and taking the limit as $\delta t \to 0$, we obtain from the above

$$\left(\frac{du}{dt}\right)_{\mathrm{Co}} = 2\Omega a \frac{d\phi}{dt} \sin \phi_0 = 2\Omega v \sin \phi$$

where $v = a\, d\phi/dt$ is the northward velocity component.

Similarly, it is easy to show that if the particle is launched vertically at latitude ϕ_0, conservation of the absolute angular momentum will require an acceleration in the zonal direction equal to $-2\Omega w \cos \phi_0$, where w is

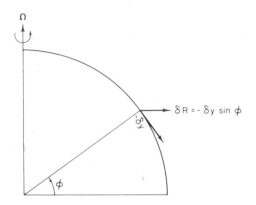

Fig. 1.9 Relationship of δR and $\delta y = a\, \delta\phi$ for an equatorward displacement.

the vertical velocity. Thus, in the general case where both horizontal and vertical relative motions are included

$$\left(\frac{du}{dt}\right)_{\text{Co}} = 2\Omega v \sin \phi - 2\Omega w \cos \phi \tag{1.12}$$

Again, the effect of the horizontal relative velocity is to deflect the particle to the right in the Northern Hemisphere. This deflection force is negligible for motions with time scales that are very short compared to the period of the earth's rotation (a point that is illustrated by several problems at the end of the chapter). Thus, the Coriolis force is not important for the dynamics of individual cumulus clouds but is essential to the understanding of longer time scale phenomena such as synoptic scale systems. The Coriolis force must also be taken into account when computing long-range missile or artillery trajectories. As an example, suppose that a ballistic missile is fired due eastward at 43°N latitude ($2\Omega \sin \phi = 10^{-4}\,\text{s}^{-1}$ at 43°N). If the missile travels 1000 km at a horizontal speed $u_0 = 1000\ \text{m s}^{-1}$, by how much is the missile deflected from its eastward path by the Coriolis force? Integrating (1.10) with respect to time, we find that

$$v = -2\Omega u_0 t \sin \phi \tag{1.13}$$

where it is assumed that the deflection is sufficiently small so that we may let $u = u_0$ be constant. To find the total southward displacement we must integrate (1.13) with respect to time:

$$\int_0^t v\, dt = \int_{y_0}^{y_0 + \delta y} dy = -2\Omega u_0 \int_0^t t\, dt \sin \phi$$

Thus, the total displacement is

$$\delta y = -\Omega u_0 t^2 \sin \phi \approx -50 \text{ km}$$

Therefore, the missile is deflected southward by 50 km owing to the Coriolis effect. Further examples of the deflection of particles by the Coriolis force are given in some of the problems at the end of the chapter.

1.6 Structure of the Static Atmosphere

The thermodynamic state of the atmosphere at any point is determined by the values of pressure, temperature, and density (or specific volume) at that point. These field variables are related to each other by the equation of state for an ideal gas. Letting p, T, ρ, and α denote pressure, temperature, density, and specific volume respectively, we can express the equation of state for dry air as

$$p\alpha = RT \quad \text{or} \quad p = \rho RT \tag{1.14}$$

where R is the gas constant for dry air ($R = 287 \text{ J kg}^{-1} \text{ K}^{-1}$).

1.6.1 THE HYDROSTATIC EQUATION

In the absence of atmospheric motions the gravity force must be exactly balanced by the vertical component of the pressure gradient force. Thus, as illustrated in Fig. 1.10,

$$dp/dz = -\rho g \tag{1.15}$$

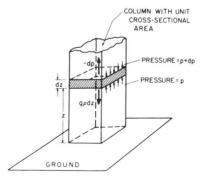

Fig. 1.10 Balance of forces for hydrostatic equilibrium. Note that dp is negative since pressure decreases with height. (After Wallace and Hobbs, 1977.)

COLUMN WITH UNIT
CROSS-SECTIONAL
AREA

PRESSURE = p+dp

PRESSURE = p

GROUND

This condition of *hydrostatic balance* provides an excellent approximation for the vertical dependence of the pressure field in the real atmosphere. Only for intense small-scale systems such as squall lines and tornadoes is it necessary to consider departures from hydrostatic balance. Integrating (1.15) from a height z to the top of the atmosphere we find that

$$p(z) = \int_z^\infty \rho g \, dz$$

so that the pressure at any point is simply equal to the weight of the unit cross section column of air overlying the point. Thus, mean sea-level pressure $p(0) = 101.325 \text{ kPa}$ (1013.25 mb) is simply the average weight per square meter of the total atmospheric column.[4] It is often useful to express the hydrostatic equation in terms of the geopotential rather than the geometric height. Noting from (1.8) that $d\Phi = g \, dz$ and from (1.14) that $\alpha = RT/p$, we can express the hydrostatic equation in the form

$$d\Phi = -(RT/p) \, dp = -RT \, d \ln p \qquad (1.16)$$

Thus, the variation of geopotential with respect to pressure depends only on temperature. Integration of (1.16) in the vertical yields a form of the *hypsometric equation*:

$$\Phi(z_2) - \Phi(z_1) = R \int_{p_2}^{p_1} T \, d \ln p \qquad (1.17)$$

Meteorologists often prefer to replace $\Phi(z)$ in (1.17) by a quantity called *geopotential height*, which is defined as $Z \equiv \Phi(z)/g_0$, where $g_0 \equiv 9.80665 \text{ m s}^{-2}$ is the global average of gravity at mean sea level. Thus in the troposphere and lower stratosphere Z is numerically almost identical to the geometric height z. In terms of Z the hypsometric equation becomes

$$Z_T \equiv Z_2 - Z_1 = \frac{R}{g_0} \int_{p_2}^{p_1} T \, d \ln p \qquad (1.18)$$

where Z_T is the *thickness* of the atmospheric layer between the pressure surfaces p_2 and p_1. Defining a layer mean temperature

$$\langle T \rangle = \int_{p_2}^{p_1} T \, d \ln p \left[\int_{p_2}^{p_1} d \ln p \right]^{-1}$$

---

[4] For computational convenience the mean surface pressure is often assumed to equal 100 kPa (1000 mb).

and a layer mean scale height $H \equiv R\langle T\rangle / g_0$, we have from (1.18)

$$Z_T = H \ln(p_1/p_2) \tag{1.19}$$

Thus the thickness of a layer bounded by isobaric surfaces is proportional to the mean temperature of the layer. Pressure decreases more rapidly with height in a cold layer than in a warm layer. It also follows immediately from (1.19) that in an isothermal atmosphere of temperature T the geopotential height is proportional to the natural logarithm of pressure normalized by the surface pressure;

$$Z = -H \ln(p/p_0)$$

where p_0 is the pressure at $Z = 0$. Thus, the pressure decreases exponentially with geopotential height by a factor of e^{-1} per scale height,

$$p(Z) = p(0)e^{-Z/H}$$

1.6.2   Pressure as a Vertical Coordinate

From the hydrostatic equation (1.15) it is clear that there exists a single-valued monotonic relationship between pressure and height in each vertical column of the atmosphere. Thus we may use pressure as the independent vertical coordinate and height (or geopotential) as a dependent variable. The thermodynamic state of the atmosphere is then specified by the fields of $\Phi(x, y, p, t)$ and $T(x, y, p, t)$.

Now the horizontal components of the pressure gradient force given by (1.1) are evaluated by partial differentiation holding z constant. However, when pressure is used as the vertical coordinate horizontal partial derivatives must be evaluated holding p constant. Transformation of the horizontal pressure gradient force from height to pressure coordinates may be carried out with the aid of Fig. 1.11. Considering only the x, z plane, we see from the figure that

$$\left[\frac{(p_0 + \delta p) - p_0}{\delta x}\right]_z = \left[\frac{(p_0 + \delta p) - p_0}{\delta z}\right]_x \left(\frac{\delta z}{\delta x}\right)_p$$

where the subscripts indicate variables that remain constant in evaluating the differentials. Thus, for example, in the limit $\delta z \to 0$

$$\left[\frac{(p_0 + \delta p) - p_0}{\delta z}\right]_x \to \left(-\frac{\partial p}{\partial z}\right)_x$$

where the minus sign is included because $\delta z < 0$ for $\delta p > 0$.

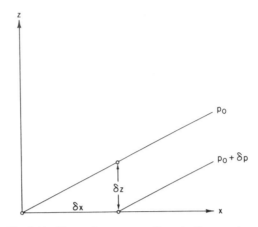

Fig. 1.11   Slope of pressure surfaces in the x, z plane.

Taking the limits $\delta x, \delta z \to 0$ we obtain[5]

$$\left(\frac{\partial p}{\partial x}\right)_z = -\left(\frac{\partial p}{\partial z}\right)_x \left(\frac{\partial z}{\partial x}\right)_p$$

which after substitution from the hydrostatic equation (1.15) yields

$$-\frac{1}{\rho}\left(\frac{\partial p}{\partial x}\right)_z = -g\left(\frac{\partial z}{\partial x}\right)_p = -\left(\frac{\partial \Phi}{\partial x}\right)_p \qquad (1.20)$$

Similarly, it is easy to show that

$$-\frac{1}{\rho}\left(\frac{\partial p}{\partial y}\right)_z = -\left(\frac{\partial \Phi}{\partial y}\right)_p \qquad (1.21)$$

Thus in the *isobaric* coordinate system the horizontal pressure gradient force is measured by the gradient of geopotential at constant pressure. Density no longer appears explicitly in the pressure gradient force; this is a distinct advantage of the isobaric system.

1.6.3   A Generalized Vertical Coordinate

Any single-valued monotonic function of pressure or height may be used as the independent vertical coordinate. For example, in many numerical

---

[5] It is important to note the minus sign on the right in this expression!

weather prediction models pressure normalized by the pressure at the ground $[\sigma \equiv p(x, y, z, t)/p_s(x, y, t)]$ is used as a vertical coordinate. This choice guarantees that the ground is a coordinate surface ($\sigma \equiv 1$) even in the presence of spatial and temporal surface pressure variations. Thus this so-called σ-coordinate system is particularly useful in regions of strong topographic variations.

We now obtain a general expression for the horizontal pressure gradient, which is applicable to any vertical coordinate $s = s(x, y, z, t)$ that is a single-valued monotonic function of height. Referring to Fig. 1.12, we see that for a horizontal distance δx the pressure difference evaluated along the surface $s = $ const. is related to that evaluated at $z = $ const. as

$$\frac{p_C - p_A}{\delta x} = \frac{p_C - p_B}{\delta z}\frac{\delta z}{\delta x} + \frac{p_B - p_A}{\delta x}$$

Taking the limits as $\delta x, \delta z \to 0$ we obtain

$$\left(\frac{\partial p}{\partial x}\right)_s = \frac{\partial p}{\partial z}\left(\frac{\partial z}{\partial x}\right)_s + \left(\frac{\partial p}{\partial x}\right)_z \tag{1.22}$$

Using the identity $\partial p/\partial z \equiv (\partial s/\partial z)(\partial p/\partial s)$ we can express (1.22) in the alternate form

$$\left(\frac{\partial p}{\partial x}\right)_s = \left(\frac{\partial p}{\partial x}\right)_z + \frac{\partial s}{\partial z}\left(\frac{\partial z}{\partial x}\right)_s\left(\frac{\partial p}{\partial s}\right) \tag{1.23}$$

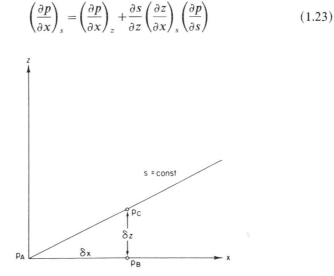

Fig. 1.12   Transformation of the pressure gradient force to s coordinates.

In later chapters we will apply (1.22) or (1.23) and similar expressions for other fields to transform the dynamical equations to several different vertical coordinate systems.

Problems

1.1. Neglecting the latitudinal variation in the radius of the earth, calculate the angle between the gravitational force and gravity vectors at the surface of the earth as a function of latitude.

1.2. Calculate the altitude at which an artificial satellite orbiting in the equatorial plane can be a synchronous satellite (i.e., can remain above the same spot on the surface of the earth).

1.3. An artificial satellite is placed into a natural synchronous orbit above the equator and is attached to the earth below by a wire. A second satellite is attached to the first by a wire of the same length and is placed in orbit directly above the first at the same angular velocity. Assuming that the wires have zero mass, calculate the tension in the wires per unit mass of satellite. Could this tension be used to lift objects into orbit with no additional expenditure of energy?

1.4. A train is running smoothly along a curved track at the rate of 50 m s^{-1}. A passenger standing on a set of scales observes that his weight is 10% greater than when the train is at rest. The track is banked so that the force acting on the passenger is normal to the floor of the train. What is the radius of curvature of the track?

1.5. If a baseball player throws a ball a horizontal distance of 100 m at 30° latitude in 4 s, by how much is it deflected laterally as a result of the rotation of the earth?

1.6. Two balls 4 cm in diameter are placed 100 m apart on a frictionless horizontal plane at 43°N. If the balls are impulsively propelled directly at each other with equal speeds, at what speed must they travel so that they just miss each other?

1.7. A locomotive of 2×10^5 kg mass travels 50 m s^{-1} along a straight horizontal track at 43°N. What lateral force is exerted on the rails? Compare the magnitudes of the upward reaction force exerted by the rails for cases where the locomotive is traveling eastward and westward, respectively.

1.8. Find the horizontal displacement of a body dropped from a fixed platform at a height h at the equator neglecting the effects of air resistance. What is the numerical value of the displacement for $h = 5$ km?

1.9. A bullet is fired vertically upward with initial speed w_0, at latitude ϕ. Neglecting air resistance, by what distance will it be displaced horizontally when it returns to the ground? (Neglect $2\Omega u \cos \phi$ compared to g in the vertical equation.)

1.10. A block of mass $M = 1$ kg is suspended from the end of a weightless string. The other end of the string is passed through a small hole in a horizontal platform and a ball of mass $m = 10$ kg is attached. At what angular velocity must the ball rotate on the horizontal platform to balance the weight of the block if the horizontal distance of the ball from the hole is 1 m? While the ball is rotating, the block is pulled down 10 cm. What is the new angular velocity of the ball? How much work is done in pulling down the block?

1.11. A particle is free to slide on a horizontal frictionless plane located at a latitude ϕ on the earth. Find the equation governing the path of the particle if it is given an impulsive eastward velocity $u = u_0$ at $t = 0$. Give the solution for the position of the particle as a function of time. Assume that f is constant.

1.12. Calculate the 100–50 kPa (1000–500 mb) thicknesses for isothermal conditions with temperatures of 273 K and 250 K, respectively.

1.13. Isolines of 100–50 kPa (1000–500 mb) thickness are drawn on a weather map using a contour interval of 60 m. What is the corresponding layer mean temperature interval?

1.14. Show that a homogeneous atmosphere (density independent of height) has a finite height that depends only on the temperature at the lower boundary. Compute the height of a homogeneous atmosphere with surface temperature $T_0 = 273$ K and surface pressure 100 kPa. (Use the ideal gas law and hydrostatic balance.)

1.15. For the conditions of Problem 1.14 compute the variation of the temperature with respect to height.

1.16. Show that in an atmosphere with uniform *lapse rate* γ (where $\gamma \equiv -dT/dz$) the geopotential height at pressure level p_1 is given by

$$Z = \frac{T_0}{\gamma}\left[1 - \left(\frac{p_0}{p_1}\right)^{-R\gamma/g}\right]$$

where T_0 and p_0 are the sea-level temperature and pressure, respectively.

1.17. Calculate the 100–50 kPa (1000–500 mb) thickness for a constant lapse rate atmosphere with $\gamma = 6.5$ K km^{-1} and $T_0 = 273$ K. Compare your results with the results in Problem 1.12.

1.18. Derive an expression for the variation of density with respect to height in a constant lapse rate atmosphere.

1.19. Derive an expression for the altitude variation of the pressure change δp that occurs when an atmosphere with constant lapse rate is subjected to a height-independent temperature change δT while surface pressure remains constant. At what height is the magnitude of the pressure change a maximum?

Suggested References

Complete reference information is provided in the Bibliography at the end of the book.

Wallace and Hobbs, *Atmospheric Science: An Introductory Survey,* discusses much of the material in this chapter at an elementary level.

Iribarne and Godson, *Atmospheric Thermodynamics,* contains a more advanced discussion of atmospheric statics.

Chapter

2 | The Basic Conservation Laws

Atmospheric motions are governed by three fundamental physical principles: conservation of mass, conservation of momentum, and conservation of energy. The mathematical relations that express these laws may be derived by considering the budgets of mass, momentum, and energy for an infinitesimal *control volume* in the fluid. Two types of control volume are commonly used in fluid dynamics. In the *Eulerian* frame of reference the control volume consists of a parallelepiped of sides δx, δy, δz, whose position is fixed relative to the coordinate axes. Mass, momentum, and energy budgets will depend on fluxes caused by the flow of fluid through the boundaries of the control volume. (This type of control volume was used in Section 1.4.1.) In the *Lagrangian* frame, however, the control volume consists of an infinitesimal mass of "tagged" fluid particles; thus, the control volume moves about following the motion of the fluid, always containing the same fluid particles.

The Lagrangian frame is particularly useful for deriving conservation laws since such laws may be stated most simply in terms of a particular mass element of the fluid. The Eulerian system is, however, more convenient for solving most problems because in that system the field variables are related by a set of partial differential equations in which the independent variables are the coordinates x, y, z, t. In the Lagrangian system, on the other hand, it is necessary to follow the time evolution of the fields for various individual fluid parcels. Thus the independent variables are x_0, y_0, z_0, and t, where x_0, y_0, z_0 designate the position that a particular parcel passed through at a reference time t_0.

2.1 Total Differentiation

The conservation laws to be derived in this chapter contain expressions for the rates of change of density, momentum, and thermodynamic energy following the motion of particular fluid parcels. In order to apply these laws in the Eulerian frame it is necessary to derive a relationship between the rate of change of a field variable following the motion and its rate of change at a fixed point. The former is called the *substantial*, the *total*, or the *material* derivative (it will be denoted by D/Dt). The latter is called the *local* derivative (it is merely the partial derivative with respect to time). To derive a relationship between the total derivative and the local derivative it is convenient to refer to a particular field variable (temperature, for example). For a given air parcel the location (x, y, z) is a function of t so that $x = x(t)$, $y = y(t)$, $z = z(t)$. Following the parcel T may then be considered as truly a function only of time, and its rate of change is just the total derivative DT/Dt. In order to relate the total derivative to the local rate of change at a fixed point we consider the temperature measured on a balloon that moves with the wind. Suppose that this temperature is T_0 at the point x_0, y_0, z_0 and time t_0. If the balloon moves to the point $x_0 + \delta x$, $y_0 + \delta y$, $z_0 + \delta z$ in a time increment δt, then the temperature change recorded on the balloon, δT can be expressed in a Taylor series expansion as

$$\delta T = \left(\frac{\partial T}{\partial t}\right)\delta t + \left(\frac{\partial T}{\partial x}\right)\delta x + \left(\frac{\partial T}{\partial y}\right)\delta y + \left(\frac{\partial T}{\partial z}\right)\delta z + (\text{higher-order terms})$$

Dividing through by δt and noting that δT is the change in temperature following the motion so that

$$\frac{DT}{Dt} \equiv \lim_{\delta t \to 0} \frac{\delta T}{\delta t}$$

we find that in the limit $\delta t \to 0$

$$\frac{DT}{Dt} = \frac{\partial T}{\partial t} + \left(\frac{\partial T}{\partial x}\right)\frac{Dx}{Dt} + \left(\frac{\partial T}{\partial y}\right)\frac{Dy}{Dt} + \left(\frac{\partial T}{\partial z}\right)\frac{Dz}{Dt}$$

is the rate of change of T following the motion.
   If we now let

$$\frac{Dx}{Dt} \equiv u, \qquad \frac{Dy}{Dt} \equiv v, \qquad \frac{Dz}{Dt} \equiv w$$

then u, v, w are the velocity components in the x, y, z directions, respectively, and

$$\frac{DT}{Dt} = \frac{\partial T}{\partial t} + \left(u\frac{\partial T}{\partial x} + v\frac{\partial T}{\partial y} + w\frac{\partial T}{\partial z}\right) \tag{2.1}$$

Using vector notation, this expression may be rewritten as

$$\frac{\partial T}{\partial t} = \frac{DT}{Dt} - \mathbf{U} \cdot \mathbf{\nabla} T$$

where $\mathbf{U} = \mathbf{i}u + \mathbf{j}v + \mathbf{k}w$ is the velocity vector. The term $-\mathbf{U} \cdot \mathbf{\nabla} T$ is called the temperature *advection*. It gives the contribution to the local temperature change owing to the air motion. For example, if the wind is blowing from a cold region toward a warm region $-\mathbf{U} \cdot \mathbf{\nabla} T$ will be negative (cold advection) and the advection term will contribute negatively to the local temperature change. Thus, the local rate of change of temperature equals the rate of change of temperature following the motion (that is, the heating or cooling of individual air parcels) plus the advective rate of change of temperature.

The relationship given for temperature in (2.1) holds for any of the field variables. Furthermore, the total derivative can be defined following a motion field other than the actual wind field. For example, we may wish to relate the pressure change measured by a barometer on a moving ship to the local pressure change.

Example.   The surface pressure decreases by 0.3 kPa/180 km in the eastward direction. A ship steaming eastward at 10 km/h measures a pressure fall of 0.1 kPa/3 h. What is the pressure change on an island that the ship is passing? If we take the x axis oriented eastward, then the local rate of change of pressure on the island is

$$\frac{\partial p}{\partial t} = \frac{Dp}{Dt} - u\frac{\partial p}{\partial x}$$

where Dp/Dt is the pressure change observed by the ship and u the velocity of the ship. Thus,

$$\frac{\partial p}{\partial t} = \frac{-0.1 \text{ kPa}}{3 \text{ h}} - \left(10 \frac{\text{km}}{\text{h}}\right)\left(\frac{-0.3 \text{ kPa}}{180 \text{ km}}\right) = -\frac{0.1 \text{ kPa}}{6 \text{ h}}$$

so that the rate of pressure fall on the island is only half the rate measured on the moving ship.

If the total derivative of a field variable is zero, then that variable is a conservative quantity following the motion. The local change is then entirely owing to advection. As we shall see later, field variables that are approximately conserved following the motion play an important role in dynamic meteorology.

2.1.1 TOTAL DIFFERENTIATION OF A VECTOR IN A ROTATING SYSTEM

The conservation law for momentum (Newton's second law of motion) relates the rate of change of the absolute momentum following the motion in an inertial reference frame to the sum of the forces acting on the fluid. For most applications in meteorology it is desirable to refer the motion to a reference frame rotating with the earth. The transformation of the momentum equation to a rotating coordinate system requires a relationship between the total derivative of a vector in an inertial reference frame and the corresponding total derivative in a rotating system.

To derive this relationship we let $\mathbf{A}$ be an arbitrary vector whose Cartesian components in an inertial frame are given by

$$\mathbf{A} = \mathbf{i}A_x + \mathbf{j}A_y + \mathbf{k}A_z$$

and whose components in a frame rotating with an angular velocity $\mathbf{\Omega}$ are

$$\mathbf{A} = \mathbf{i}'A_x' + \mathbf{j}'A_y' + \mathbf{k}'A_z'$$

Letting $D_a\mathbf{A}/Dt$ be the total derivative of $\mathbf{A}$ in the inertial frame we can write

$$\frac{D_a\mathbf{A}}{Dt} = \mathbf{i}\frac{DA_x}{Dt} + \mathbf{j}\frac{DA_y}{Dt} + \mathbf{k}\frac{DA_z}{Dt}$$

$$= \mathbf{i}'\frac{DA_x'}{Dt} + \mathbf{j}'\frac{DA_y'}{Dt} + \mathbf{k}'\frac{DA_z'}{Dt} + \frac{D_a\mathbf{i}'}{Dt}A_x' + \frac{D_a\mathbf{j}'}{Dt}A_y' + \frac{D_a\mathbf{k}'}{Dt}A_z'$$

Now,

$$\mathbf{i}' \frac{DA_x'}{Dt} + \mathbf{j}' \frac{DA_y'}{Dt} + \mathbf{k}' \frac{DA_z'}{Dt} \equiv \frac{D\mathbf{A}}{Dt}$$

is just the total derivative of $\mathbf{A}$ as viewed in the rotating coordinates (that is, the rate of change of $\mathbf{A}$ following the relative motion). Furthermore, since $\mathbf{i}'$ may be regarded as a position vector of unit length, $D_a\mathbf{i}'/Dt$ is the velocity of $\mathbf{i}'$ owing to its rotation. Thus $D_a\mathbf{i}'/Dt = \mathbf{\Omega} \times \mathbf{i}'$ and similarly $D_a\mathbf{j}'/Dt = \mathbf{\Omega} \times \mathbf{j}'$ and $D_a\mathbf{k}'/Dt = \mathbf{\Omega} \times \mathbf{k}'$. Therefore the above can be rewritten as

$$\frac{D_a\mathbf{A}}{Dt} = \frac{D\mathbf{A}}{Dt} + \mathbf{\Omega} \times \mathbf{A} \tag{2.2}$$

which is the desired relationship.

2.2 The Vectorial Form of the Momentum Equation in Rotating Coordinates

In an inertial reference frame Newton's second law of motion may be written symbolically as

$$\frac{D_a\mathbf{U}_a}{Dt} = \sum \mathbf{F} \tag{2.3}$$

The left-hand side represent the rate of change of the absolute velocity $\mathbf{U}_a$, following the motion as viewed in an inertial system. The right-hand side represents the sum of the real forces acting per unit mass. In Section 1.5 we found through simple physical reasoning that when the motion is viewed in a rotating coordinate system certain additional apparent forces must be included if Newton's second law is to be valid. The same result may be obtained by a formal transformation of coordinates in (2.3).

In order to transform this expression to rotating coordinates we must first find a relationship between $\mathbf{U}_a$ and the velocity relative to the rotating system, which we will designate by $\mathbf{U}$. This relationship is obtained by applying (2.2) to the position vector $\mathbf{r}$ for an air parcel on the rotating earth:

$$\frac{D_a\mathbf{r}}{Dt} = \frac{D\mathbf{r}}{Dt} + \mathbf{\Omega} \times \mathbf{r} \tag{2.4}$$

But $D_a\mathbf{r}/Dt \equiv \mathbf{U}_a$ and $D\mathbf{r}/Dt \equiv \mathbf{U}$; therefore (2.4) may be written as

$$\mathbf{U}_a = \mathbf{U} + \mathbf{\Omega} \times \mathbf{r} \qquad (2.5)$$

which states simply that the absolute velocity of an object on the rotating earth is equal to its velocity relative to the earth plus the velocity owing to the rotation of the earth.

Now we apply (2.2) to the velocity vector $\mathbf{U}_a$ and obtain

$$\frac{D_a\mathbf{U}_a}{Dt} = \frac{D\mathbf{U}_a}{Dt} + \mathbf{\Omega} \times \mathbf{U}_a \qquad (2.6)$$

Substituting from (2.5) into the right-hand side of (2.6) gives

$$\frac{D_a\mathbf{U}_a}{Dt} = \frac{D}{Dt}(\mathbf{U} + \mathbf{\Omega} \times \mathbf{r}) + \mathbf{\Omega} \times (\mathbf{U} + \mathbf{\Omega} \times \mathbf{r})$$

$$= \frac{D\mathbf{U}}{Dt} + 2\mathbf{\Omega} \times \mathbf{U} - \Omega^2 \mathbf{R} \qquad (2.7)$$

where $\mathbf{\Omega}$ is assumed to be constant. Here $\mathbf{R}$ is a vector perpendicular to the axis of rotation, with magnitude equal to the distance to the axis of rotation, so that with the aid of a vector identity,

$$\mathbf{\Omega} \times (\mathbf{\Omega} \times \mathbf{r}) = \mathbf{\Omega} \times (\mathbf{\Omega} \times \mathbf{R}) = -\Omega^2 \mathbf{R}$$

Equation (2.7) states that the acceleration following the motion in an inertial system equals the rate of change of relative velocity following the relative motion in the rotating frame plus the Coriolis acceleration owing to relative motion in the rotating frame plus the centripetal acceleration caused by the rotation of the coordinates.

If we assume that the only real forces acting on the atmosphere are the pressure gradient force, gravitation, and friction, we can rewrite Newton's second law (2.3) with the aid of (2.7) as

$$\frac{D\mathbf{U}}{Dt} = -2\mathbf{\Omega} \times \mathbf{U} - \frac{1}{\rho}\nabla p + \mathbf{g} + \mathbf{F}_r \qquad (2.8)$$

where F_r designates the frictional force (see Section 1.4.3), and the centrifugal force has been combined with gravitation in the gravity term $\mathbf{g}$ (see Section 1.5.2). Equation (2.8) is the statement of Newton's second law of motion relative to a rotating coordinate frame. It states that the acceleration following the relative motion in the rotating frame equals the sum of the Coriolis force, the pressure gradient force, effective gravity, and friction. It is this form of the momentum equation that is basic to most work in dynamic meteorology.

2.3 The Component Equations in Spherical Coordinates

For purposes of theoretical analysis and numerical prediction, it is necessary to expand the vectorial momentum equation (2.8) into its scalar components. Since the departure of the shape of the earth from sphericity is entirely negligible for meteorological purposes, it is convenient to expand (2.8) in spherical coordinates so that the (level) surface of the earth corresponds to a coordinate surface. The coordinate axes are then (λ, ϕ, z), where λ is longitude, ϕ is latitude, and z is the vertical distance above the surface of the earth. If the unit vectors $\mathbf{i}, \mathbf{j}, \mathbf{k}$ are now taken to be directed eastward, northward, and upward, respectively, the relative velocity becomes

$$\mathbf{U} \equiv \mathbf{i}u + \mathbf{j}v + \mathbf{k}w$$

where the components u, v, and w are defined as follows:

$$u \equiv r \cos \phi \frac{D\lambda}{Dt}, \qquad v \equiv r \frac{D\phi}{Dt}, \qquad w \equiv \frac{Dz}{Dt} \qquad (2.9)$$

Here, r is the distance to the center of the earth, which is related to z by $r = a + z$, where a is the radius of the earth. Traditionally, the variable r in (2.9) is replaced by the constant a. This is a very good approximation since $z \ll a$ for the regions of the atmosphere with which meteorologists are concerned. For notational simplicity, it is conventional to define x and y as eastward and northward distance, such that $Dx = a \cos \phi \, D\lambda$ and $Dy = a \, D\phi$. Thus, the horizontal velocity components are $u \equiv Dx/Dt$ and $v \equiv Dy/Dt$ in the eastward and northward directions, respectively. The (x, y, z) coordinate system defined in this way is not, however, a Cartesian coordinate system because the directions of the $\mathbf{i}, \mathbf{j}, \mathbf{k}$ unit vectors are not constant, but are functions of position on the spherical earth. This position dependence of the unit vectors must be taken into account when the acceleration vector is expanded into its components on the sphere. Thus, we write

$$\frac{D\mathbf{U}}{Dt} = \mathbf{i}\frac{Du}{Dt} + \mathbf{j}\frac{Dv}{Dt} + \mathbf{k}\frac{Dw}{Dt} + u\frac{D\mathbf{i}}{Dt} + v\frac{D\mathbf{j}}{Dt} + w\frac{D\mathbf{k}}{Dt} \qquad (2.10)$$

In order to obtain the component equations, it is necessary first to evaluate the rates of change of the unit vectors following the motion.

We first consider $D\mathbf{i}/Dt$. Expanding the total derivative as in (2.1) and noting that $\mathbf{i}$ is a function only of x (that is, an eastward-directed vector does not change its orientation if the motion is in the north–south or vertical directions), we get

$$\frac{D\mathbf{i}}{Dt} = u\frac{\partial\mathbf{i}}{\partial x}$$

From Fig. 2.1 we see that

$$\lim_{\delta x \to 0} \frac{|\delta\mathbf{i}|}{\delta x} = \left|\frac{\partial\mathbf{i}}{\partial x}\right| = \frac{1}{a\cos\phi}$$

and that the vector $\partial\mathbf{i}/\partial x$ is directed toward the axis of rotation. Thus, as illustrated in Fig. 2.2,

$$\frac{\partial\mathbf{i}}{\partial x} = \frac{1}{a\cos\phi}(\mathbf{j}\sin\phi - \mathbf{k}\cos\phi)$$

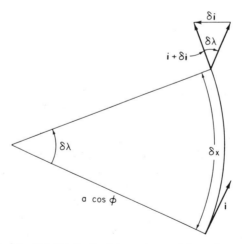

Fig. 2.1   The longitudinal dependence of the unit vector **i**.

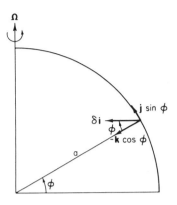

Fig. 2.2   Resolution of δ**i** into northward and
vertical components.

Therefore

$$\frac{D\mathbf{i}}{Dt} = \frac{u}{a\cos\phi}\,(\mathbf{j}\sin\phi - \mathbf{k}\cos\phi) \tag{2.11}$$

Considering now $D\mathbf{j}/Dt$, we note that $\mathbf{j}$ is a function only of x and y.
Thus, with the aid of Fig. 2.3 we see that for eastward motion $|\delta\mathbf{j}| = \delta x/(a/\tan\phi)$. Since the vector $\partial\mathbf{j}/\partial x$ is directed in the negative x direction,

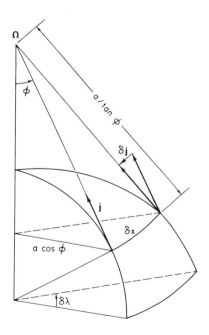

Fig. 2.3   The dependence of unit vector **j**
on longitude.

we have then

$$\frac{\partial \mathbf{j}}{\partial x} = -\frac{\tan \phi}{a}\mathbf{i}$$

From Fig. 2.4 it is clear that for northward motion $|\delta \mathbf{j}| = \delta \phi$. But $\delta y = a\,\delta \phi$, and $\delta \mathbf{j}$ is directed downward, so that

$$\frac{\partial \mathbf{j}}{\partial y} = -\frac{\mathbf{k}}{a}$$

Hence,

$$\frac{D\mathbf{j}}{Dt} = -\frac{u \tan \phi}{a}\mathbf{i} - \frac{v}{a}\mathbf{k} \tag{2.12}$$

Finally, by similar arguments it can be shown that

$$\frac{D\mathbf{k}}{Dt} = \mathbf{i}\frac{u}{a} + \mathbf{j}\frac{v}{a} \tag{2.13}$$

Substituting (2.11)–(2.13) into (2.10) and rearranging the terms, we obtain the spherical polar coordinate expansion of the acceleration following the relative motion

$$\frac{D\mathbf{U}}{Dt} = \left(\frac{Du}{Dt} - \frac{uv \tan \phi}{a} + \frac{uw}{a}\right)\mathbf{i} + \left(\frac{Dv}{Dt} + \frac{u^2 \tan \phi}{a} + \frac{vw}{a}\right)\mathbf{j}$$

$$+ \left(\frac{Dw}{Dt} - \frac{u^2 + v^2}{a}\right)\mathbf{k} \tag{2.14}$$

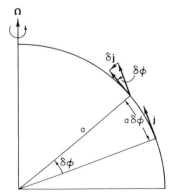

Fig. 2.4   The dependence of unit vector $\mathbf{j}$ on latitude.

We turn next to the component expansion of the force terms in (2.8). The Coriolis force is expanded by noting that $\mathbf{\Omega}$ has no component parallel to $\mathbf{i}$ and that its components parallel to $\mathbf{j}$ and $\mathbf{k}$ are $2\Omega \cos \phi$ and $2\Omega \sin \phi$, respectively. Thus, using the definition of the vector cross product,

$$-2\mathbf{\Omega} \times \mathbf{U} = -2\Omega \begin{vmatrix} \mathbf{i} & \mathbf{j} & \mathbf{k} \\ 0 & \cos \phi & \sin \phi \\ u & v & w \end{vmatrix}$$

$$= -(2\Omega w \cos \phi - 2\Omega v \sin \phi)\mathbf{i} - 2\Omega u \sin \phi\, \mathbf{j} + 2\Omega u \cos \phi\, \mathbf{k} \quad (2.15)$$

The pressure gradient force may be expressed as

$$\nabla p = \mathbf{i}\frac{\partial p}{\partial x} + \mathbf{j}\frac{\partial p}{\partial y} + \mathbf{k}\frac{\partial p}{\partial z} \tag{2.16}$$

and gravity is conveniently represented as

$$\mathbf{g} = -g\mathbf{k} \tag{2.17}$$

where g is a positive scalar ($g \cong 9.8 \text{ m s}^{-2}$ at the earth's surface). Finally, recall from (1.5) that

$$\mathbf{F}_r = \mathbf{i}F_{rx} + \mathbf{j}F_{ry} + \mathbf{k}F_{rz} \tag{2.18}$$

Substituting (2.14)–(2.18) into the equation of motion (2.8) and equating all terms in the $\mathbf{i}, \mathbf{j}$, and $\mathbf{k}$ directions, respectively, we obtain

$$\frac{Du}{Dt} - \frac{uv \tan \phi}{a} + \frac{uw}{a} = -\frac{1}{\rho}\frac{\partial p}{\partial x} + 2\Omega v \sin \phi - 2\Omega w \cos \phi + F_{rx} \tag{2.19}$$

$$\frac{Dv}{Dt} + \frac{u^2 \tan \phi}{a} + \frac{vw}{a} = -\frac{1}{\rho}\frac{\partial p}{\partial y} - 2\Omega u \sin \phi + F_{ry} \tag{2.20}$$

$$\frac{Dw}{Dt} - \frac{u^2 + v^2}{a} = -\frac{1}{\rho}\frac{\partial p}{\partial z} - g + 2\Omega u \cos \phi + F_{rz} \tag{2.21}$$

which are the eastward, northward, and vertical component momentum equations, respectively. The terms proportional to $1/a$ on the left-hand sides in (2.19)-(2.21) are called the *curvature* terms; they arise owing to the curvature of the earth.[1] Because they are nonlinear (that is, they are quadratic in the dependent variables) they are difficult to handle in theoretical analyses. Fortunately, as will be shown in the next section, the curvature terms are unimportant for midlatitude synoptic-scale motions. However, even when the curvature terms are neglected (2.19)-(2.21) are still nonlinear partial differential equations, as can be seen by expanding the total derivatives into their local and advective parts:

$$\frac{Du}{Dt} = \frac{\partial u}{\partial t} + u\frac{\partial u}{\partial x} + v\frac{\partial u}{\partial y} + w\frac{\partial u}{\partial z}$$

with similar expressions for Dv/Dt and Dw/Dt. In general the advective acceleration terms are comparable in magnitude to the local acceleration. The presence of nonlinear advection processes is one reason that dynamic meteorology is an interesting and challenging subject.

2.4 Scale Analysis of the Equations of Motion

In Section 1.3 we discussed the basic notion of scaling the equations of motion in order to determine whether some terms in the equations are negligible for motions of meteorological concern. Elimination of terms on scaling considerations not only has the advantage of simplifying the mathematics, but, as shown in later chapters, the elimination of small terms in some cases has the very important property of completely eliminating, or *filtering*, an unwanted type of motion. The complete equations of motion (2.19)-(2.21) describe all types and scales of atmospheric motions. Sound waves, for example, are a perfectly valid class of solutions to these equations. However, sound waves are of negligible importance in dynamical meteorology. Therefore, it will be a distinct advantage if, as turns out to be true, we can neglect the terms that lead to the production of sound waves and filter out this unwanted class of motions.

In order to simplify (2.19)-(2.21) for synoptic-scale motions we define the following characteristic scales of the field variables based on observed values for midlatitude synoptic systems.

---

[1] It can be shown that when r is replaced by a as done here (the traditional approximation) the Coriolis terms proportional to cos ϕ in (2.19) and (2.21) must be neglected if the equations are to satisfy angular momentum conservation.

$$U \sim 10 \text{ m s}^{-1}$$ horizontal velocity scale
$$W \sim 1 \text{ cm s}^{-1}$$ vertical velocity scale
$$L \sim 10^6 \text{ m}$$ length scale [$\sim 1/(2\pi)$ wavelength]
$$H \sim 10^4 \text{ m}$$ depth scale
$$\delta P/\rho \sim 10^3 \text{ m}^2 \text{ s}^{-2}$$ horizontal pressure fluctuation scale
$$L/U \sim 10^5 \text{ s}$$ time scale

The horizontal pressure fluctuation δP is normalized by the density ρ in order to produce a scale estimate that is valid at all heights in the troposphere despite the approximate exponential decrease with height of both δP and ρ. Note that $\delta P/\rho$ has units of geopotential. Referring back to (1.21), we see that indeed the magnitude of the fluctuation of $\delta P/\rho$ on a surface of constant height must equal the magnitude of the fluctuation of the geopotential of an isobaric surface. The time scale here is an advective time scale, which is appropriate for pressure systems that move at approximately the speed of the horizontal wind, as is observed for synoptic-scale motions. Thus, L/U is the time required to travel a distance L at a speed U, and the substantial differential operator $D/Dt \sim U/L$ for such motions.

It should be pointed out here that the synoptic-scale vertical velocity is not a directly measurable quantity. However, as is shown in Chaper 3, the magnitude of w can be deduced from knowledge of the horizontal velocity field.

We can now estimate the magnitude of each term in (2.19) and (2.20) for synoptic-scale motions at a given latitude. It is convenient to consider a disturbance centered at latitude $\phi_0 = 45°$ and introduce the notation

$$f_0 = 2\Omega \sin \phi_0 = 2\Omega \cos \phi_0 \cong 10^{-4} \text{ s}^{-1}$$

Table 2.1 shows the characteristic magnitude of each term in (2.19) and (2.20) based on the above scaling considerations. The molecular friction

Table 2.1 *Scale Analysis of the Horizontal Momentum Equations*

	A	B	C	D	E	F	G
x-Eq.	$\dfrac{Du}{Dt}$	$-2\Omega v \sin \phi$	$+2\Omega w \cos \phi$	$+\dfrac{uw}{a}$	$-\dfrac{uv \tan \phi}{a}$	$=-\dfrac{1}{\rho}\dfrac{\partial p}{\partial x}$	$+F_{rx}$
y-Eq.	$\dfrac{Dv}{Dt}$	$+2\Omega u \sin \phi$		$+\dfrac{vw}{a}$	$+\dfrac{u^2 \tan \phi}{a}$	$=-\dfrac{1}{\rho}\dfrac{\partial p}{\partial y}$	$+F_{ry}$
Scales	U^2/L	$f_0 U$	$f_0 W$	$\dfrac{UW}{a}$	$\dfrac{U^2}{a}$	$\dfrac{\delta P}{\rho L}$	$\dfrac{vU}{H^2}$
(m s^{-2})	10^{-4}	10^{-3}	10^{-6}	10^{-8}	10^{-5}	10^{-3}	10^{-12}

term is so small that it may be neglected for all motions except the smallest-scale turbulent motions near the ground, where vertical wind shears can become very large and the molecular friction term must be retained, as will be discussed in Chapter 5.

2.4.1   THE GEOSTROPHIC APPROXIMATION AND THE GEOSTROPHIC WIND

It is apparent from Table 2.1 that for midlatitude synoptic-scale disturbances the Coriolis force (term B) and the pressure gradient force (term F) are in approximate balance. Retaining only these two terms in (2.19) and (2.20), gives as a first approximation the *geostrophic* relationship

$$-fv \approx -\frac{1}{\rho}\frac{\partial p}{\partial x}, \qquad fu \approx -\frac{1}{\rho}\frac{\partial p}{\partial y} \qquad (2.22)$$

where $f \equiv 2\Omega \sin \phi$ is called the *Coriolis parameter*. The geostrophic balance is a *diagnostic* expression that gives the approximate relationship between the pressure field and horizontal velocity in large-scale extratropical systems. The approximation (2.22) contains no reference to time and therefore cannot be used to predict the evolution of the velocity field. It is for this reason that the geostrophic relationship is called a diagnostic relationship.

By analogy to the geostrophic approximation (2.22) it is possible to define a horizontal velocity field $\mathbf{V}_g \equiv \mathbf{i}u_g + \mathbf{j}v_g$, called the *geostrophic wind*, which satisfies (2.22) identically. In vectorial form

$$\mathbf{V}_g \equiv \mathbf{k} \times \frac{1}{\rho f}\nabla p \qquad (2.23)$$

Thus, knowledge of the pressure distribution at any time determines the geostrophic wind. It should be kept clearly in mind that (2.23) always defines the geostrophic wind; but only for large-scale motions away from the equator should the geostrophic wind be used as an approximation to the actual horizontal wind field. For the scales used in Table 2.1 the geostrophic wind approximates the true horizontal velocity to within 10–15% in midlatitudes.

2.4.2   APPROXIMATE PROGNOSTIC EQUATIONS: THE ROSSBY NUMBER

To obtain prediction equations it is necessary to retain the acceleration (term A) in (2.19) and (2.20). The resulting approximate horizontal momentum equations are

$$\frac{Du}{Dt} = fv - \frac{1}{\rho}\frac{\partial p}{\partial x} = f(v - v_g) \qquad (2.24)$$

$$\frac{Dv}{Dt} = -fu - \frac{1}{\rho}\frac{\partial p}{\partial y} = -f(u - u_g) \tag{2.25}$$

where (2.23) is used to rewrite the pressure gradient force in terms of the geostrophic wind. Since the acceleration terms in (2.24) and (2.25) are proportional to the difference between the actual wind and the geostrophic wind, they are about an order of magnitude smaller than the Coriolis force and the pressure gradient force, in agreement with our scale analysis. The fact that the horizontal flow is in approximate geostrophic balance is helpful for diagnostic analysis. However, it makes actual applications of these equations in weather prognosis difficult because acceleration (which must be measured accurately) is given by the small difference between two large terms. Thus, a small error in measurement of either velocity or pressure gradient will lead to very large errors in estimating the acceleration. This problem is discussed in some detail in Chapter 13.

A convenient measure of the magnitude of the acceleration compared to the Coriolis force may be obtained by forming the ratio of the characteristic scales for the acceleration and the Coriolis force terms: $(U^2/L)/(f_0 U)$. This ratio is a nondimensional number called the *Rossby number* after the Swedish meteorologist C. G. Rossby (1898–1957) and is designated by

$$\text{Ro} \equiv U/(f_0 L)$$

Thus, the smallness of the Rossby number is a measure of the validity of the geostrophic approximation.

2.4.3 THE HYDROSTATIC APPROXIMATION

A similar scale analysis can be applied to the vertical component of the momentum equation (2.21). Since pressure decreases by about an order of magnitude from the ground to the tropopause, the vertical pressure gradient may be scaled by P_0/H, where P_0 is the surface pressure and H is the depth of the troposphere. The terms in (2.21) may then be estimated for synoptic-scale motions and are shown in Table 2.2. As with the horizontal component equations, we consider motions centered at 45° latitude and neglect friction.

Table 2.2 *Scale Analysis of the Vertical Momentum Equation*

z-Eq.	Dw/Dt	$-2\Omega u \cos\phi$	$-(u^2 + v^2)/a$	$= -\rho^{-1}\partial p/\partial z$	$-g$	$+F_{rz}$
Scales	UW/L	$f_0 U$	U^2/a	$P_0/(\rho H)$	g	νWH^{-2}
m s^{-2}	10^{-7}	10^{-3}	10^{-5}	10	10	10^{-15}

The scaling indicates that to a high degree of accuracy the pressure field is in *hydrostatic equilibrium*; that is, the pressure at any point is simply equal to the weight of a unit cross-section column of air above that point.

The above analysis of the vertical momentum equation is, however, somewhat misleading. It is not sufficient to show merely that the vertical acceleration is small compared to g. Since only that part of the pressure field that varies horizontally is directly coupled to the horizontal velocity field, it is actually necessary to show that the horizontally varying pressure component is itself in hydrostatic equilibrium with the horizontally varying density field. To do this it is convenient first to define a standard pressure $p_0(z)$, which is the horizontally averaged pressure at each height, and a corresponding standard density $\rho_0(z)$, defined so that $p_0(z)$ and $\rho_0(z)$ are in *exact* hydrostatic balance:

$$\frac{1}{\rho_0}\frac{dp_0}{dz}\equiv -g \tag{2.26}$$

We may then write the total pressure and density fields as

$$p(x, y, z, t) = p_0(z) + p'(x, y, z, t)$$

$$\rho(x, y, z, t) = \rho_0(z) + \rho'(x, y, z, t) \tag{2.27}$$

where p' and ρ' are deviations from the standard values of pressure and density. For an atmosphere at rest, p' and ρ' would thus be zero. Using the definitions (2.26) and (2.27) and assuming that ρ'/ρ_0 is much less than unity in magnitude so that $(\rho_0 + \rho')^{-1} \cong \rho_0^{-1}(1 - \rho'/\rho_0)$, we find that

$$-\frac{1}{\rho}\frac{\partial p}{\partial z} - g = -\frac{1}{\rho_0 + \rho'}\frac{\partial}{\partial z}(p_0 + p') - g$$

$$\approx \frac{1}{\rho_0}\left(\frac{\rho'}{\rho_0}\frac{dp_0}{dz} - \frac{\partial p'}{\partial z}\right) = -\frac{1}{\rho_0}\left(\rho' g + \frac{\partial p'}{\partial z}\right) \tag{2.28}$$

For synoptic-scale motions, the terms in (2.28) have the magnitudes

$$\frac{1}{\rho_0}\frac{\partial p'}{\partial z} \sim \frac{\delta P}{\rho_0 H} \sim 10^{-1}\,\mathrm{m\ s^{-2}}, \qquad \frac{\rho' g}{\rho_0} \sim 10^{-1}\,\mathrm{m\ s^{-2}}$$

Comparing these with the magnitudes of other terms in the vertical momentum equation (Table 2.2), we see that to a very good approximation the perturbation pressure field is in hydrostatic equilibrium with the perturbation density field so that

$$\frac{\partial p'}{\partial z} + \rho' g = 0 \qquad (2.29)$$

Therefore, for synoptic-scale motions, vertical accelerations are negligible and the vertical velocity cannot be determined from the vertical momentum equation. However, we show in Chapter 3 that it is, nevertheless, possible to deduce the vertical motion field indirectly.

2.5 The Continuity Equation

We turn now to the second of the three fundamental conservation principles, conservation of mass. The mathematical relationship that expresses conservation of mass for a fluid is called the *continuity equation*. In this section we develop the continuity equation using two alternative methods. The first method is based on an Eulerian control volume, while the second is based on a Lagrangian control volume.

2.5.1 AN EULERIAN DERIVATION

We consider a volume element $\delta x \, \delta y \, \delta z$ that is fixed in a Cartesian coordinate frame as shown in Fig. 2.5. For such a fixed control volume the

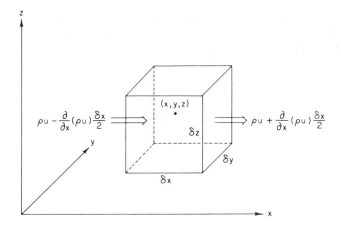

Fig. 2.5 Mass inflow into a fixed (Eulerian) control volume, owing to motion parallel to the *x* axis.

net rate of mass inflow through the sides must equal the rate of accumulation of mass within the volume. The rate of inflow of mass through the left-hand face per unit area is

$$\left[\rho u - \frac{\partial}{\partial x} (\rho u) \frac{\delta x}{2} \right]$$

whereas the rate of outflow per unit area through the right-hand face is

$$\left[\rho u + \frac{\partial}{\partial x} (\rho u) \frac{\delta x}{2} \right]$$

Since the area of each of these faces is $\delta y \, \delta z$, the net rate of flow into the volume owing to the x velocity component is

$$\left[\rho u - \frac{\partial}{\partial x} (\rho u) \frac{\delta x}{2} \right] \delta y \, \delta z - \left[\rho u + \frac{\partial}{\partial x} (\rho u) \frac{\delta x}{2} \right] \delta y \, \delta z = - \frac{\partial}{\partial x} (\rho u) \, \delta x \, \delta y \, \delta z$$

Similar expressions obviously hold for the y and z directions. Thus, the net rate of mass inflow is

$$- \left[\frac{\partial}{\partial x} (\rho u) + \frac{\partial}{\partial y} (\rho v) + \frac{\partial}{\partial z} (\rho w) \right] \delta x \, \delta y \, \delta z$$

and the mass inflow per unit volume is just $-\nabla \cdot (\rho \mathbf{U})$, which must equal the rate of mass increase per unit volume. Now the increase of mass per unit volume is just the local density change $\partial \rho / \partial t$. Therefore,

$$\frac{\partial \rho}{\partial t} + \nabla \cdot (\rho \mathbf{U}) = 0 \qquad\qquad (2.30)$$

Equation (2.30) is the mass divergence form of the continuity equation.

An alternative form of the continuity equation is obtained by applying the vector identity

$$\nabla \cdot (\rho \mathbf{U}) \equiv \rho \nabla \cdot \mathbf{U} + \mathbf{U} \cdot \nabla \rho$$

and the relationship

$$\frac{D}{Dt} \equiv \frac{\partial}{\partial t} + \mathbf{U} \cdot \nabla$$

to get

$$\frac{1}{\rho}\frac{D\rho}{Dt}+\mathbf{\nabla}\cdot\mathbf{U}=0 \qquad (2.31)$$

Equation (2.31) is the velocity divergence form of the continuity equation. It states that the fractional rate of increase of the density *following the motion* of an air parcel is equal to minus the velocity divergence. This should be clearly distinguished from (2.30), which states that the *local* rate of change of density is equal to minus the mass divergence.

2.5.2   A LAGRANGIAN DERIVATION

The physical meaning of divergence can be illustrated by the following alternative derivation of (2.31). Consider a control volume of fixed mass δM that moves with the fluid. Letting $\delta V = \delta x\, \delta y\, \delta z$ be the volume, we find that, since $\delta M = \rho\, \delta V = \rho\, \delta x\, \delta y\, \delta z$ is conserved following the motion, we can write

$$\frac{1}{\delta M}\frac{D}{Dt}(\delta M)=\frac{1}{\rho\,\delta V}\frac{D}{Dt}(\rho\delta V)=\frac{1}{\rho}\frac{D\rho}{Dt}+\frac{1}{\delta V}\frac{D}{Dt}(\delta V)=0 \qquad (2.32)$$

But

$$\frac{1}{\delta V}\frac{D}{Dt}(\delta V)=\frac{1}{\delta x}\frac{D}{Dt}(\delta x)+\frac{1}{\delta y}\frac{D}{Dt}(\delta y)+\frac{1}{\delta z}\frac{D}{Dt}(\delta z)$$

Referring to Fig. 2.6, we see that the faces of the control volume in the y, z plane (designated A and B) are advected with the flow in the x direction at the speeds $u_A = Dx/Dt$ and $u_B = D(x+\delta x)/Dt$, respectively.

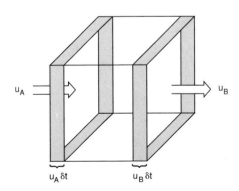

Fig. 2.6   Change in Lagrangian control volume owing to fluid motion parallel to the x axis.

Thus, the difference in speeds of the two faces is $\delta u = u_B - u_A = D(x + \delta x)/Dt - Dx/Dt$, or $\delta u = D(\delta x)/Dt$. Similarly, $\delta v = D(\delta y)/Dt$ and $\delta w = D(\delta z)/Dt$. Therefore,

$$\lim_{\delta x, \delta y, \delta z \to 0} \left[\frac{1}{\delta V} \frac{D}{Dt} (\delta V) \right] = \frac{\partial u}{\partial x} + \frac{\partial v}{\partial y} + \frac{\partial w}{\partial z} = \nabla \cdot \mathbf{U}$$

so that in the limit $\delta V \to 0$ (2.32) reduces to the continuity equation (2.31); the divergence of the three-dimensional velocity field is equal to the fractional rate of change of volume of a fluid parcel in the limit $\delta V \to 0$. It is left as a problem for the student to show that the divergence of the *horizontal* velocity field is equal to the fractional rate of change of the horizontal area δA of a fluid parcel in the limit $\delta A \to 0$.

2.5.3 SCALE ANALYSIS OF THE CONTINUITY EQUATION

Following the technique developed in Section 2.4.3 we can rewrite the continuity equation (2.31) as

$$\underbrace{\frac{1}{\rho_0} \left(\frac{\partial \rho'}{\partial t} + \mathbf{U} \cdot \nabla \rho' \right)}_{A} + \underbrace{\frac{w}{\rho_0} \frac{d\rho_0}{dz}}_{B} + \underbrace{\nabla \cdot \mathbf{U}}_{C} \approx 0 \qquad (2.33)$$

where ρ' designates the local deviation of density from its horizontally averaged value, $\rho_0(z)$. For synoptic-scale motions $\rho'/\rho_0 \sim 10^{-2}$, so using the characteristic scales given in Section 2.4 we find that term A has magnitude

$$\frac{1}{\rho_0} \left(\frac{\partial \rho'}{\partial t} + \mathbf{U} \cdot \nabla \rho' \right) \sim \frac{\rho'}{\rho_0} \frac{U}{L} \approx 10^{-7} \, \mathrm{s}^{-1}$$

For motions in which the depth scale H is comparable to the density scale height H, $d \ln \rho_0 / dz \sim H^{-1}$, so term B scales as

$$\frac{w}{\rho_0} \frac{d\rho_0}{dz} \sim \frac{W}{H} \approx 10^{-6} \, \mathrm{s}^{-1}$$

Expanding term C in Cartesian coordinates, we have

$$\nabla \cdot \mathbf{U} = \frac{\partial u}{\partial x} + \frac{\partial v}{\partial y} + \frac{\partial w}{\partial z}$$

For synoptic-scale motions the terms $\partial u/\partial x$ and $\partial v/\partial y$ tend to be of equal magnitude but opposite sign. Thus, they tend to balance so that

$$\left(\frac{\partial u}{\partial x}+\frac{\partial v}{\partial y}\right) \sim 10^{-1}\frac{U}{L} \approx 10^{-6}\,\text{s}^{-1}$$

and in addition

$$\frac{\partial w}{\partial z} \sim \frac{W}{H} \approx 10^{-6}\,\text{s}^{-1}$$

Thus, terms B and C are each an order of magnitude greater than term A, and to a first approximation terms B and C balance in the continuity equation. Thus,

$$\frac{\partial u}{\partial x}+\frac{\partial v}{\partial y}+\frac{\partial w}{\partial z}+ w\frac{d}{dz}(\ln \rho_0) = 0$$

or, alternatively in vector form

$$\nabla \cdot (\rho_0\mathbf{U}) = 0 \qquad\qquad (2.34)$$

Thus for synoptic-scale motions the mass flux computed using the basic state density ρ_0 is nondivergent. This approximation is similar to the idealization of incompressibility, which is often used in fluid mechanics. However, an *incompressible* fluid has density constant following the motion:

$$\frac{D\rho}{Dt}=0$$

Thus by (2.31) the velocity divergence vanishes ($\nabla \cdot \mathbf{U}=0$) in an incompressible fluid, which is not the same as (2.34). Our approximation (2.34) shows that for purely horizontal flow the atmosphere behaves as though it were an incompressible fluid. However, when there is vertical motion the compressibility associated with the height dependence of ρ_0 must be taken into account.

2.6   The Thermodynamic Energy Equation

We now turn to the third fundamental conservation principle, the conservation of energy, as applied to a moving fluid element. The first law of thermodynamics is usually derived by considering a system in thermo-dynamic equilibrium, i.e., a system that is initially at rest and after exchang-ing heat with its surroundings and doing work on the surroundings is again at rest. For such a system the first law states that *the change in internal energy of the system is equal to the difference between the heat added to the system and the work done by the system.* A Lagrangian control volume consisting of a specified mass of fluid may be regarded as a thermodynamic system. However, unless the fluid is at rest it will not be in thermodynamic equilibrium. Nevertheless, the first law of thermodynamics still applies. To show that this is the case, we note that the total thermodynamic energy of the control volume is considered to consist of the sum of the internal energy (owing to the kinetic energy of the individual molecules) and the kinetic energy owing to the macroscopic motion of the fluid. The rate of change of this total thermodynamic energy is equal to the rate of diabatic heating plus the rate at which work is done on the fluid parcel by external forces.

If we let e designate the internal energy per unit mass, then the total thermodynamic energy contained in a Lagrangian fluid element of density ρ and volume δV is $\rho[e + (1/2)\mathbf{U} \cdot \mathbf{U}] \, \delta V$. The external forces that act on a fluid element may be divided into surface forces, such as pressure and viscosity, and body forces, such as gravity or the Coriolis force. The rate at which work is done on the fluid element by the x component of the pressure force is illustrated in Fig. 2.7. Recalling that pressure is a force per unit area and that the rate at which a force does work is given by the dot product of the force and velocity vectors, we see that the rate at which the surrounding fluid does work on the element owing to the pressure force on the two boundary surfaces in the y, z plane is given by

$$(pu)_A \, \delta y \, \delta z - (pu)_B \, \delta y \, \delta z$$

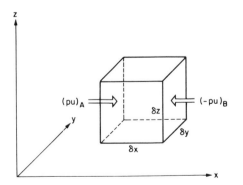

Fig. 2.7   Rate of working on a fluid element owing to the x component of the pressure force.

(The negative sign is needed before the second term because the work done on the fluid element is positive if u is negative across face B.) Now by expanding in a Taylor series we can write

$$(pu)_B = (pu)_A + \left[\frac{\partial}{\partial x}(pu)\right]_A \delta x + \cdots$$

Thus the net rate at which the pressure force does work owing to the x component of motion is

$$[(pu)_A - (pu)_B]\,\delta y\,\delta z = -\left[\frac{\partial}{\partial x}(pu)\right]_A \delta V$$

where $\delta V = \delta x\,\delta y\,\delta z$.

Similarly, we can show that the net rates at which the pressure force does work owing to the y and z components of motion are

$$-\left[\frac{\partial}{\partial y}(pv)\right]\delta V \quad \text{and} \quad -\left[\frac{\partial}{\partial z}(pw)\right]\delta V$$

respectively. Hence, the total rate at which work is done by the pressure force is simply

$$-\boldsymbol{\nabla} \cdot (p\mathbf{U})\,\delta V$$

The only body forces of meteorological significance that act on an element of mass in the atmosphere are the Coriolis force and gravity. However, since the Coriolis force, $-2\boldsymbol{\Omega} \times \mathbf{U}$, is perpendicular to the velocity vector it can do no work. Thus the rate at which body forces do work on the mass element is just $\rho\mathbf{g} \cdot \mathbf{U}\,\delta V$.

Applying the principle of energy conservation to our Lagrangian control volume (neglecting effects of molecular viscosity), we thus obtain

$$\frac{D}{Dt}\left[\rho\left(e + \frac{1}{2}\mathbf{U} \cdot \mathbf{U}\right)\delta V\right] = -\boldsymbol{\nabla} \cdot (p\mathbf{U})\,\delta V + \rho\mathbf{g} \cdot \mathbf{U}\,\delta V + \rho J\,\delta V$$

(2.35)

Here J is the rate of heating per unit mass owing to radiation, conduction, and latent heat release. With the aid of the chain rule of differentiation we can rewrite (2.35) as

$$\rho\,\delta V\frac{D}{Dt}\left(e + \frac{1}{2}\mathbf{U} \cdot \mathbf{U}\right) + \left(e + \frac{1}{2}\mathbf{U} \cdot \mathbf{U}\right)\frac{D(\rho\,\delta V)}{Dt}$$

$$= -\mathbf{U} \cdot \boldsymbol{\nabla}p\,\delta V - p\boldsymbol{\nabla} \cdot \mathbf{U}\,\delta V - \rho g w\,\delta V + \rho J\,\delta V \qquad (2.36)$$

where we have used $\mathbf{g} = -g\mathbf{k}$. Now from (2.32) the second term on the left in (2.36) vanishes so that

$$\rho \frac{De}{Dt} + \rho \frac{D}{Dt}\left(\frac{1}{2}\mathbf{U}\cdot\mathbf{U}\right) = -\mathbf{U}\cdot\nabla p - p\nabla\cdot\mathbf{U} - \rho g w + \rho J \qquad (2.37)$$

This equation can be simplified by noting that if we take the dot product of $\mathbf{U}$ with the momentum equation (2.8) we obtain (neglecting friction)

$$\rho \frac{D}{Dt}\left(\frac{1}{2}\mathbf{U}\cdot\mathbf{U}\right) = -\mathbf{U}\cdot\nabla p - \rho g w \qquad (2.38)$$

Subtracting (2.38) from (2.37) we obtain

$$\rho \frac{De}{Dt} = -p\nabla\cdot\mathbf{U} + \rho J \qquad (2.39)$$

The terms in (2.37) that were eliminated by subtracting (2.38) represent the balance of mechanical energy owing to the motion of the fluid element; the remaining terms represent the thermal energy balance.

Using the definition of geopotential (1.15) we have

$$gw = g\frac{Dz}{Dt} = \frac{D\Phi}{Dt}$$

so that (2.38) can be rewritten as

$$\rho \frac{D}{Dt}\left(\frac{1}{2}\mathbf{U}\cdot\mathbf{U} + \Phi\right) = -\mathbf{U}\cdot\nabla p \qquad (2.40)$$

which is referred to as the *mechanical energy equation*. The sum of the kinetic energy plus the gravitational potential energy is called the *mechanical energy*. Thus (2.40) states that following the motion, the rate of change of mechanical energy per unit volume equals the rate at which work is done by the pressure gradient force.

The thermal energy equation (2.39) can be written in more familiar form by noting from (2.31) that

$$\frac{1}{\rho}\nabla\cdot\mathbf{U} = -\frac{1}{\rho^2}\frac{D\rho}{Dt} = \frac{D\alpha}{Dt}$$

and that for dry air the internal energy per unit mass is given by $e = c_v T$, where c_v ($=717$ J kg^{-1} K^{-1}) is the specific heat at constant volume. We then obtain

$$c_v \frac{DT}{Dt} + p \frac{D\alpha}{Dt} = J \tag{2.41}$$

which is the usual form of the thermodynamic energy equation. Thus the first law of thermodynamics indeed is applicable to a fluid in motion. The second term on the left, representing the rate of working by the fluid system (per unit mass), represents a conversion between thermal and mechanical energy. It is this conversion process that enables the solar heat energy to drive the motions of the atmosphere.

2.7 Thermodynamics of the Dry Atmosphere

Taking the total derivative of the equation of state (1.14) we obtain

$$p \frac{D\alpha}{Dt} + \alpha \frac{Dp}{Dt} = R \frac{DT}{Dt}$$

Substituting for $p \, D\alpha/Dt$ in (2.41) and using $c_p = c_v + R$, where c_p ($=1004$ J kg^{-1} K^{-1}) is the specific heat at constant pressure, we can rewrite the first law of thermodynamics as

$$c_p \frac{DT}{Dt} - \alpha \frac{Dp}{Dt} = J \tag{2.42}$$

Dividing through by T and again the equation of state, we obtain the entropy form of the first law of thermodynamics:

$$c_p \frac{D \ln T}{Dt} - R \frac{D \ln p}{Dt} = \frac{J}{T} \equiv \frac{Ds}{Dt} \tag{2.43}$$

Equation (2.43) gives the rate of change of entropy per unit mass following the motion for a thermodynamically *reversible* process. A reversible process is one in which a system changes its thermodynamic state and then returns to the original state without changing its surroundings. For such a process the entropy s defined by (2.43) is a field variable that depends only on the state of the fluid. Thus Ds is a perfect differential, and Ds/Dt is to be regarded as a total derivative. However, "heat" is not a field variable, so that the heating rate J is not a total derivative.[2]

2.7.1 POTENTIAL TEMPERATURE

For an ideal gas undergoing an *adiabatic* process (i.e., a reversible process in which no heat is exchanged with the surroundings) the first law of thermodynamics can be written in the form

$$c_p \, D \ln T - R \, D \ln p = 0$$

Integrating this expression from a state at pressure p and temperature T to a state in which the pressure is p_s and the temperature is θ, we obtain after taking the antilogarithm

$$\theta = T(p_s/p)^{R/c_p} \tag{2.44}$$

This relationship is referred to as Poisson's equation, and the temperature θ defined by (2.44) is called the *potential temperature*. θ is simply the temperature that a parcel of dry air at pressure p and temperature T would have if it were expanded or compressed adiabatically to a standard pressure p_s [usually taken to be 100 kPa (1000 mb)]. Thus, every air parcel has a unique value of potential temperature, and this value is conserved for dry adiabatic motion. Since synoptic-scale motions are approximately adiabatic outside regions of active precipitation, θ is a quasi-conserved quantity for such motions.

Taking the logarithm of (2.44) and differentiating, we find that

$$c_p \frac{D \ln \theta}{Dt} = c_p \frac{D \ln T}{Dt} - R \frac{D \ln p}{Dt} \tag{2.45}$$

Comparing (2.43) and (2.45) we obtain

$$c_p \frac{D \ln \theta}{Dt} = \frac{Ds}{Dt} \tag{2.46}$$

Thus, for reversible processes, fractional potential temperature changes are indeed proportional to entropy changes. A parcel that conserves entropy following the motion must move along an isentropic (constant θ) surface.

---

[2] For a discussion of entropy and its role in the second law of thermodynamics see Wallace and Hobbs (1977), for example.

2.7.2 THE ADIABATIC LAPSE RATE

A relationship between the *lapse rate* of temperature (i.e., the rate of *decrease* of temperature with respect to height) and the rate of change of potential temperature with respect to height can be obtained by taking the logarithm of (2.44) and differentiating with respect to height. Using the hydrostatic equation and the ideal gas law to simplify the result gives

$$\frac{T}{\theta}\frac{\partial \theta}{\partial z} = \frac{\partial T}{\partial z} + \frac{g}{c_p} \tag{2.47}$$

For an atmosphere in which the potential temperature is constant with respect to height the lapse rate is thus

$$-\frac{dT}{dz} = \frac{g}{c_p} \equiv \Gamma_d \tag{2.48}$$

Hence, the dry adiabatic lapse rate is approximately constant throughout the lower atmosphere.

2.7.3 STATIC STABILITY

If potential temperature is a function of height, the atmospheric lapse rate, $\Gamma \equiv -\partial T/\partial z$, will differ from the adiabatic lapse rate and

$$\frac{T}{\theta}\frac{\partial \theta}{\partial z} = \Gamma_d - \Gamma \tag{2.49}$$

If $\Gamma < \Gamma_d$ so that θ increases with height, an air parcel that undergoes an adiabatic displacement from its equilibrium level will be positively (negatively) buoyant when displaced vertically downward (upward) so that it will tend to return to its equilibrium level and the atmosphere is said to be statically stable or *stably stratified*.

Adiabatic oscillations of a fluid parcel about its equilibrium level in a stably stratified atmosphere are referred to as *buoyancy oscillations*. The characteristic frequency of such oscillations can be derived by considering a parcel that is displaced vertically small distance δz without disturbing its environment. If the environment is in hydrostatic balance, $\rho_0 g = -dp_0/dz$, where p_0 and ρ_0 are the pressure and density of the environment. The vertical acceleration of the parcel is

$$\frac{Dw}{Dt} = \frac{D^2}{Dt^2}(\delta z) = -g - \frac{1}{\rho}\frac{\partial p}{\partial z} \tag{2.50}$$

where p and ρ are the pressure and density of the parcel. In the parcel method it is assumed that the pressure of the parcel instantaneously adjusts to the environmental pressure during the displacement: $p = p_0$. This condition must be true if the parcel is to leave the environment undisturbed. Thus with the aid of the hydrostatic relationship pressure can be eliminated in (2.50) to give

$$\frac{D^2}{Dt^2}(\delta z) = g\left(\frac{\rho_0 - \rho}{\rho}\right) = g\frac{\theta}{\theta_0} \tag{2.51}$$

where (2.44) and the ideal gas law have been used to express the buoyancy force in terms of potential temperature. Here θ designates the deviation of the potential temperature of the parcel from its basic state (environmental) value $\theta_0(z)$. If the parcel is initially at level $z = 0$ where the potential temperature is $\theta_0(0)$, then for a small displacement δz we can represent the environmental potential temperature as

$$\theta_0(\delta z) \approx \theta_0(0) + (d\theta_0/dz)\,\delta z$$

If the parcel displacement is adiabatic, the potential temperature of the parcel is conserved. Thus, $\theta(\delta z) = \theta_0(0) - \theta_0(\delta z) = -(d\theta_0/dz)\,\delta z$, and (2.51) becomes

$$\frac{D^2}{Dt^2}(\delta z) = -N^2\,\delta z \tag{2.52}$$

where

$$N^2 = g\frac{d\ln\theta_0}{dz}$$

is a measure of the static stability of the environment. Equation (2.52) has a general solution of the form $\delta z = A\exp(iNt)$. Therefore, if $N^2 > 0$ the parcel will oscillate about its initial level with a period $\tau = 2\pi/N$. The corresponding frequency N is the *buoyancy frequency*.[3] For average tropospheric conditions $N \approx 1.2 \times 10^{-2}\,\mathrm{s}^{-1}$, so that the period of a buoyancy oscillation is about 8 min. In the case of $N = 0$, examination of (2.52) indicates that no accelerating force will exist and the parcel will be in neutral equilibrium at its new level. On the other hand, if $N^2 < 0$ (potential

---

[3] N is often referred to as the Brunt–Väisälä frequency.

temperature decreasing with height) the displacement will increase exponentially in time. We thus arrive at the familiar gravitational or static stability criteria for dry air:

$$d\theta_0/dz > 0 \quad \text{statically stable}$$

$$d\theta_0/dz = 0 \quad \text{statically neutral}$$

$$d\theta_0/dz < 0 \quad \text{statically unstable}$$

On the synoptic scale the atmosphere is always stably stratified because any unstable regions that develop are quickly stabilized by convective overturning. For a moist atmosphere, the situation is more complicated and discussion of that situation will be deferred until Chapter 9.

2.7.4 SCALE ANALYSIS OF THE THERMODYNAMIC ENERGY EQUATION

If potential temperature is divided into a basic state $\theta_0(z)$ and a deviation $\theta(x, y, z, t)$ so that the total potential temperature at any point is given by $\theta_{tot} \equiv \theta_0(z) + \theta(x, y, z, t)$, the first law of thermodynamics (2.46) can be written approximately as

$$\frac{1}{\theta_0}\left(\frac{\partial\theta}{\partial t} + \mathbf{U}\cdot\nabla\theta\right) + w\frac{d\ln\theta_0}{dz} = \frac{J}{c_p T} \tag{2.53}$$

where we have used the fact that for $|\theta/\theta_0| \ll 1$

$$\ln\theta_{tot} \approx \ln\theta_0 + \theta/\theta_0$$

Outside regions of active precipitation, the diabatic heating is due primarily to net radiative heating. In the troposphere radiative heating is quite weak so that typically $J/c_p \leq 1°C\ d^{-1}$ (except near cloud tops, where substantially larger cooling can occur owing to thermal emission by the cloud particles). The typical amplitude of horizontal potential temperature fluctuations in a midlatitude synoptic system (above the boundary layer) is $\theta \sim 4°C$. Thus,

$$\frac{T}{\theta_0}\left(\frac{\partial\theta}{\partial t} + \mathbf{U}\cdot\nabla\theta\right) \sim \frac{\theta U}{L} \sim 4°C\ d^{-1}$$

The cooling owing to vertical advection of the basic state potential temperature (usually called the adiabatic cooling) has a typical magnitude of

$$w\left(\frac{T}{\theta_0}\frac{d\theta_0}{dz}\right) = w(\Gamma_d - \Gamma) \sim 4°C\ d^{-1}$$

where $w \sim 1\ cm\ s^{-1}$ and $\Gamma_d - \Gamma$, the difference between the dry adiabatic and actual lapse rates, is $\sim 4°C\ km^{-1}$.

Thus, in the absence of strong diabatic heating, the rate of change of the perturbation potential temperature is equal to the adiabatic heating or cooling owing to vertical motion in the statically stable basic state, and (2.53) can be approximated as

$$\frac{D\theta}{Dt} + w\frac{d\theta_0}{dz} \approx 0 \tag{2.54}$$

Problems

2.1. A ship is steaming northward at a rate of $10\ km\ h^{-1}$. The surface pressure increases towards the northwest at the rate of $5\ Pa\ km^{-1}$. What is the pressure tendency recorded at a nearby island station if the pressure aboard the ship decreases at a rate of $100\ Pa/3\ h$?

2.2. The temperature at a point 50 km north of a station is 3°C cooler than at the station. If the wind is blowing from the northeast at $20\ m\ s^{-1}$ and the air is being heated by radiation at the rate of $1°C\ h^{-1}$, what is the local temperature change at the station?

2.3. Derive the relationship

$$\mathbf{\Omega} \times (\mathbf{\Omega} \times \mathbf{r}) = -\Omega^2 \mathbf{R}$$

which was used in Eq. (2.7).

2.4. Derive the expression given in Eq. (2.13) for the rate of change of $\mathbf{k}$ following the motion.

2.5. Compare the magnitudes of the curvature term $u^2 \tan \phi/a$ and the Coriolis force for a ballistic missile fired eastward with a velocity of $1000\ m\ s^{-1}$ at 45° latitude. If the missile travels 1000 km by how much is it deflected from its eastward path owing to both these terms? Can the curvature term be neglected in this case?

2.6. Suppose a 1-kg parcel of dry air is rising at a constant vertical velocity. If the parcel is being heated by radiation at the rate of $10^{-1}\ W\ kg^{-1}$, what must the speed of rise be to maintain the parcel at a constant temperature?

2.7. Derive an expression for the density ρ that results when an air parcel initially at pressure p_s and density ρ_s expands adiabatically to pressure p.

2.8. An air parcel that has a temperature of 20°C at the 1000-mb level is lifted dry adiabatically. What is its density when it reaches the 500-mb level?

2.9. Suppose an air parcel starts from rest at the 80-kPa (800-mb) level and rises vertically to 50 kPa (500 mb) while maintaining a constant 1°C temperature excess over the environment. Assuming that the mean temperature of the 80–50-kPa layer is 260 K, compute the energy released owing to the work of the buoyancy force. Assuming that all the released energy is realized as kinetic energy of the parcel, what will the vertical velocity of the parcel be at 50 kPa?

2.10. Show that for an atmosphere with an adiabatic lapse rate (i.e., constant potential temperature) the geopotential height is given by

$$Z = H_\theta \left[1 - \left(\frac{p}{p_0} \right)^{R/c_p} \right]$$

where p_0 is the pressure at $Z = 0$ and $H_\theta \equiv c_p \theta / g_0$ is the total geopotential height of the atmosphere.

2.11. In the *isentropic* coordinate system (see Section 4.6) potential temperature is used as the vertical coordinate. Since in adiabatic flow potential temperature is conserved following the motion, isentropic coordinates are useful for tracing the actual paths of travel of individual air parcels. Show that the transformation of the horizontal pressure gradient force from z to θ coordinates is given by

$$\frac{1}{\rho} \nabla_z p = \nabla_\theta M$$

where $M \equiv c_p T + \Phi$ is the *Montgomery streamfunction*.

Suggested References

Brown, *Fluid Mechanics of the Atmosphere*, contains a thorough development of the basic conservation laws at the graduate level.

Haltiner and Williams, *Numerical Weather Prediction and Dynamic Meteorology*, contains a more complete discussion of scale analysis for synoptic scale motions.

Iribarne and Godson, *Atmospheric Thermodynamics*, contains a thorough discussion of the thermodynamics of both dry and moist atmospheres.

3 Elementary Applications of the Basic Equations

In addition to the geostrophic wind, which was discussed in Chapter 2, there are other approximate expressions for the relationships among the velocity, pressure, and temperature fields, which are useful in the analysis of weather systems. These are most conveniently discussed using a coordinate system in which pressure is the vertical coordinate. Thus, before introducing the elementary applications of the present chapter it is useful to present the dynamical equations in isobaric coordinates.

3.1 The Basic Equations in Isobaric Coordinates

3.1.1 THE HORIZONTAL MOMENTUM EQUATION

The approximate horizontal momentum equations (2.24) and (2.25) may be written in vectorial form as

$$\frac{D\mathbf{V}}{Dt} + f\mathbf{k} \times \mathbf{V} = -\frac{1}{\rho}\boldsymbol{\nabla}p \tag{3.1}$$

where $\mathbf{V} = \mathbf{i}u + \mathbf{j}v$ is the *horizontal* velocity vector. In order to express (3.1) in isobaric coordinate form we transform the pressure gradient force using (1.20) and (1.21) to obtain

$$\frac{D\mathbf{V}}{Dt} + f\mathbf{k} \times \mathbf{V} = -\nabla_p \Phi \tag{3.2}$$

where ∇_p is the horizontal gradient operator applied with pressure held constant.

Since p is the independent vertical coordinate we must expand the total derivative as follows:

$$\frac{D}{Dt} \equiv \frac{\partial}{\partial t} + \frac{Dx}{Dt}\frac{\partial}{\partial x} + \frac{Dy}{Dt}\frac{\partial}{\partial y} + \frac{Dp}{Dt}\frac{\partial}{\partial p}$$

$$= \frac{\partial}{\partial t} + u\frac{\partial}{\partial x} + v\frac{\partial}{\partial y} + \omega\frac{\partial}{\partial p} \tag{3.3}$$

Here $\omega \equiv Dp/Dt$ (usually called the "omega" vertical motion) is the pressure change following the motion, which plays the same role in the isobaric coordinate system that $w \equiv Dz/Dt$ plays in height coordinates.

From (3.2) we see that the isobaric coordinate form of the geostrophic relationship is

$$f\mathbf{V}_g = \mathbf{k} \times \nabla_p \Phi \tag{3.4}$$

One advantage of isobaric coordinates is easily seen by comparing (2.23) and (3.4). In the latter equation density does not appear. Thus, a given geopotential gradient implies the same geostrophic wind at any height, whereas a given horizontal pressure gradient implies different values of the geostrophic wind depending on the density. Furthermore, if f is regarded as a constant, the horizontal divergence of the geostrophic wind at constant pressure is zero:

$$\nabla_p \cdot \mathbf{V}_g = 0$$

3.1.2 THE CONTINUITY EQUATION

It is possible to transform the continuity equation (2.31) from height coordinates to pressure coordinates. However, it is simpler to directly derive the isobaric form by considering a Lagrangian control volume $\delta V = \delta x\, \delta y\, \delta z$ and applying the hydrostatic equation $\delta p = -\rho g\, \delta z$ (note that $\delta p < 0$) to express the volume element as $\delta V = -\delta x \, \delta y \, \delta p/(\rho g)$. The mass of this fluid element, which is conserved following the motion, is then $\delta M = \rho \, \delta V = -\delta x \, \delta y \, \delta p/g$:

$$\frac{1}{\delta M} \frac{D}{Dt} (\delta M) = \frac{g}{\delta x \, \delta y \, \delta p} \frac{D}{Dt} \left(\frac{\delta x \, \delta y \, \delta p}{g} \right) = 0$$

After differentiating, using the chain rule, and changing the order of the differential operators we obtain[1]

$$\frac{1}{\delta x} \delta \left(\frac{Dx}{Dt} \right) + \frac{1}{\delta y} \delta \left(\frac{Dy}{Dt} \right) + \frac{1}{\delta p} \delta \left(\frac{Dp}{Dt} \right) = 0$$

or

$$\frac{\delta u}{\delta x} + \frac{\delta v}{\delta y} + \frac{\delta \omega}{\delta p} = 0$$

Taking the limit $\delta x, \delta y, \delta p \to 0$, and observing that δx and δy are evaluated at constant pressure, we obtain the continuity equation in the isobaric system:

$$\left(\frac{\partial u}{\partial x} + \frac{\partial v}{\partial y} \right)_p + \frac{\partial \omega}{\partial p} = 0 \tag{3.5}$$

This form of the continuity equation contains no reference to the density field and does not involve time derivatives. The simplicity of (3.5) is one of the chief advantages of the isobaric coordinate system.

3.1.3 THE THERMODYNAMIC ENERGY EQUATION

The first law of thermodynamics (2.42) can be expressed in the isobaric system by letting $Dp/Dt = \omega$ and expanding DT/Dt by using (3.3):

$$c_p \left(\frac{\partial T}{\partial t} + u \frac{\partial T}{\partial x} + v \frac{\partial T}{\partial y} + \omega \frac{\partial T}{\partial p} \right) - \alpha \omega = J$$

This may be rewritten as

$$\left(\frac{\partial T}{\partial t} + u \frac{\partial T}{\partial x} + v \frac{\partial T}{\partial y} \right) - S_p \omega = \frac{J}{c_p} \tag{3.6}$$

[1] From now on g will be regarded as a constant.

where, with the aid of the equation of state and Poisson's equation (2.44), we have

$$S_p \equiv \frac{RT}{c_p p} - \frac{\partial T}{\partial p} = -\frac{T}{\theta} \frac{\partial \theta}{\partial p} \qquad (3.7)$$

which is the static stability parameter for the isobaric system. Using (2.49) and the hydrostatic equation, (3.7) may be rewritten as

$$S_p = (\Gamma_d - \Gamma)/\rho g$$

Thus, S_p is positive provided that the lapse rate is less than dry adiabatic. However, since density decreases approximately exponentially with height, S_p increases rapidly with height. This strong height dependence of the stability measure S_p is a minor disadvantage of isobaric coordinates.

3.2 Balanced Flow

Despite the apparent complexity of atmospheric motion systems as depicted on synoptic weather charts, the pressure (or geopotential height) and velocity distributions in meteorological disturbances are actually related by rather simple approximate force balances. In order to gain a qualitative understanding of the horizontal balance of forces in atmospheric motions we idealize by considering flows that are steady state (i.e., time independent) and have no vertical component of velocity. Furthermore it is useful to describe the flow field by expanding the isobaric form of the horizontal momentum equation (3.2) into its components in a so-called *natural* coordinate system.

3.2.1 NATURAL COORDINATES

The natural coordinate system is defined by the orthogonal set of unit vectors $\mathbf{t}$, $\mathbf{n}$, and $\mathbf{k}$. Unit vector $\mathbf{t}$ is oriented parallel to the horizontal velocity at each point; unit vector $\mathbf{n}$ is normal to the horizontal velocity and is oriented so that it is positive to the left of the flow direction; unit vector $\mathbf{k}$ is directed vertically upward. In this system the horizontal velocity may be written $\mathbf{V} = V\mathbf{t}$, where V, the horizontal speed, is a nonnegative scalar defined by $V \equiv Ds/Dt$, where $s(x, y, t)$ is the curve followed by a parcel moving in the horizontal plane. The acceleration following the motion is thus

$$\frac{D\mathbf{V}}{Dt} = \mathbf{t}\frac{DV}{Dt} + V\frac{D\mathbf{t}}{Dt}$$

The rate of change of **t** following the motion may be derived from geometrical considerations with the aid of Fig. 3.1:

$$\delta\psi = \frac{\delta s}{R} = \frac{|\delta\mathbf{t}|}{|\mathbf{t}|} = |\delta\mathbf{t}|$$

Here R is the *radius of curvature* following the parcel motion, and we have used the fact that $|\mathbf{t}| = 1$. By convention R is taken to be positive when the center of curvature is in the positive **n** direction. Thus, for $R > 0$ the air parcels turn toward the left following the motion, and for $R < 0$ the air parcels turn toward the right following the motion.

Noting that in the limit $\delta s \to 0$, $\delta\mathbf{t}$ is directed parallel to **n**, the above relationship yields $D\mathbf{t}/Ds = \mathbf{n}/R$. Thus,

$$\frac{D\mathbf{t}}{Dt} = \frac{D\mathbf{t}}{Ds}\frac{Ds}{Dt} = \frac{\mathbf{n}}{R} V$$

and

$$\frac{D\mathbf{V}}{Dt} = \mathbf{t}\frac{DV}{Dt} + \mathbf{n}\frac{V^2}{R} \tag{3.8}$$

Therefore, the acceleration following the motion is the sum of the rate of change of speed of the air parcel and its centripetal acceleration due to the

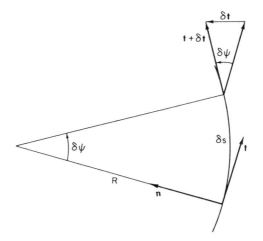

Fig. 3.1 Rate of change of the unit tangent vector **t** following the motion.

curvature of the trajectory. Since the Coriolis force always acts normal to the direction of motion, its natural coordinate form is simply

$$-f\mathbf{k} \times \mathbf{V} = -fV\mathbf{n}$$

while the pressure gradient force can be expressed as

$$-\nabla_p\Phi = -\left(\mathbf{t}\frac{\partial\Phi}{\partial s} + \mathbf{n}\frac{\partial\Phi}{\partial n}\right)$$

The horizontal momentum equation may thus be expanded into the following component equations in the natural coordinate system:

$$\frac{DV}{Dt} = -\frac{\partial\Phi}{\partial s} \tag{3.9}$$

$$\frac{V^2}{R} + fV = -\frac{\partial\Phi}{\partial n} \tag{3.10}$$

Equations (3.9) and (3.10) express the force balances parallel to and normal to the direction of flow, respectively. For motion parallel to the geopotential height contours, $\partial\Phi/\partial s = 0$ and the speed is constant following the motion. If, in addition, the geopotential gradient normal to the direction of motion is constant along a trajectory, (3.10) implies that the radius of curvature of the trajectory is also constant. In that case the flow can be classified into several simple categories depending on the relative contributions of the three terms in (3.10) to the net force balance.

3.2.2 GEOSTROPHIC FLOW

Flow in a straight line ($R \to \pm\infty$) parallel to the height contours is referred to as *geostrophic motion*. In geostrophic motion the horizontal components of the Coriolis force and pressure gradient force are in exact balance so that $V = V_g$, where the geostrophic wind V_g is defined by[2]

$$fV_g = -\partial\Phi/\partial n \tag{3.11}$$

This balance is indicated schematically in Fig. 3.2. The actual wind can be in exact geostrophic motion only if the height contours are parallel to latitude circles. As discussed in Section 2.4.1, the geostrophic wind is generally a good approximation to the actual wind in extratropical synoptic-scale disturbances. However, in some of the special cases to be treated below this is not true.

[2] Note that although the actual speed V must always be positive in the natural coordinates, V_g, which is proportional to the height gradient normal to the direction of flow may be negative, as in the "anomalous" low shown in Fig. 3.5c.

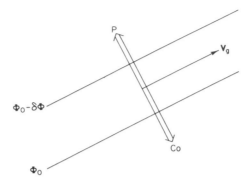

Fig. 3.2 Balance of forces for geostrophic equilibrium. The pressure gradient force is designated by P and the Coriolis force by Co.

3.2.3 INERTIAL FLOW

If the geopotential field is uniform on an isobaric surface so that the horizontal pressure gradient vanishes, (3.10) reduces to a balance between the Coriolis force and centrifugal force:

$$V^2/R + fV = 0 \qquad (3.12)$$

Equation (3.12) may be solved for the radius of curvature

$$R = -V/f$$

Since from (3.9), the speed must be constant in this case, the radius of curvature is also constant (neglecting the latitudinal dependence of f). Thus the air parcels follow circular paths in an anticyclonic sense.[3] The period of this oscillation is

$$P = \left| \frac{2\pi R}{V} \right| = \frac{2\pi}{|f|} = \frac{\frac{1}{2}\,\text{day}}{|\sin \phi|} \qquad (3.13)$$

P is equivalent to the time that is required for a Foucault pendulum to turn through an angle of 180°. Hence, it is often referred to as one-half *pendulum day*.

---

[3] Anticyclonic flow is a clockwise rotation in the Northern Hemisphere and counterclockwise in the Southern Hemisphere. Cyclonic flow has the opposite sense of rotation in each hemisphere.

Since both the Coriolis force and the centrifugal force owing to the relative motion are caused by the inertia of the fluid, this type of motion is referred to as an *inertial oscillation,* and the circle of radius $|R|$ is called the inertia circle. It is important to realize, however, that the "inertial flow" governed by (3.12) is not the same as inertial motion in an absolute reference frame. In the former, the force of gravity, acting orthogonal to the plane of motion, keeps the oscillation on a horizontal surface. In the latter, all forces vanish and the motion maintains a uniform absolute velocity.

In the atmosphere, motions are nearly always generated and maintained by pressure gradient forces; the conditions of uniform pressure required for pure inertial flow rarely exist. However, in the oceans, currents are often generated by transient winds blowing across the surface, rather than by internal pressure gradients. As a result, significant amounts of energy occur in currents that oscillate with near inertial periods. An example recorded by a current meter near the island of Barbados is shown in Fig. 3.3.

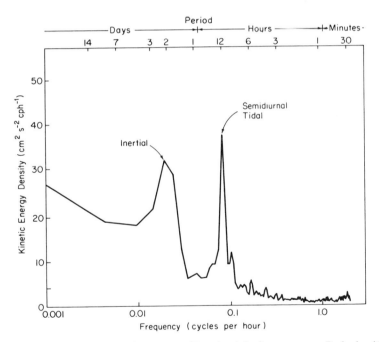

Fig. 3.3 Power spectrum of kinetic energy at 30-m depth in the ocean near Barbados (13°N). Ordinate shows kinetic energy density per unit frequency interval (cph^{-1} designates cycles per hour). This type of plot indicates the manner in which the total kinetic energy is partitioned among oscillations of different periods. Note the strong peak at 53 h, which is the period of an inertial oscillation at 13° latitude. (After Warsh *et al.,* 1971. Reproduced with permission of the American Meteorological Society.)

3.2.4 CYCLOSTROPHIC FLOW

If the horizontal scale of a disturbance is small enough, the Coriolis force may be neglected in (3.10) compared to the pressure gradient force and the centrifugal force. The force balance normal to the direction of flow is then

$$\frac{V^2}{R} = -\frac{\partial \Phi}{\partial n}$$

If this equation is solved for V, we obtain the speed of the *cyclostrophic wind*

$$V = \left(-R\frac{\partial \Phi}{\partial n}\right)^{1/2} \tag{3.14}$$

As indicated in Fig. 3.4, the cyclostrophic flow may be either cyclonic or anticyclonic. In both cases the pressure gradient force is directed toward the center of curvature and the centrifugal force away from the center of curvature.

The cyclostrophic balance approximation is valid provided that the ratio of the centrifugal force to the Coriolis force is large. This ratio V/fR is equivalent to the Rossby number discussed in Section 2.4.2. As an example of cyclostrophic scale motion we consider a typical tornado. Suppose that

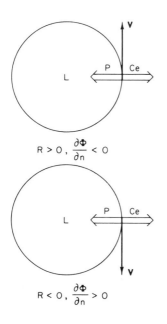

Fig. 3.4 Force balance in cyclostrophic flow; P designates the pressure gradient, Ce the centrifugal force.

the tangential velocity is 30 m s^{-1} at a distance of 300 m from the center of the vortex. Assuming that $f = 10^{-4}$ s^{-1}, we obtain a Rossby number of Ro $= V/|fR| \approx 10^3$, which implies that the Coriolis force can be neglected in computing the balance of forces for a tornado. However, the majority of tornadoes in the Northern Hemisphere are observed to rotate in a cyclonic (counterclockwise) sense. This is apparently because they are imbedded in environments that favor cyclonic rotation (see Section 9.6.1). Smaller-scale vortices, on the other hand, such as dust devils and water spouts, do not have a preferred direction of rotation. According to data collected by Sinclair (1965), they are observed to be anticyclonic as often as cyclonic.

3.2.5 THE GRADIENT WIND APPROXIMATION

Horizontal frictionless flow that is parallel to the height contours so that the tangential acceleration vanishes ($DV/Dt = 0$) is called *gradient flow*. Gradient flow is a three-way balance between the Coriolis force, the centrifugal force, and the horizontal pressure gradient force. Like geostrophic flow, pure gradient flow can exist only under very special circumstances. It is always possible, however, to define a gradient wind, which at any point is just the wind component parallel to the height contours that satisfies (3.10). For this reason (3.10) is commonly referred to as the gradient wind equation. Because (3.10) takes account of the centrifugal force owing to the curvature of parcel trajectories, the gradient wind is often a better approximation to the actual wind than is the geostrophic wind.

The gradient wind speed is obtained by solving (3.10) for V to yield

$$V = -\frac{fR}{2} \pm \left(\frac{f^2 R^2}{4} - R \frac{\partial \Phi}{\partial n} \right)^{1/2} \tag{3.15}$$

Not all the mathematically possible roots of (3.15) correspond to physically possible solutions since it is required that V be real and nonnegative. In Table 3.1 the various roots of (3.15) are classified according to the signs of R and $\partial \Phi / \partial n$ in order to isolate the physically meaningful solutions.

The force balances for the four permitted solutions are illustrated in Fig. 3.5. Equation (3.15) shows that in the cases of both the regular and anomalous highs the pressure gradient is limited by the requirement that the quantity under the radical be nonnegative; that is,

$$\left| \frac{\partial \Phi}{\partial n} \right| < \frac{|R| f^2}{4} \tag{3.16}$$

Table 3.1   *Classification of Roots of the Gradient Wind Equation in the Northern Hemisphere[a]*

Sign $\partial\Phi/\partial n$	$R > 0$	$R < 0$
Positive	Positive root: unphysical	Positive root: antibaric flow (anomalous low)
	Negative root: unphysical	Negative root: unphysical
Negative	Positive root: cyclonic flow (regular low)	Positive root: ($V > -fR/2$): anticyclonic flow (anomalous high)
	Negative root: unphysical	Negative root: ($V < -fR/2$): anticyclonic flow (regular high)

[a] The terms *positive root* and *negative root* in columns 2 and 3 refer to the sign taken in the final term in (3.15).

Thus, the pressure gradient in a high must approach zero as $|R| \rightarrow 0$. It is for this reason that the pressure field near the center of a high is always flat and the wind gentle compared to the region near the center of a low.

The absolute angular momentum about the axis of rotation for the circularly symmetric motions shown in Fig. 3.5 is given by $VR + fR^2/2$.

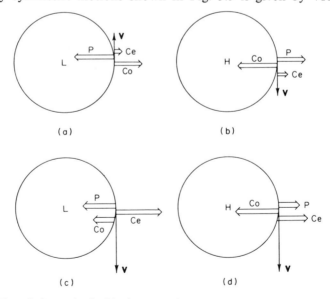

Fig. 3.5   Force balances in the Northern Hemisphere for the four types of gradient flow: (a) regular low; (b) regular high; (c) anomalous low; (d) anomalous high. *P* designates the pressure gradient, *Ce* the centrifugal force, and *Co* the Coriolis force.

From (3.15) it is readily verified that the regular gradient wind balances have positive absolute angular momentum in the Northern Hemisphere, while the anomalous cases have negative absolute angular momentum. Since the only source of negative absolute angular momentum is the Southern Hemisphere, the anomalous cases are unlikely to occur except perhaps close to the equator.

In all cases except the anomalous low (Fig. 3.5c) the horizontal components of the Coriolis and pressure gradient forces are oppositely directed. Such flow is called *baric*. The anomalous low is antibaric; the geostrophic wind V_g defined in (3.11) is negative for an anomalous low and clearly not a useful approximation to the actual speed.[4] Furthermore, as shown in Fig. 3.5, gradient flow is cyclonic only when the centrifugal force and the horizontal component of the Coriolis force have the same sense $(Rf > 0)$; it is anticyclonic when these forces have the opposite sense $(Rf < 0)$. Since the direction of anticyclonic and cyclonic flow is reversed in the Southern Hemisphere, the requirement that $Rf > 0$ for cyclonic flow holds regardless of the hemisphere considered.

The definition of the geostrophic wind (3.11) can be used to rewrite the force balance normal to the direction of flow (3.10) in the form

$$V^2/R + fV - fV_g = 0$$

Dividing through by fV shows that the ratio of the geostrophic wind to the gradient wind is

$$\frac{V_g}{V} = 1 + \frac{V}{fR} \tag{3.17}$$

For normal cyclonic flow $(fR > 0)$ V_g is larger than V, while for anticyclonic flow $(fR < 0)$ V_g is smaller than V. Therefore, the geostrophic wind is an overestimate of the balanced wind in a region of cyclonic curvature and an underestimate in a region of anticyclonic curvature. For midlatitude synoptic systems, the difference between the gradient and geostrophic wind speeds generally does not exceed 10–20%. (Note that the magnitude of $V/(fR)$ is just the Rossby number.) For tropical disturbances, the Rossby number is in the range of 1–10, and the gradient wind formula must be applied rather than the geostrophic wind. Equation (3.17) also shows that the antibaric anomalous low, which has $V_g < 0$, can exist only when $V/(fR) < -1$. Thus, antibaric flow is associated with small-scale intense vortices such as tornadoes.

---

[4] Remember that in the natural coordinate system the speed V is positive definite.

3.3 Trajectories and Streamlines

In the natural coordinate system used in the previous section to discuss balanced flow $s(x, y, t)$ was defined as the curve in the horizontal plane traced out by the path of an air parcel. The path followed by a particular air parcel over a finite period of time is called the *trajectory* of the parcel. Thus, the radius of curvature R of the path s referred to in the gradient wind equation is the radius of curvature for a parcel trajectory. In practice R is often estimated by using the radius of curvature of a geopotential height contour, since this can easily be estimated from a synoptic chart. However, the height contours are actually *streamlines* of the gradient wind (that is, lines that are everywhere parallel to the instantaneous wind velocity).

It is important to distinguish clearly between streamlines, which give a "snapshot" of the velocity field at any instant, and trajectories, which trace the motion of individual fluid parcels over a finite time interval. In Cartesian coordinates horizontal trajectories are determined by integration of

$$\frac{Ds}{Dt} = V(x, y, t) \tag{3.18}$$

over a finite time span for each parcel to be followed, whereas streamlines are determined by the integration of

$$\frac{dy}{dx} = \frac{v(x, y, t_0)}{u(x, y, t_0)} \tag{3.19}$$

with respect to x at time t_0. (Note that since a streamline is parallel to the velocity field its slope in the horizontal plane is just the ratio of the horizontal velocity components.) Only for *steady-state* motion fields (i.e., fields in which the local rate of change of velocity vanishes) do the streamlines and trajectories coincide. However, synoptic disturbances are not steady-state motions. They generally move at speeds of the same order as the winds that circulate about them. In order to gain an appreciation for the possible errors involved in using the curvature of the streamlines instead of the curvature of the trajectories in the gradient wind equation, it is necessary to investigate the relationship between the curvature of the trajectories and the curvature of the streamlines for a moving pressure system.

We let $\beta(x, y, t)$ designate the angular direction of the wind at each point on an isobaric surface, and R_t and R_s designate the radii of curvature of the trajectories and streamlines, respectively. Then, from Fig. 3.6, $\delta s = R\,\delta\beta$ so that in the limit $\delta s \to 0$

$$\frac{D\beta}{Ds} = \frac{1}{R_t} \quad \text{and} \quad \frac{\partial\beta}{\partial s} = \frac{1}{R_s} \tag{3.20}$$

where $D\beta/Ds$ means the rate of change of wind direction along a trajectory (positive for counterclockwise turning) and $\partial\beta/\partial s$ is the rate of change of wind direction along a streamline at any instant. Thus, the rate of change of wind direction following the motion is

$$\frac{D\beta}{Dt} = \frac{D\beta}{Ds}\frac{Ds}{Dt} = \frac{V}{R_t} \tag{3.21}$$

or, after expanding the total derivative,

$$\frac{D\beta}{Dt} = \frac{\partial\beta}{\partial t} + V\frac{\partial\beta}{\partial s} = \frac{\partial\beta}{\partial t} + \frac{V}{R_s} \tag{3.22}$$

Combining (3.21) and (3.22) we obtain a formula for the local turning of the wind:

$$\frac{\partial\beta}{\partial t} = V\left(\frac{1}{R_t} - \frac{1}{R_s}\right) \tag{3.23}$$

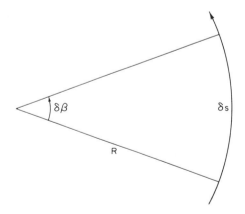

Fig. 3.6 Relationship between the change in angular direction of the wind $\delta\beta$ and the radius of curvature R.

Equation (3.23) indicates that the trajectories and streamlines will coincide only when the local rate of change of the wind direction vanishes.

In general, midlatitude synoptic systems move eastward as a result of advection by the upper-level westerly winds. In such cases there is a local turning of the wind due to the motion of the system even if the shape of the height contour pattern remains constant as it moves. The relationship between R_t and R_s in such a situation can easily be determined for an idealized circular pattern of height contours moving at a constant velocity **C**. In this case the local turning of the wind is entirely due to the motion of the streamline pattern, so

$$\frac{\partial \beta}{\partial t} = -\mathbf{C} \cdot \nabla \beta = -C \frac{\partial \beta}{\partial s} \cos \gamma$$

where γ is the angle between the streamlines (height contours) and the direction of motion of the system. Substituting the above into (3.23) and solving for R_s with the aid of (3.20), we obtain the desired relationship between the curvature of the streamlines and the curvature of the trajectories:

$$R_s = R_t \left(1 - \frac{C \cos \gamma}{V} \right) \tag{3.24}$$

Equation (3.24) can be used to compute the curvature of the trajectory anywhere on a moving pattern of streamlines. In Fig. 3.7 the curvatures of the trajectories for parcels initially located due north, east, south, and west of the center of a cyclonic system are shown both for the case of a wind speed greater than the speed of movement of the height contours and for the case of a wind speed less than the speed of movement of the height contours. In these examples the plotted trajectories are based on a geo-strophic balance so that the height contours are equivalent to streamlines. It is also assumed for simplicity that the wind speed does not depend on the distance from the center of the system. In the case shown in Fig. 3.7b there is a region south of the low center where the curvature of the trajectories is opposite that of the streamlines. Since synoptic-scale pressure systems usually move at speeds comparable to the wind speed, the gradient wind speed computed on the basis of the curvature of the height contours is often no better an approximation to the actual wind speed than is the geostrophic wind. In fact, the actual gradient wind speed will vary along a height contour with the variation of the trajectory curvature.

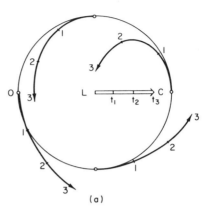

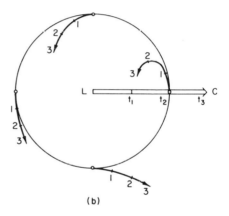

Fig. 3.7 Trajectories for moving circular cyclonic circulation systems in the Northern Hemisphere with (a) $V = 2C$ and (b) $2V = C$. Numbers indicate positions at successive times. The L designates a pressure minimum.

3.4 The Thermal Wind

The geostrophic wind must have vertical shear in the presence of a horizontal temperature gradient, as can easily be shown from simple physical considerations based on hydrostatic equilibrium. Since the geostrophic wind (3.4) is proportional to the geopotential gradient on an isobaric surface, a geostrophic wind directed along the positive y axis that increases in magnitude with height requires the slope of the isobars along the x axis to increase with height as well, as shown in Fig. 3.8. According to the hypsometric equation (1.18) the thickness δz corresponding to a pressure interval δp is

$$\delta z \approx -g^{-1} RT \delta \ln p \qquad (3.25)$$

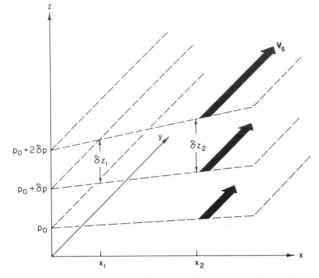

Fig. 3.8 Relationship between vertical shear of the geostrophic wind and horizontal temperature gradients. (Note: $\delta p < 0$.)

Therefore, referring to Fig. 3.8, since the interval $\delta \ln p$ is a constant, (3.25) implies that for $\delta z_1 < \delta z_2$, T_1 must be less than T_2. Thus, the increase with height of the positive x-directed pressure gradient must be associated with a positive x-directed temperature gradient. The air in a vertical column at x_2, since it is warmer (less dense), must occupy a greater depth for a given pressure drop than the air at x_1.

The equations for the rate of change of the geostrophic wind components with height are most easily derived using the isobaric coordinate system. In isobaric coordinates the geostrophic wind (3.4) has components given by

$$v_g = \frac{1}{f} \frac{\partial \Phi}{\partial x} \quad \text{and} \quad u_g = -\frac{1}{f} \frac{\partial \Phi}{\partial y} \tag{3.26}$$

where the derivatives are evaluated with pressure held constant. Also, with the aid of the ideal gas law we can write the hydrostatic equation as

$$\frac{\partial \Phi}{\partial p} = -\alpha = -\frac{RT}{p} \tag{3.27}$$

Differentiating (3.26) with respect to pressure and applying (3.27) we obtain

$$p \frac{\partial v_g}{\partial p} \equiv \frac{\partial v_g}{\partial \ln p} = -\frac{R}{f} \left(\frac{\partial T}{\partial x} \right)_p \tag{3.28}$$

$$p \frac{\partial u_g}{\partial p} \equiv \frac{\partial u_g}{\partial \ln p} = \frac{R}{f}\left(\frac{\partial T}{\partial y}\right)_p \qquad (3.29)$$

or in vectorial form

$$\frac{\partial \mathbf{V}_g}{\partial \ln p} = -\frac{R}{f}\mathbf{k} \times \nabla_p T \qquad (3.30)$$

Equation (3.30) is often referred to as the *thermal wind* equation. However, it is actually a relationship for the vertical wind *shear* (that is, the rate of change of the geostrophic wind with respect to ln p). Strictly speaking, the term thermal wind refers to the vector difference between the geostrophic winds at two levels. Designating the thermal wind vector by $\mathbf{V}_T$ we may integrate (3.30) from pressure level p_0 to level p_1 ($p_1 < p_0$) to get

$$\mathbf{V}_T \equiv \mathbf{V}_g(p_1) - \mathbf{V}_g(p_0) = -\frac{R}{f}\int_{p_0}^{p_1} (\mathbf{k} \times \nabla_p T)\, d \ln p \qquad (3.31)$$

Letting $\langle T \rangle$ denote the mean temperature in the layer between pressure p_0 and p_1, the x and y components of the thermal wind are thus given by

$$u_T = -\frac{R}{f}\left(\frac{\partial \langle T \rangle}{\partial y}\right)_p \ln\left(\frac{p_0}{p_1}\right), \qquad v_T = \frac{R}{f}\left(\frac{\partial \langle T \rangle}{\partial x}\right)_p \ln\left(\frac{p_0}{p_1}\right) \qquad (3.32)$$

Alternatively, we may express the thermal wind for a given layer in terms of the horizontal gradient of the geopotential difference between the top and bottom of the layer:

$$u_T = -\frac{1}{f}\frac{\partial}{\partial y}(\Phi_1 - \Phi_0), \qquad v_T = \frac{1}{f}\frac{\partial}{\partial x}(\Phi_1 - \Phi_0) \qquad (3.33)$$

The equivalence of (3.32) and (3.33) can be readily verified by integrating the hydrostatic equation (3.27) vertically from p_0 to p_1 after replacing T by the mean $\langle T \rangle$. The result is the hypsometric equation (1.17):

$$\delta\Phi \equiv \Phi_1 - \Phi_0 = R\langle T \rangle \ln\left(\frac{p_0}{p_1}\right) \qquad (3.34)$$

The quantity $\delta\Phi$ is the *thickness* of the layer between p_0 and p_1 measured in units of geopotential. From (3.34) we see that the thickness is proportional to the mean temperature in the layer. Hence, lines of equal $\delta\Phi$ (isolines of thickness) are equivalent to the isotherms of mean temperature in the layer. Note also that $\delta z \equiv \delta\Phi/g_0$ gives the approximate thickness in geopotential height.

The thermal wind equation is an extremely useful diagnostic tool, which is often used to check analyses of the observed wind and temperature fields for consistency. It can also be used to estimate the mean horizontal temperature advection in a layer as shown in Fig. 3.9. It is clear from the vector form of (3.33),

$$\mathbf{V}_T = \frac{1}{f} \mathbf{k} \times \nabla(\Phi_1 - \Phi_0)$$

that the thermal wind blows parallel to the isotherms (lines of constant thickness) with the warm air to the right facing downstream in the Northern Hemisphere. Thus, as is illustrated in Fig. 3.9a, a geostrophic wind that

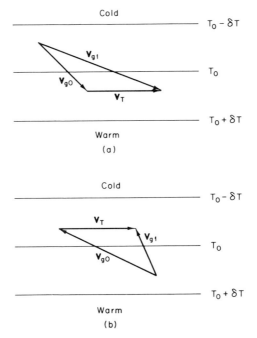

Fig. 3.9 Relationship between turning of the geostrophic wind and temperature advection: (a) backing of the wind with height; (b) veering of the wind with height.

turns counterclockwise with height (backs) is associated with cold-air advection. Conversely, as shown in Fig. 3.9b, clockwise turning (veering) of the geostrophic wind with height implies warm advection by the geostrophic wind in the layer. It is therefore possible to obtain a reasonable estimate of the horizontal temperature advection and its vertical dependence at a given location solely from data on the vertical profile of the wind given by a single sounding. Alternatively, the geostrophic wind at any level can be estimated from the mean temperature field provided that the geostrophic velocity is known at a single level. Thus, for example, if the geostrophic wind at 850 mb is known and the mean horizontal temperature gradient in the layer 850–500 mb is also known, the thermal wind equation can be applied to obtain the geostrophic wind at 500 mb.

3.4.1 BAROTROPIC AND BAROCLINIC ATMOSPHERES

A *barotropic* atmosphere is one in which the density depends only on the pressure, $\rho = \rho(p)$, so that isobaric surfaces are also surfaces of constant density. For an ideal gas, the isobaric surfaces will also be isothermal if the atmosphere is barotropic. Thus, $\nabla_p T = 0$ in a barotropic atmosphere, and the thermal wind equation (3.30) becomes $\partial \mathbf{V}_g / \partial \ln p = 0$, which states that the geostrophic wind is independent of height in a barotropic atmosphere. Thus, barotropy provides a very strong constraint on the motions in a rotating fluid; the large-scale motion can depend only on horizontal position and time, not on height.

An atmosphere in which density depends on both the temperature and the pressure, $\rho = \rho(p,T)$, is referred to as a *baroclinic* atmosphere. In a baroclinic atmosphere the geostrophic wind generally has vertical shear, and this shear is related to the horizontal temperature gradient by the thermal wind equation (3.30). Obviously, the baroclinic atmosphere is of primary importance in dynamic meteorology. However, as will be seen in later chapters, much can be learned by study of the simpler barotropic atmosphere.

3.5 Vertical Motion

As previously mentioned, for synoptic-scale motions the vertical velocity component is typically of the order of a few centimeters per second. Routine meteorological soundings however, give the wind speed to an accuracy of only about a meter per second. Thus, in general, the vertical velocity is not measured directly but must be inferred from the fields that are directly measured.

Two commonly used method for inferring the vertical motion field are the kinematic method, based on the equation of continuity, and the adiabatic method, based on the thermodynamic energy equation. Both methods are usually applied using the isobaric coordinate system so that $\omega(p)$ is inferred rather than $w(z)$. These two measures of vertical motion can be related to each other with the aid of the hydrostatic approximation.

Expanding Dp/Dt in the (x, y, z) coordinate system yields

$$\omega \equiv \frac{Dp}{Dt} = \frac{\partial p}{\partial t} + \mathbf{V} \cdot \boldsymbol{\nabla} p + w \left(\frac{\partial p}{\partial z} \right) \tag{3.35}$$

Now, for synoptic-scale motions, the horizontal velocity is geostrophic to a first approximation. Therefore, we can write $\mathbf{V} = \mathbf{V}_g + \mathbf{V}_a$, where $\mathbf{V}_a$ is the *ageostrophic* wind and $|\mathbf{V}_a| \ll |\mathbf{V}_g|$. But $\mathbf{V}_g = (\rho f)^{-1} \mathbf{k} \times \boldsymbol{\nabla} p$, so that $\mathbf{V}_g \cdot \boldsymbol{\nabla} p = 0$. Using this result plus the hydrostatic approximation, (3.35) may be rewritten as

$$\omega = \frac{\partial p}{\partial t} + \mathbf{V}_a \cdot \boldsymbol{\nabla} p - g\rho w \tag{3.36}$$

Comparing the magnitudes of the three terms on the right in (3.36) we find that for synoptic-scale motions

$$\partial p / \partial t \sim 1 \text{ kPa d}^{-1}$$

$$\mathbf{V}_a \cdot \boldsymbol{\nabla} p \sim (1 \text{ m s}^{-1})(1 \text{ Pa km}^{-1}) \sim 0.1 \text{ kPa d}^{-1}$$

$$g\rho w \sim 10 \text{ kPa d}^{-1}$$

Thus, it is quite a good approximation to let

$$\omega = -\rho g w \tag{3.37}$$

3.5.1   THE KINEMATIC METHOD

One method of deducing the vertical velocity is based on integrating the continuity equation in the vertical. Integration of (3.5) with respect to pressure from a reference level p_s to any level p yields

$$\omega(p) = \omega(p_s) - \int_{p_s}^{p} \left(\frac{\partial u}{\partial x} + \frac{\partial v}{\partial y} \right)_p dp$$

$$= \omega(p_s) + (p_s - p) \left(\frac{\partial \langle u \rangle}{\partial x} + \frac{\partial \langle v \rangle}{\partial y} \right)_p \tag{3.38}$$

Here the angle brackets denote a pressure-weighted vertical average:

$$\langle\ \rangle \equiv (p - p_s)^{-1} \int_{p_s}^{p} (\ )\, dp$$

With the aid of (3.37) and the hydrostatic equation, (3.38) can be rewritten as

$$w(z) = \frac{\rho(z_s)w(z_s)}{\rho(z)} - \frac{p_s - p}{\rho(z)g}\left(\frac{\partial\langle u\rangle}{\partial x} + \frac{\partial\langle v\rangle}{\partial y}\right) \qquad (3.39)$$

where z and z_s are the heights of pressure levels p and p_s, respectively.

Application of (3.39) to infer the vertical velocity field requires knowledge of the horizontal divergence. In order to determine the horizontal divergence the partial derivatives $\partial u/\partial x$ and $\partial v/\partial y$ are generally estimated from the fields of u and v by using *finite-difference* approximations (see Section 13.3.1). For example, to determine the divergence of the horizontal velocity at the point x_0, y_0 in Fig. 3.10 we write

$$\frac{\partial u}{\partial x} + \frac{\partial v}{\partial y} \approx \frac{u(x_0 + d) - u(x_0 - d)}{2d} + \frac{v(y_0 + d) - v(y_0 - d)}{2d} \qquad (3.40)$$

However, for synoptic-scale motions in midlatitudes the horizontal velocity is nearly in geostrophic equilibrium. Except for the small effect owing to the variation of the Coriolis parameter (see Problem 3.19) the geostrophic wind is nondivergent; that is, $\partial u/\partial x$ and $\partial v/\partial y$ are nearly equal in magnitude but opposite in sign. Thus, the horizontal divergence is due primarily to

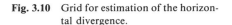

Fig. 3.10   Grid for estimation of the horizontal divergence.

the small departures of the wind from geostrophic balance (i.e., the ageostrophic wind). A 10% error in evaluating one of the wind components in (3.40) can easily cause the estimated divergence to be in error by 100%. For this reason, the continuity equation method is not recommended for estimating the vertical motion field from observed horizontal winds.

3.5.2   THE ADIABATIC METHOD

A second method for inferring vertical velocities, which is not so sensitive to errors in the measured horizontal velocities, is based on the thermodynamic energy equation. If we assume that the diabatic heating Q is small compared to the other terms in the heat balance (local change, horizontal advection, and adiabatic expansion owing to vertical motion) we can solve (3.6) for ω to yield

$$\omega = S_p^{-1}\left(\frac{\partial T}{\partial t} + u\frac{\partial T}{\partial x} + v\frac{\partial T}{\partial y}\right) \tag{3.41}$$

Since temperature advection can usually be estimated quite accurately in midlatitudes by using the geostrophic winds, the adiabatic method can be applied only when geopotential and temperature data are available. A disadvantage of the adiabatic method is that the local rate of change of temperature is required. Unless observations are taken at close intervals in time it may be difficult to estimate $\partial T/\partial t$ accurately over a wide area. This method will also be rather inaccurate in situations where strong diabatic heating is present, such as storms in which heavy rainfall occurs over a large area. In Chapter 6 we develop an alternative method, based on the so-called omega equation, that does not suffer from these difficulties.

3.6   Surface Pressure Tendency

The development of a negative surface *pressure tendency* is a classic warning of an approaching cyclonic weather disturbance. A simple expression that relates the surface pressure tendency to the wind field, and hence in theory might be used as the basis for short-range forecasts, can be obtained by taking the limit $p \to 0$ in (3.38) to get

$$\omega(p_s) = -\int_0^{p_s} (\nabla \cdot \mathbf{V})\, dp \tag{3.42}$$

followed by substituting from (3.36) to yield

$$\frac{\partial p_s}{\partial t} \approx -\int_0^{p_s} (\mathbf{\nabla} \cdot \mathbf{V}) \, dp \tag{3.43}$$

Here we have assumed that the surface is horizontal so that $w_s = 0$ and have neglected advection by the ageostrophic surface velocity in accord with the scaling arguments in Section 3.5.1.

According to (3.43) the surface pressure tendency at a given point is determined by the total convergence of mass into the vertical column of atmosphere above that point. This result is a direct consequence of the hydrostatic assumption, which implies that the pressure at a point is determined solely by the weight of the column of air above that point.

Although, as stated above, the tendency equation might appear to have potential as a forecasting aid, its utility is severely limited owing to the fact that, as discussed in Section 3.5.1, $\mathbf{\nabla} \cdot \mathbf{V}$ is difficult to compute accurately from observations since it depends on the ageostrophic wind field. In addition, there is a strong tendency for vertical compensation. Thus, when there is convergence in the lower troposphere there is divergence aloft, and vice versa. The net integrated convergence or divergence is then a small residual in the vertical integral of a poorly determined quantity.

Nevertheless, (3.43) does have qualitative value as an aid in understanding the origin of surface pressure changes and the relationship of such changes to the horizontal divergence. This can be illustrated by considering (as one possible example) the development of a thermal cyclone. We suppose that a heat source generates a local warm anomaly in the midtroposphere (Fig. 3.11a). Then according to the hypsometric equation (3.34) the heights of the upper-level pressure surfaces are raised above the warm anomaly, and as a result there is a horizontal pressure gradient force at the upper levels,

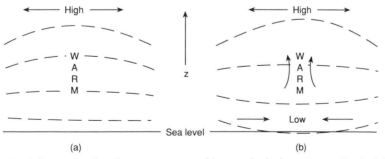

Fig. 3.11 Adjustment of surface pressure to midtropospheric heat source. Dashed lines indicate isobars.

which drives a divergent upper-level wind. By (3.43) this upper-level divergence will initially cause the surface pressure to decrease, thus generating a surface low below the warm anomaly (Fig. 3.11b). The horizontal pressure gradient associated with the surface low then drives a low-level convergence and vertical circulation which tends to compensate the upper-level divergence. The degree of compensation between the upper divergence and lower convergence will determine whether the surface pressure continues to fall, remains steady, or rises.

The thermally driven circulation of the foregoing example is by no means the only type of circulation possible (for example, cold core cyclones are important synoptic-scale features). However, it does provide insight into how dynamical processes at upper levels are communicated to the surface and how the surface and upper troposphere are dynamically connected through the divergent circulation. This subject will be considered in detail in Chapter 6.

Equation (3.43) is a lower boundary condition that determines the evolution of pressure at constant height. If the isobaric coordinate system of dynamical equations (3.2), (3.5), (3.6), and (3.27) is used as the set of governing equations, the lower boundary condition should be expressed in terms of the evolution of geopotential (or geopotential height) at constant pressure. Such an expression can be obtained simply by expanding $D\Phi/Dt$ in isobaric coordinates

$$\frac{\partial \Phi}{\partial t} = -\mathbf{V}_a \cdot \boldsymbol{\nabla}\Phi - \omega \frac{\partial \Phi}{\partial p}$$

and substituting from (3.27) and (3.42) to get

$$\frac{\partial \Phi_s}{\partial t} \approx -\frac{RT_s}{p_s} \int_0^{p_s} (\boldsymbol{\nabla} \cdot \mathbf{V}) \, dp \tag{3.44}$$

where we have again neglected advection by the ageostrophic wind.

In practice, the boundary condition (3.44) is difficult to use because it should be applied at a pressure p_s which is itself changing in time and space. In simple models it is usual to assume that p_s is constant (usually 1000 mb) and to let $\omega = 0$ at p_s. For modern forecast models an alternative coordinate system is generally employed in which the lower boundary is always a coordinate surface. This approach will be described in Section 10.3.1.

Problems

3.1. An aircraft flying a heading of 60° (i.e., 60° to the east of north) at air speed 200 m s^{-1} moves relative to the ground due east (90°) at 225 m s^{-1}. If the plane is flying at constant pressure, what is its rate of change in altitude (in meters per kilometer horizontal distance) assuming a steady pressure field, geostrophic winds, and $f = 10^{-4} \text{ s}^{-1}$?

3.2. The actual wind is directed 30° to the right of the geostrophic wind. If the geostrophic wind is 20 m s^{-1}, what is the rate of change of wind speed? Let $f = 10^{-4} \text{ s}^{-1}$.

3.3. A tornado rotates with constant angular velocity ω. Show that the surface pressure at the center of the tornado is given by

$$p = p_0 \exp\left(\frac{-\omega^2 r_0^2}{2RT}\right)$$

where p_0 is the surface pressure at a distance r_0 from the center and T is the temperature (assumed constant). If the temperature is 288 K and pressure and wind speed at 100 m from the center are 10^2 kPa and 100 m s^{-1}, respectively, what is the central pressure?

3.4. Calculate the geostrophic wind speed in meters per second for a pressure gradient of $1 \text{ kPa}/10^3 \text{ km}$ and compare with all possible gradient wind speeds for the same pressure gradient and a radius of curvature of $\pm 500 \text{ km}$. Let $\rho = 1 \text{ kg m}^{-3}$ and $f = 10^{-4} \text{ s}^{-1}$.

3.5. Determine the maximum possible ratio of the normal anticyclonic gradient wind speed to the geostrophic wind speed for the same pressure gradient.

3.6. Show that the geostrophic balance in isothermal coordinates may be written

$$f\mathbf{V}_g = \mathbf{k} \times \nabla_T(RT \ln p + \Phi)$$

3.7. Determine the radii of curvature for the trajectories of air parcels located 500 km to the east, north, south, and west of the center of a circular low-pressure system, respectively. The system is moving eastward at 15 m s^{-1}. Assume geostrophic flow with a uniform tangential wind speed of 15 m s^{-1}.

3.8. Determine the gradient wind speeds for the four air parcels of Problem 3.7 and compare these speeds with the geostrophic speed. (Let $f = 10^{-4}\,\text{s}^{-1}$.)

3.9. Show that as the pressure gradient approaches zero the gradient wind reduces to the geostrophic wind for a normal anticyclone and to the inertia circle for an anomalous anticyclone.

3.10. The mean temperature in the layer between 75 and 50 kPa decreases eastward by 3°C per 100 km. If the 75-kPa geostrophic wind is from the southeast at $20\,\text{m s}^{-1}$, what is the geostrophic wind speed and direction at 50 kPa? Let $f = 10^{-4}\,\text{s}^{-1}$.

3.11. What is mean temperature advection in the 75–50-kPa layer in Problem 3.10.

3.12. Suppose that a vertical column of the atmosphere at 43°N is initially isothermal from 90 to 50 kPa. The geostrophic wind is $10\,\text{m s}^{-1}$ from the south at 90 kPa, $10\,\text{m s}^{-1}$ from the west at 70 kPa, and $20\,\text{m s}^{-1}$ from the west at 50 kPa. Calculate the mean horizontal temperature gradients in the two layers 90–70 kPa and 70–50 kPa. Compute the rate of advective temperature change in each layer. How long would this advection pattern have to persist in order to establish a dry adiabatic lapse rate between 60 and 80 kPa? (Assume that the lapse rate is constant between 90 and 50 kPa and that the 80–60-kPa layer thickness is 2.25 km.)

3.13. An airplane pilot crossing the ocean at 45°N latitude has both a pressure altimeter and a radar altimeter, the latter measuring his absolute height above the sea. Flying at an air speed of $100\,\text{m s}^{-1}$ the pilot maintains altitude by referring to his pressure altimeter set at an altimeter setting of 101.3 kPa, holding an indicated 6000 m altitude. At the beginning of a 1-h period he notes that the radar altimeter reads 5700 m and at the end of the hour its reads 5950 m. In what direction and approximately how far has the pilot drifted from his heading?

3.14. Work out a gradient wind classification scheme equivalent to Table 3.1 for the Southern Hemisphere ($f < 0$) case.

3.15. In the geostrophic momentum approximation (Hoskins, 1975) the gradient wind formula for steady circular flow (3.17) is replaced by the approximation

$$VV_g R^{-1} + fV = fV_g$$

Compare the speeds V computed using this approximation with those obtained in Problem 3.8 using the gradient wind formula.

3.16. How large can the ratio $V_g/(fR)$ be before the geostrophic momentum approximation differs from the gradient wind approximation by 10% for cyclonic flow?

3.17. The planet Venus rotates about its axis so slowly that to a reasonable approximation the Coriolis parameter may be set equal to zero. For steady, frictionless motion parallel to latitude circles the momentum equation (2.20) then reduces to a type of cyclostrophic balance:

$$\frac{u^2 \tan \phi}{a} = -\frac{1}{\rho}\frac{\partial p}{\partial y}$$

By transforming this expression to isobaric coordinates, show that the thermal wind equation in this case can be expressed in the form

$$\omega_r^2(p_1) - \omega_r^2(p_0) = \frac{-R \ln(p_0/p_1)}{(a \sin \phi \cos \phi)}\frac{\partial \langle T \rangle}{\partial y}$$

where R is the gas constant, a the radius of the planet, and $\omega_r \equiv u/(a \cos \phi)$ is the relative angular velocity. How must $\langle T \rangle$ (the vertically averaged temperature) vary with respect to latitude in order that ω_r be a function only of pressure? If the zonal velocity at about 60 km height above the equator ($p_1 = 2.9 \times 10^5$ Pa) is 100 m s^{-1} and the zonal velocity vanishes at the surface of the planet ($p_0 = 9.5 \times 10^6$ Pa), what is the vertically averaged temperature difference between the equator and pole assuming that ω_r depends only on pressure? The planetary radius is $a = 6100$ km, and the gas constant is $R = 187$ J kg^{-1} K^{-1}.

3.18. Suppose that during the passage of a cyclonic storm the radius of curvature of the isobars is observed to be +800 km at a station where the wind is veering (turning clockwise) at a rate of 10° per hour. What is the radius of curvature of the trajectory for an air parcel that is passing over the station? (The wind speed is 20 m s^{-1}.)

3.19. Show that the divergence of the geostrophic wind in isobaric coordinates on the spherical earth is given by

$$\mathbf{\nabla} \cdot \mathbf{V}_g = -\frac{1}{fa}\frac{\partial \Phi}{\partial x}\left(\frac{\cos \phi}{\sin \phi}\right) = -v_g\left(\frac{\cot \phi}{a}\right)$$

(Use the spherical coordinate expression for the divergence operator given in Appendix C.)

3.20. The following wind data were received from 50 km to the east, north, west, and south of a station, respectively: 90°, 10 m s^{-1}; 120°, 4 m s^{-1}; 90°, 8 m s^{-1}; 60°, 4 m s^{-1}. Calculate the approximate horizontal divergence at the station.

3.21. Suppose that the wind speeds given in Problem 3.20 are each in error by ±10%. What would be the percent error in the calculated horizontal divergence in the worst case?

3.22. The divergence of the horizontal wind at various pressure levels above a given station is shown in the following table.

Pressure (kPa)	$\nabla \cdot \mathbf{V}$ ($\times 10^{-5}$ s^{-1})
100	+0.9
85	+0.6
70	+0.3
50	0.0
30	−0.6
10	−1.0

Compute the vertical velocity at each level assuming an isothermal atmosphere with temperature 260 K and letting $w = 0$ at 100 kPa (1000 mb).

3.23. Suppose that the lapse rate at the 85-kPa (850-mb) level is 4 K km^{-1}. If the temperature at a given location is decreasing at a rate of 2 K h^{-1}, the wind is westerly at 10 m s^{-1}, and the temperature decreases toward the west at a rate of 5 K/100 km, compute the vertical velocity at the 85-kPa level using the adiabatic method.

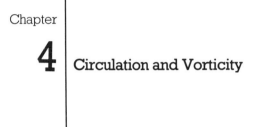

Chapter

4 | Circulation and Vorticity

In classical mechanics the principle of conservation of angular momentum is often invoked in the analysis of motions that involve rotation. This principle provides a powerful constraint on the behavior of rotating objects. Analogous conservation laws also apply to the rotational field of a fluid. However, it should be obvious that in a continuous medium such as the atmosphere the definition of "rotation" is subtler than for a solid object.

Circulation and vorticity are the two primary measures of rotation in a fluid. Circulation, which is a scalar integral quantity, is a *macroscopic* measure of rotation for a finite area of the fluid. Vorticity, however, is a vector field that gives a *microscopic* measure of the rotation at any point in the fluid.

4.1 The Circulation Theorem

The *circulation*, C, about a closed contour in a fluid is defined as the line integral about the contour of the component of the velocity vector that is locally tangent to the contour:

$$C \equiv \oint \mathbf{U} \cdot d\mathbf{l} = \oint |\mathbf{U}| \cos \alpha \, dl$$

where $\mathbf{l}(s)$ is a position vector extending from the origin to the point $s(x, y, z)$ on the contour C, and $d\mathbf{l}$ represents the limit of $\delta\mathbf{l} = \mathbf{l}(s + \delta s) - \mathbf{l}(s)$ as $\delta s \rightarrow 0$. Hence, as indicated in Fig. 4.1, $d\mathbf{l}$ is a displacement vector locally tangent to the contour. By convention the circulation is taken to be positive if $C > 0$ for counterclockwise integration around the contour.

That circulation is a measure of rotation is readily demonstrated by considering a circular ring of fluid of radius R in solid-body rotation at angular velocity Ω about the z axis. In this case $\mathbf{U} = \Omega \times \mathbf{R}$ where $\mathbf{R}$ is the distance from the axis of rotation to the ring of fluid. Thus the circulation about the ring is given by

$$C \equiv \oint \mathbf{U} \cdot d\mathbf{l} = \int_0^{2\pi} \Omega R^2 \, d\lambda = 2\Omega \pi R^2$$

In this case the circulation is just 2π times the angular momentum of the fluid ring about the axis of rotation. Alternatively, note that $C/(\pi R^2) = 2\Omega$, so that the circulation divided by the area enclosed by the loop is just twice the angular speed of rotation of the ring. Unlike angular momentum or angular velocity, circulation can be computed without reference to an axis of rotation; it can thus be used to characterize fluid rotation in situations where "angular velocity" is not easily defined.

The circulation theorem is obtained by taking the line integral of Newton's second law for a closed chain of fluid particles. In the absolute coordinate system the result (neglecting viscous forces) is

$$\oint \frac{D_a \mathbf{U}_a}{Dt} \cdot d\mathbf{l} = -\oint \frac{\nabla p \cdot d\mathbf{l}}{\rho} - \oint \nabla \Phi \cdot d\mathbf{l} \qquad (4.1)$$

Fig. 4.1   Circulation about a closed contour.

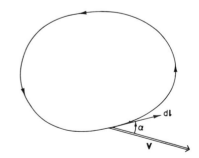

where the gravitational force $\mathbf{g}$ is represented as the gradient of the geopotential $\Phi(-\nabla\Phi = \mathbf{g} = -g\mathbf{k})$. The integrand on the left-hand side can be rewritten as[1]

$$\frac{D_a\mathbf{U}_a}{Dt} \cdot d\mathbf{l} = \frac{D}{Dt}(\mathbf{U}_a \cdot d\mathbf{l}) - \mathbf{U}_a \cdot \frac{D_a}{Dt}(d\mathbf{l})$$

or after observing that, since $\mathbf{l}$ is a position vector, $D_a\mathbf{l}/Dt \equiv \mathbf{U}_a$,

$$\frac{D_a\mathbf{U}_a}{Dt} \cdot d\mathbf{l} = \frac{D}{Dt}(\mathbf{U}_a \cdot d\mathbf{l}) - \mathbf{U}_a \cdot d\mathbf{U}_a \tag{4.2}$$

Substituting (4.2) into (4.1) and using the fact that the line integral about a closed loop of a perfect differential is zero, so that

$$\oint \nabla\Phi \cdot d\mathbf{l} = \oint d\Phi = 0$$

and noting that

$$\oint \mathbf{U}_a \cdot d\mathbf{U}_a = \tfrac{1}{2}\oint d(\mathbf{U}_a \cdot \mathbf{U}_a) = 0$$

we obtain the circulation theorem:

$$\frac{DC_a}{Dt} = \frac{D}{Dt}\oint \mathbf{U}_a \cdot d\mathbf{l} = -\oint \rho^{-1}\, dp \tag{4.3}$$

The term on the right-hand side in (4.3) is called the solenoidal term. For a barotropic fluid, the density is a function only of pressure and the solenoidal term is zero. Thus, in a barotropic fluid the absolute circulation is conserved following the motion. This result, called *Kelvin's circulation theorem*, is a fluid analog of angular momentum conservation in solid-body mechanics.

For meteorological analysis, it is more convenient to work with the relative circulation C rather than the absolute circulation because a portion of the absolute circulation, C_e, is due to the rotation of the earth about its axis. To compute C_e we apply Stokes' theorem to the vector $\mathbf{U}_e$, where $\mathbf{U}_e = \mathbf{\Omega} \times \mathbf{r}$ is the velocity of the earth at the position $\mathbf{r}$:

$$C_e = \oint \mathbf{U}_e \cdot d\mathbf{l} = \int_A \int (\nabla \times \mathbf{U}_e) \cdot \mathbf{n}\, dA$$

[1] Note that for a scalar $D_a/Dt = D/Dt$ (i.e., the rate of change following the motion does not depend on the reference system). For a vector, however, this is not the case, as was shown in Section 2.2.1.

where A is the area enclosed by the contour, and the unit normal $\mathbf{n}$ is defined by the counterclockwise sense of the line integration using the "right-hand screw rule." Thus, for the contour of Fig. 4.1 $\mathbf{n}$ would be directed out of the page. If the line integral is computed in the horizontal plane, $\mathbf{n}$ is directed along the local vertical (see Fig. 4.2). Now, by a vector identity (see Appendix C)

$$\mathbf{\nabla} \times \mathbf{U}_e = \mathbf{\nabla} \times (\mathbf{\Omega} \times \mathbf{r}) = \mathbf{\nabla} \times (\mathbf{\Omega} \times \mathbf{R}) = \mathbf{\Omega}\mathbf{\nabla} \cdot \mathbf{R} = 2\mathbf{\Omega}$$

so that $(\mathbf{\nabla} \times \mathbf{U}_e) \cdot \mathbf{n} = 2\Omega \sin \phi \equiv f$ is just the Coriolis parameter. Hence, the circulation in the horizontal plane owing to the rotation of the earth is

$$C_e = 2\Omega\langle\sin \phi\rangle A = 2\Omega A_e$$

where $\langle\sin \phi\rangle$ denotes an average over the area element A, and A_e is the projection of A in the equatorial plane as illustrated in Fig. 4.2. Thus, the relative circulation may be expressed as

$$C = C_a - C_e = C_a - 2\Omega A_e \tag{4.4}$$

Differentiating (4.4) following the motion and substituting from (4.3), we obtain the Bjerknes circulation theorem:

$$\frac{DC}{Dt} = -\oint \frac{dp}{\rho} - 2\Omega \frac{DA_e}{Dt} \tag{4.5}$$

For a barotropic fluid, (4.5) can be integrated following the motion from an initial state (designated by subscript 1) to a final state (designated by subscript 2), yielding the circulation change

$$C_2 - C_1 = -2\Omega(A_2 \sin \phi_2 - A_1 \sin \phi_1) \tag{4.6}$$

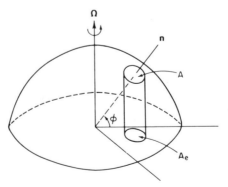

Fig. 4.2 Area A_e subtended on the equatorial plane by horizontal area A centered at latitude ϕ.

Equation (4.6) indicates that in a barotropic fluid the relative circulation for a closed chain of fluid particles will be changed if either the horizontal area enclosed by the loop changes or the latitude changes. Furthermore, a negative absolute circulation in the Northern Hemisphere can develop only if a closed chain of fluid particles is advected across the equator from the Southern Hemisphere. The anomalous gradient wind balances discussed in Section 3.2.5 are examples of systems with negative absolute circulations (see Problem 4.6).

Example. Suppose that the air within a circular region of radius 100 km centered at the equator is initially motionless with respect to the earth. If this circular air mass were moved to the North Pole along an isobaric surface preserving its area, the circulation about the circumference would be

$$C = -2\Omega \pi r^2 [\sin(\pi/2) - \sin(0)]$$

Thus the mean tangential velocity at the radius $r = 100$ km would be

$$V = C/(2\pi r) = -\Omega r \approx -7 \text{ m s}^{-1}$$

The negative sign here indicates that the air has acquired anticyclonic relative circulation.

In a baroclinic fluid, circulation may be generated by the pressure–density solenoid term in (4.3). This process can be effectively illustrated by considering the development of a sea breeze circulation, as shown in Fig. 4.3. For the situation depicted the mean temperature in the air over the ocean is colder than the mean temperature over the adjoining land. Thus, if the

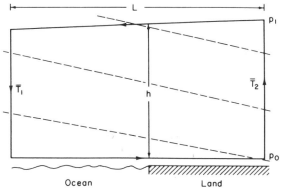

Fig. 4.3 Application of the circulation theorem to the sea breeze problem. The closed heavy solid line is the loop about which the circulation is to be evaluated. Dashed lines indicate surfaces of constant density.

pressure is uniform at ground level, the isobaric surfaces above the ground will slope downward toward the ocean while the isopycnal surfaces (isolines of constant density) will slope downward toward the land. To compute the acceleration as a result of the intersection of the pressure–density surfaces we apply the circulation theorem by integrating around a circuit in a vertical plane perpendicular to the coastline. Substituting the ideal gas law into (4.3) we obtain

$$\frac{DC_a}{Dt} = -\oint RT\, d\ln p$$

For the circuit shown in Fig. 4.3 there is a contribution to the line integral only for the vertical segments of the loop because the horizontal segments are taken at constant pressure. The resulting rate of increase in the circulation is

$$\frac{DC_a}{Dt} = R\ln\left(\frac{p_0}{p_1}\right)(\bar{T}_2 - \bar{T}_1) > 0$$

Letting $\langle v \rangle$ be the mean tangential velocity along the circuit, we find that

$$\frac{D\langle v \rangle}{Dt} = \frac{R\ln(p_0/p_1)}{2(h+L)}(\bar{T}_2 - \bar{T}_1) \tag{4.7}$$

If we let $p_0 = 100$ kPa, $p_1 = 90$ kPa, $\bar{T}_2 - \bar{T}_1 = 10°C$, $L = 20$ km, and $h = 1$ km, (4.7) yields an acceleration of about 7×10^{-3} m s^{-2}. In the absence of frictional retarding forces this would produce a wind speed of 25 m s^{-1} in about 1 h. In reality, as the wind speed increases the frictional force reduces the acceleration rate, and temperature advection reduces the land–sea temperature contrast, so that a balance is obtained between the generation of kinetic energy by the pressure–density solenoids and frictional dissipation.

4.2 Vorticity

Vorticity, the microscopic measure of rotation in a fluid, is a vector field defined as the curl of velocity. The absolute vorticity $\boldsymbol{\omega}_a$ is the curl of the absolute velocity, while the relative vorticity $\boldsymbol{\omega}$ is the curl of the relative velocity:

$$\boldsymbol{\omega}_a \equiv \nabla \times \mathbf{U}_a, \qquad \boldsymbol{\omega} \equiv \nabla \times \mathbf{U}$$

so that in Cartesian coordinates

$$\boldsymbol{\omega} = \left(\frac{\partial w}{\partial y} - \frac{\partial v}{\partial z}, \frac{\partial u}{\partial z} - \frac{\partial w}{\partial x}, \frac{\partial v}{\partial x} - \frac{\partial u}{\partial y} \right)$$

However, in large-scale dynamic meteorology we are in general concerned only with the vertical components of absolute and relative vorticity, which are designated by η and ζ, respectively.

$$\eta \equiv \mathbf{k} \cdot (\nabla \times \mathbf{U}_a), \qquad \zeta \equiv \mathbf{k} \cdot (\nabla \times \mathbf{U})$$

In the remainder of this book η and ζ will be referred to as the absolute and relative vorticities, respectively, without adding the explicit modifier "vertical component of." Regions of large positive (negative) ζ tend to develop in association with cyclonic storms in the Northern (Southern) Hemisphere. Thus, the distribution of relative vorticity is an excellent diagnostic for weather analysis. Absolute vorticity tends to be conserved following the motion at midtropospheric levels; this conservation property is the basis for the simplest dynamical forecast scheme to be discussed in Chapter 13.

The difference between absolute and relative vorticity is the *planetary vorticity*, which is just the local vertical component of the vorticity of the earth owing to its rotation; $\mathbf{k} \cdot \nabla \times \mathbf{U}_e = 2\Omega \sin \phi \equiv f$. Thus, $\eta = \zeta + f$, or in Cartesian coordinates

$$\zeta = \frac{\partial v}{\partial x} - \frac{\partial u}{\partial y}, \qquad \eta = \frac{\partial v}{\partial x} - \frac{\partial u}{\partial y} + f$$

The relationship between relative vorticity and the relative circulation C discussed in the previous section can be clearly seen by considering an alternative approach in which the vertical component of vorticity is defined as the circulation about a closed contour in the horizontal plane divided by the area enclosed, in the limit where the area approaches zero:

$$\zeta \equiv \lim_{A \to 0} \left(\oint \mathbf{V} \cdot d\mathbf{l} \right) A^{-1} \tag{4.8}$$

This latter definition makes explicit the relationship between circulation and vorticity discussed in the introduction to this chapter. The equivalence of these two definitions of ζ is easily shown by considering the circulation about a rectangular element of area $\delta x \, \delta y$ in the x, y plane as shown in

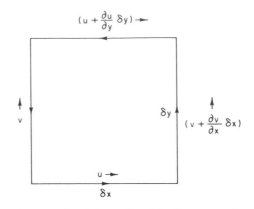

Fig. 4.4 Relationship between circulation and vorticity for an area element in the horizontal plane.

Fig. 4.4. Evaluating $\mathbf{V} \cdot d\mathbf{l}$ for each side of the rectangle in the figure yields the circulation

$$\delta C = u\,\delta x + \left(v + \frac{\partial v}{\partial x}\,\delta x\right)\delta y - \left(u + \frac{\partial u}{\partial y}\,\delta y\right)\delta x - v\,\delta y$$

$$= \left(\frac{\partial v}{\partial x} - \frac{\partial u}{\partial y}\right)\delta x\,\delta y$$

Dividing through by the area $\delta A = \delta x\,\delta y$ gives

$$\frac{\delta C}{\delta A} = \left(\frac{\partial v}{\partial x} - \frac{\partial u}{\partial y}\right) \equiv \zeta$$

In more general terms the relationship between vorticity and circulation is given simply by Stokes' theorem applied to the velocity vector:

$$\oint \mathbf{U} \cdot d\mathbf{l} = \iint_A (\nabla \times \mathbf{U}) \cdot \mathbf{n}\,dA$$

Here A is the area enclosed by the contour and $\mathbf{n}$ is a unit normal to the area element dA (positive in the right-hand sense). Thus, Stokes' theorem states that the circulation about any closed loop is equal to the integral of the normal component of vorticity over the area enclosed by the contour. Hence, for a finite area, the circulation divided by the area gives the *average* normal component of vorticity in the region. As a consequence, the vorticity of a fluid in solid-body rotation is just twice the angular velocity of rotation. Vorticity may thus be regarded as a measure of the local angular velocity of the fluid.

4.2.1 VORTICITY IN NATURAL COORDINATES

Physical interpretation of vorticity is facilitated by considering the vertical component of vorticity in the natural coordinate system (see Section 3.2.1). If we compute the circulation about the infinitesimal contour shown in Fig. 4.5 we obtain[2]

$$\delta C = V[\delta s + d(\delta s)] - \left(V + \frac{\partial V}{\partial n} \delta n \right) \delta s$$

But from Fig. 4.5 we see that $d(\delta s) = \delta\beta \, \delta n$, where $\delta\beta$ is the angular change in the wind direction in the distance δs. Hence,

$$\delta C = \left(-\frac{\partial V}{\partial n} + V \frac{\delta\beta}{\delta s} \right) \delta n \, \delta s$$

or in the limit $\delta n, \delta s \to 0$

$$\zeta = \lim_{\delta n, \delta s \to 0} \frac{\delta C}{(\delta n \, \delta s)} = -\frac{\partial V}{\partial n} + \frac{V}{R_s} \tag{4.9}$$

where R_s is the radius of curvature of the streamlines [Eq. (3.20)]. It is now apparent that the net vertical vorticity component is the result of the sum of two parts: (1) the rate of change of wind speed normal to the direction of flow, $-\partial V/\partial n$, called the *shear* vorticity; and (2) the turning of the wind

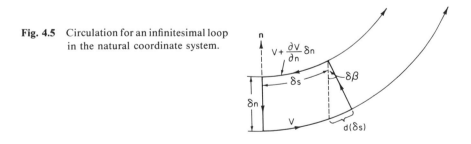

Fig. 4.5 Circulation for an infinitesimal loop in the natural coordinate system.

[2] Recall that n is a coordinate in the horizontal plane perpendicular to the local flow direction with positive values to the left of an observer facing downstream.

along a streamline V/R_s, called the *curvature* vorticity. Thus, even straight-line motion may have vorticity if the speed changes normal to the flow axis. For example, in the jet stream shown schematically in Fig. 4.6a there will be cyclonic relative vorticity north of the velocity maximum and anticyclonic relative vorticity to the south (Northern Hemisphere conditions), as is easily recognized when the turning of a small paddle wheel placed in the flow is considered. The lower of the two paddle wheels in Fig. 4.6a will turn in a clockwise direction (anticyclonically) since the wind force on the blades north of its axis of rotation is stronger than the force on the blades to the south of the axis. The upper wheel will, of course, experience a counterclockwise (cyclonic) turning. Thus, the poleward and equatorward sides of a westerly jet stream are referred to as the cyclonic and anticyclonic shear sides, respectively.

Conversely, curved flow may have zero vorticity provided that the shear vorticity is equal and opposite to the curvature vorticity. This will be the case in the example shown in Fig. 4.6b, where a frictionless fluid with zero

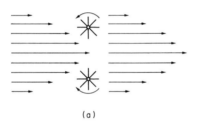

(a)

Fig. 4.6 Two types of two-dimensional flow: (a) linear shear flow with vorticity and (b) curved flow with zero vorticity.

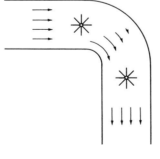

(b)

relative vorticity upstream flows around a bend in a canal. The fluid along the inner boundary on the curve flows faster in just the right proportion so that the paddle wheel does not turn.

4.3 Potential Vorticity

With the aid of the ideal gas law, potential temperature (2.44) can be expressed as a relationship between pressure and density for a surface of constant θ:

$$\rho = p^{c_v/c_p}(R\theta)^{-1}(p_s)^{R/c_p}$$

Hence, on an isentropic surface density is a function of pressure alone, and the solenoidal term in the circulation theorem (4.3) vanishes:

$$\oint \frac{dp}{\rho} \propto \oint dp^{(1-c_v/c_p)} = 0$$

Thus, for adiabatic flow the circulation computed for a closed chain of fluid parcels on a constant-θ surface reduces to the same form as in a barotropic fluid, i.e., it satisfies Kelvin's circulation theorem, which may be expressed as

$$\frac{D}{Dt}(C + 2\Omega \, \delta A \sin \phi) = 0 \tag{4.10}$$

where C is evaluated for a closed loop encompassing the area δA on an isentropic surface. If the isentropic surface is approximately horizontal and it is recalled that the vertical component of relative vorticity is given approximately by

$$\zeta = \lim_{\delta A \to 0} \frac{C}{\delta A}$$

then for an infinitesimal parcel of air (4.10) can be expressed as

$$\delta A(\zeta_\theta + f) = \text{const} \tag{4.11}$$

where ζ_θ designates the vertical component of relative vorticity evaluated on an isentropic surface and $f = 2\Omega \sin \phi$ is the Coriolis parameter.

Suppose that the parcel of (4.11) is confined between potential temperature surfaces θ_0 and $\theta_0 + \delta\theta$, which are separated by a pressure interval $-\delta p$ as shown in Fig. 4.7. The mass of the parcel $\delta M = (-\delta p/g)\,\delta A$ must be conserved following the motion. Therefore,

$$\delta A = -\frac{\delta Mg}{\delta p} = \left(-\frac{\delta\theta}{\delta p}\right)\left(\frac{\delta Mg}{\delta\theta}\right) = \text{const} \times g\left(-\frac{\delta\theta}{\delta p}\right)$$

since $\delta\theta$ is a constant. Substituting into (4.11) to eliminate δA and taking the limit $\delta p \to 0$ we obtain

$$P \equiv (\zeta_\theta + f)\left(-g\frac{\partial\theta}{\partial p}\right) = \text{const} \tag{4.12}$$

The quantity P (units: $\text{K kg}^{-1}\,\text{m}^2\,\text{s}^{-1}$) is the isentropic coordinate form of *Ertel's potential vorticity.*[3] It is defined with a minus sign so that its value is normally positive in the Northern Hemisphere.

According to (4.12) potential vorticity is conserved following the motion in adiabatic frictionless flow. The term potential vorticity is used, as we shall see later, in connection with several other mathematical expressions. In essence, however, potential vorticity is always in some sense a measure of the ratio of the absolute vorticity to the effective depth of the vortex. In (4.12), for example, the effective depth is just the distance between potential temperature surfaces measured in pressure units $(-\partial\theta/\partial p)$.

In a homogeneous incompressible fluid, potential vorticity conservation takes a somewhat simpler form. In this case, since density is a constant,

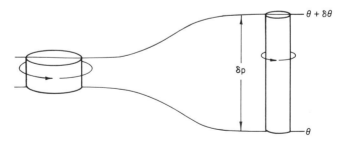

Fig. 4.7 A cylindrical column of air moving adiabatically, conserving potential vorticity.

<hr />

[3] Named for the German meteorologist Hans Ertel. A more general form of Ertel's potential vorticity is discussed, for example, in Gill (1982).

the horizontal area must be inversely proportional to the depth of the fluid parcel:

$$\delta A = M(\rho\,\delta z)^{-1} = \text{const}/\delta z$$

where δz is the depth of the parcel. Substituting to eliminate δA in (4.11) yields

$$(\zeta + f)/\delta z = \text{const} \tag{4.13}$$

where ζ is here evaluated at constant height.

The conservation of potential vorticity is a powerful constraint on the large-scale motions of the atmosphere. This can be illustrated by considering the flow of air over a large mountain barrier in which $\partial\theta/\partial p$ undergoes a substantial change along the trajectory. In order to appreciate some of the consequences of potential vorticity conservation in flow over topography it is useful to consider first a simpler situation where $-\partial\theta/\partial p$ (or δz) is constant so that the absolute vorticity $\eta = \zeta + f$ is conserved following the motion. Suppose that at a certain point (x_0, y_0) the flow is in the zonal direction and the relative vorticity vanishes so that $\eta(x_0, y_0) = f_0$. Then, if absolute vorticity is conserved, the motion at any point along a parcel trajectory that passes through (x_0, y_0) must satisfy $\zeta + f = f_0$. Since f increases toward the north, trajectories that curve northward in the downstream direction must have $\zeta = f_0 - f < 0$, while trajectories that curve southward must have $\zeta = f_0 - f > 0$. But, as indicated in Fig. 4.8, if the flow is westerly northward curvature downstream implies $\zeta > 0$, while southward curvature implies $\zeta < 0$. Thus, westerly zonal flow must remain purely zonal if absolute vorticity is to be conserved following the motion. The easterly flow case, also shown in Fig. 4.8, is just the opposite. Northward and southward curvatures are associated with negative and positive relative vorticities, respectively. Hence, an easterly current can curve either to the north or south and still conserve absolute vorticity.

When $-\partial\theta/\partial p$ (or δz) changes following the motion it is potential vorticity that is conserved. But again westerly and easterly flows behave differently.

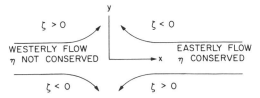

Fig. 4.8 Absolute vorticity conservation for curved flow trajectories.

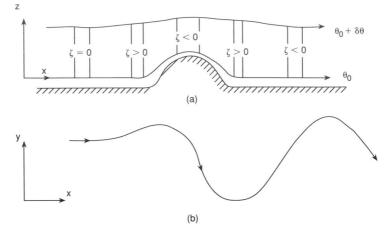

Fig. 4.9 Schematic view of westerly flow over a topographic barrier: (a) the depth of a fluid column as a function of x and (b) the trajectory of a parcel in the x, y plane.

The situation for westerly flow is shown in Fig. 4.9. In Fig. 4.9a a vertical cross section of the flow is shown. We suppose that upstream of the mountain barrier the flow is a uniform zonal flow so that $\zeta = 0$. If the flow is adiabatic, each column of air confined between the potential temperature surfaces θ_0 and $\theta_0 + \delta\theta$ remains between those surfaces as it crosses the mountain. For this reason, a potential temperature surface θ_0 near the ground must approximately follow the ground contours. A potential temperature surface $\theta_0 + \delta\theta$ several kilometers above the ground will also be deflected vertically. But, owing to pressure forces produced by interaction of the flow with the topographic barrier, the vertical displacement at upper levels is spread horizontally; it extends upstream and downstream of the barrier and has smaller amplitude in the vertical than the displacement near the ground (see Figs. 4.9 and 4.10).

As a result of the vertical displacement of the upper-level isentropes there is a vertical stretching of air columns upstream of the topographic barrier. (For motions of large horizontal scale the upstream stretching is quite small.) This stretching causes $-\partial\theta/\partial p$ to decrease, and hence from (4.12) ζ must become positive in order to conserve potential vorticity. Thus, an air column turns cyclonically as it approaches the topographic barrier. This cyclonic curvature causes a poleward drift so that f also increases, which reduces the change in ζ required for potential vorticity conservation. As the column begins to cross the barrier its vertical extent decreases; the relative vorticity must then become negative. Thus, the air column will acquire anticyclonic vorticity and move southward as shown in the x, y plane profile in Fig.

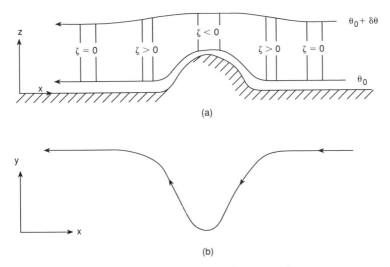

Fig. 4.10 As in Fig. 4.9, but for easterly flow

4.9b. When the air column has passed over the mountain and returned to its original depth it will be south of its original latitude so that f will be smaller and the relative vorticity must be positive. Thus, the trajectory must have cyclonic curvature so that the column will be deflected poleward. When the parcel returns to its original latitude, it will still have a poleward velocity component and will continue poleward gradually acquiring anti-cyclonic curvature until its direction is again reversed. The parcel will then move downstream conserving potential vorticity by following a wavelike trajectory in the horizontal plane. Therefore, steady westerly flow over a large-scale mountain barrier will result in a cyclonic flow pattern immediately to the east of the barrier (the lee side trough) followed by an alternating series of ridges and troughs downstream.

The situation for easterly flow impinging on a mountain barrier is quite different. As indicated schematically in Fig. 4.10b, the upstream stretching leads to a cyclonic turning of the flow, as in the westerly case. For easterly flow this cyclonic turning creates an equatorward component of motion. As the column moves westward and equatorward over the mountain its depth contracts and its absolute vorticity must then decrease so that potential vorticity can be conserved. This reduction in absolute vorticity arises both from development of anticyclonic relative vorticity and from a decrease in f owing to the equatorward motion. The anticyclonic relative vorticity gradually turns the column so that when it reaches the top of the mountain it is headed westward. As it continues westward down the mountain conserv-ing potential vorticity, the process is simply reversed with the result that

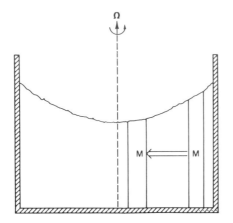

Fig. 4.11   Dependence of depth on radius in a rotating cylindrical vessel.

some distance downstream from the mountain the air column again is moving westward at its original latitude. Thus, the dependence of the Coriolis parameter on latitude creates a dramatic difference between westerly and easterly flow over large-scale topographic barriers. In the case of a westerly wind the barrier generates a wavelike disturbance in the streamlines that extends far downstream. But in the case of an easterly wind, the disturbance in the streamlines damps out away from the barrier.

Equation (4.13) also indicates that in a barotropic fluid a change in the depth is dynamically analogous to a change in the Coriolis parameter. This can easily be demonstrated in a rotating cylindrical vessel filled with water. For solid-body rotation the equilibrium shape of the free surface, determined by a balance between the radial pressure gradient and centrifugal forces, is parabolic. Thus, as shown in Fig. 4.11, if a column of fluid moves radially outward it must stretch vertically. According to (4.13) the relative vorticity must then increase to keep the ratio $(\zeta + f)/\delta z$ constant. The same result would apply if a column of fluid on a rotating sphere were moved equatorward without change in depth. In this case z would have to increase to offset the decrease of f. Therefore, in a barotropic fluid a decrease of depth with increasing latitude has the same effect on the relative vorticity as the increase of the Coriolis force with latitude.

4.4   The Vorticity Equation

In the previous section we discussed the time evolution of the vertical component of vorticity for the special case of adiabatic frictionless flow. In the present section we use the equations of motion to derive an equation for the time rate of change of vorticity without limiting the validity to adiabatic motion.

4.4.1 CARTESIAN COORDINATE FORM

For motions of synoptic scale, the vorticity equation can be derived using the approximate horizontal momentum equations (2.24) and (2.25). We differentiate the zonal component equation with respect to y and the meridional component equation with respect to x:

$$\frac{\partial}{\partial y}\left(\frac{\partial u}{\partial t}+u\frac{\partial u}{\partial x}+v\frac{\partial u}{\partial y}+w\frac{\partial u}{\partial z}-fv=-\frac{1}{\rho}\frac{\partial p}{\partial x}\right) \quad (4.14)$$

$$\frac{\partial}{\partial x}\left(\frac{\partial v}{\partial t}+u\frac{\partial v}{\partial x}+v\frac{\partial v}{\partial y}+w\frac{\partial v}{\partial z}+fu=-\frac{1}{\rho}\frac{\partial p}{\partial y}\right) \quad (4.15)$$

Subtracting (4.14) from (4.15) and recalling that $\zeta=\partial v/\partial x-\partial u/\partial y$, we obtain the vorticity equation

$$\frac{\partial \zeta}{\partial t}+u\frac{\partial \zeta}{\partial x}+v\frac{\partial \zeta}{\partial y}+w\frac{\partial \zeta}{\partial z}+(\zeta+f)\left(\frac{\partial u}{\partial x}+\frac{\partial v}{\partial y}\right)$$

$$+\left(\frac{\partial w}{\partial x}\frac{\partial v}{\partial z}-\frac{\partial w}{\partial y}\frac{\partial u}{\partial z}\right)+v\frac{df}{dy}=\frac{1}{\rho^2}\left(\frac{\partial \rho}{\partial x}\frac{\partial p}{\partial y}-\frac{\partial \rho}{\partial y}\frac{\partial p}{\partial x}\right) \quad (4.16)$$

Using the fact that the Coriolis parameter depends only on y so that $Df/Dt=v(df/dy)$, (4.16) may be rewritten in the form

$$\frac{D}{Dt}(\zeta+f)=-(\zeta+f)\left(\frac{\partial u}{\partial x}+\frac{\partial v}{\partial y}\right)$$

$$-\left(\frac{\partial w}{\partial x}\frac{\partial v}{\partial z}-\frac{\partial w}{\partial y}\frac{\partial u}{\partial z}\right)+\frac{1}{\rho^2}\left(\frac{\partial \rho}{\partial x}\frac{\partial p}{\partial y}-\frac{\partial \rho}{\partial y}\frac{\partial p}{\partial x}\right) \quad (4.17)$$

Equation (4.17) states that the rate of change of the absolute vorticity following the motion is given by the sum of the three terms on the right, called the divergence term, the tilting or twisting term, and the solenoidal term, respectively.

Generation of vorticity by horizontal divergence, the first term on the right in (4.17), is the fluid analog of the change in angular velocity resulting from a change in the moment of inertia of a solid body when angular momentum is conserved. If there is positive horizontal divergence, the area enclosed by a chain of fluid parcels will increase with time, and if circulation is to be conserved the average absolute vorticity of the enclosed fluid must decrease. This mechanism for changing vorticity is very important in synoptic-scale disturbances.

The second term on the right in (4.17) represents vertical vorticity that is generated by the tilting of horizontally oriented components of vorticity into the vertical by a nonuniform vertical motion field. This mechanism is illustrated in Fig. 4.12. The figure shows a region where the y component of velocity is increasing with height so that there is a component of shear vorticity oriented in the negative x direction as indicated by the double arrow in Fig. 4.12. If at the same time there is a vertical motion field in which w decreases with increasing x, advection by the vertical motion will tend to tilt the vorticity vector initially oriented parallel to x so that it has a component in the vertical. Thus, if $\partial v/\partial z > 0$ and $\partial w/\partial x < 0$, there will be a generation of positive vertical vorticity.

Finally, the third term on the right in (4.17) is just the microscopic equivalent of the solenoidal term in the circulation theorem (4.5). To show this equivalence we may apply Stokes' theorem to the solenoidal term to get

$$-\oint \alpha \, dp \equiv -\oint \alpha \, \nabla p \cdot d\mathbf{l} = -\int_A \int \nabla \times (\alpha \, \nabla p) \cdot \mathbf{k} \, dA$$

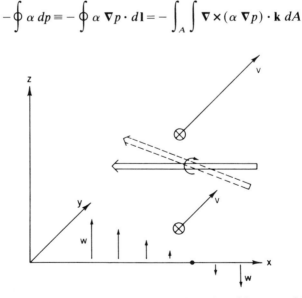

Fig. 4.12 Vorticity generation by the tilting of a horizontal vorticity vector (double arrow).

where A is the horizontal area bounded by the curve l. Applying the vector identity $\nabla \times (\alpha \, \nabla p) \equiv \nabla \alpha \times \nabla p$, this becomes

$$-\oint \alpha \, dp = -\int_A \int (\nabla \alpha \times \nabla p) \cdot \mathbf{k} \, dA$$

But the solenoidal term in the vorticity equation can be written

$$-\left(\frac{\partial \alpha}{\partial x} \frac{\partial p}{\partial y} - \frac{\partial \alpha}{\partial y} \frac{\partial p}{\partial x} \right) = -(\nabla \alpha \times \nabla p) \cdot \mathbf{k}$$

Comparing the right-hand sides of these two expressions, we see that the solenoidal term in the vorticity equation is just the limit of the solenoidal term in the circulation theorem divided by the area when the area goes to zero.

4.4.2 THE VORTICITY EQUATION IN ISOBARIC COORDINATES

A somewhat simpler form of the vorticity equation arises when the motion is referred to the isobaric coordinate system. This equation can be derived in vector form by operating on the momentum equation (3.2) with the vector operator $\mathbf{k} \cdot \nabla \times$, where ∇ now indicates the horizontal gradient on a surface of constant pressure. However, to facilitate this process it is desirable first to use the vector identity

$$(\mathbf{V} \cdot \nabla)\mathbf{V} = \nabla\left(\frac{\mathbf{V} \cdot \mathbf{V}}{2} \right) + \zeta \mathbf{k} \times \mathbf{V} \tag{4.18}$$

where $\zeta = \mathbf{k} \cdot (\nabla \times \mathbf{V})$, to rewrite (3.2) as

$$\frac{\partial \mathbf{V}}{\partial t} = -\nabla\left(\frac{\mathbf{V} \cdot \mathbf{V}}{2} + \Phi \right) - (\zeta + f)\mathbf{k} \times \mathbf{V} - \omega \frac{\partial \mathbf{V}}{\partial p} \tag{4.19}$$

We now apply the operator $\mathbf{k} \cdot \nabla \times$ to (4.19). Using the facts that for any scalar A, $\nabla \times \nabla A = 0$ and for any vectors $\mathbf{a}, \mathbf{b}$,

$$\nabla \times (\mathbf{a} \times \mathbf{b}) = (\nabla \cdot \mathbf{b})\mathbf{a} - (\mathbf{a} \cdot \nabla)\mathbf{b} - (\nabla \cdot \mathbf{a})\mathbf{b} + (\mathbf{b} \cdot \nabla)\mathbf{a} \tag{4.20}$$

we can eliminate the first term on the right and simplify the second term so that the resulting vorticity equation becomes

$$\frac{\partial \zeta}{\partial t} = -\mathbf{V} \cdot \nabla(\zeta + f) - \omega \frac{\partial \zeta}{\partial p} - (\zeta + f)\nabla \cdot \mathbf{V} + \mathbf{k} \cdot \left(\frac{\partial \mathbf{V}}{\partial p} \times \nabla \omega \right) \tag{4.21}$$

Comparing (4.17) and (4.21) we see that in the isobaric system there is no vorticity generation by pressure–density solenoids. This difference arises because in the isobaric system horizontal partial derivatives are computed with p held constant, so the vertical component of vorticity is $\zeta = (\partial v/\partial x - \partial u/\partial y)_p$, while in height coordinates it is $\zeta = (\partial v/\partial x - \partial u/\partial y)_z$. In practice the difference is generally unimportant because, as we shall see in the next section, the solenoidal term is usually sufficiently small that it can be neglected for synoptic-scale motions.

4.4.3 SCALE ANALYSIS OF THE VORTICITY EQUATION

In Section 2.4 we showed that the equations of motion could be simplified for synoptic-scale motions by evaluating the order of magnitude of various terms. The same technique can also be applied to the vorticity equation. To scale the vorticity equation we choose characteristic scales for the field variables based on typical observed magnitudes for synoptic scale motions as follows:

$$U \sim 10 \text{ m s}^{-1} \qquad \text{horizontal scale}$$
$$W \sim 1 \text{ cm s}^{-1} \qquad \text{vertical scale}$$
$$L \sim 10^6 \text{ m} \qquad \text{length scale}$$
$$H \sim 10^4 \text{ m} \qquad \text{depth scale}$$
$$\delta p \sim 1 \text{ kPa} \qquad \text{horizontal pressure scale}$$
$$\rho \sim 1 \text{ kg m}^{-3} \qquad \text{mean density}$$
$$\delta \rho / \rho \sim 10^{-2} \qquad \text{fractional density fluctuation}$$
$$L/U \sim 10^5 \text{ s} \qquad \text{time scale}$$
$$f_0 \sim 10^{-4} \text{ s}^{-1} \qquad \text{Coriolis parameter}$$
$$\beta \sim 10^{-11} \text{ m}^{-1} \text{s}^{-1} \qquad \text{"beta" parameter}$$

Again we have chosen an advective time scale because the vorticity pattern, like the pressure pattern, tends to move at a speed comparable to the horizontal wind speed. Using these scales to evaluate the magnitude of the terms in (4.16), we first note that

$$\zeta = \frac{\partial v}{\partial x} - \frac{\partial u}{\partial y} \lesssim \frac{U}{L} \sim 10^{-5} \text{ s}^{-1}$$

where the inequality in this expression means less than or equal to in order of magnitude. Thus,

$$\zeta/f_0 \lesssim U/(f_0 L) \equiv \text{Ro} \sim 10^{-1}$$

For midlatitude synoptic-scale systems the relative vorticity is small (order Rossby number) compared to the planetary vorticity.[4] Hence, ζ may be neglected compared to f in the divergence term in the vorticity equation:

$$(\zeta+f)\left(\frac{\partial u}{\partial x}+\frac{\partial v}{\partial y}\right) \approx f\left(\frac{\partial u}{\partial x}+\frac{\partial v}{\partial y}\right)$$

The magnitudes of the various terms in (4.16) can now be estimated as follows:

$$\frac{\partial \zeta}{\partial t}, u\frac{\partial \zeta}{\partial x}, v\frac{\partial \zeta}{\partial y} \sim \frac{U^2}{L^2} \sim 10^{-10}\,s^{-2}$$

$$w\frac{\partial \zeta}{\partial z} \sim \frac{WU}{HL} \sim 10^{-11}\,s^{-2}$$

$$v\frac{df}{dy} \sim U\beta \sim 10^{-10}\,s^{-2}$$

$$f\left(\frac{\partial u}{\partial x}+\frac{\partial v}{\partial y}\right) \lesssim \frac{f_0 U}{L} \sim 10^{-9}\,s^{-2}$$

$$\left(\frac{\partial w}{\partial x}\frac{\partial v}{\partial z}-\frac{\partial w}{\partial y}\frac{\partial u}{\partial z}\right) \lesssim \frac{WU}{HL} \sim 10^{-11}\,s^{-2}$$

$$\frac{1}{\rho^2}\left(\frac{\partial \rho}{\partial x}\frac{\partial p}{\partial y}-\frac{\partial \rho}{\partial y}\frac{\partial p}{\partial x}\right) \lesssim \frac{\delta\rho\,\delta p}{\rho^2 L^2} \sim 10^{-11}\,s^{-2}$$

The inequality is used in the last three terms because in each case it is possible that the two parts of the expression might partially cancel so that the actual magnitude would be less than indicated. In fact, this must be the case for the divergence term (the fourth in the list) because if $\partial u/\partial x$ and $\partial v/\partial y$ were not nearly equal and opposite, the divergence term would be an order of magnitude greater than any other term and the equation could not be satisfied. Therefore, scale analysis of the vorticity equation indicates that synoptic-scale motions must be quasi-nondivergent. The divergence term will be small enough to be balanced by the vorticity advection terms only if

$$\left(\frac{\partial u}{\partial x}+\frac{\partial v}{\partial y}\right) \lesssim 10^{-6}\,s^{-1}$$

[4] This approximation does not apply near the center of intense cyclonic storms. In such systems $|\zeta/f| \sim 1$.

so that the horizontal divergence must be small compared to the vorticity in synoptic-scale systems. From the above scalings and the definition of the Rossby number we see that

$$\left| \left(\frac{\partial u}{\partial x} + \frac{\partial v}{\partial y} \right) \Big/ f_0 \right| \lesssim \mathrm{Ro}^2$$

and

$$\left| \left(\frac{\partial u}{\partial x} + \frac{\partial v}{\partial y} \right) \Big/ \zeta \right| \lesssim \mathrm{Ro}$$

Thus the ratio of the horizontal divergence to the relative vorticity is the same magnitude as the ratio of relative vorticity to planetary vorticity.

Retaining only the terms of order 10^{-10} s^{-2} in the vorticity equation, yields the approximate form valid for synoptic-scale motions,

$$\frac{D_h(\zeta + f)}{Dt} = -f \left(\frac{\partial u}{\partial x} + \frac{\partial v}{\partial y} \right) \tag{4.22}$$

where

$$\frac{D_h}{Dt} \equiv \frac{\partial}{\partial t} + u \frac{\partial}{\partial x} + v \frac{\partial}{\partial y}$$

Equation (4.22) states that the change of absolute vorticity following the horizontal motion on the synoptic scale is approximately given by the generation (destruction) of vorticity owing to horizontal convergence (divergence). This approximation does not remain valid, however, in the vicinity of atmospheric fronts. The horizontal scale of variation in frontal zones is only ~ 100 km, and the vertical velocity scale is ~ 10 cm s^{-1}. For these scales the vertical advection, tilting, and solenoidal terms all may become as large as the divergence term.

4.5   The Barotropic (Rossby) Potential Vorticity Equation

A model that has proved useful for elucidating some aspects of the horizontal structure of large-scale atmospheric motions is the barotropic model. In this model the atmosphere is represented as a homogeneous incompressible fluid of variable depth, $h(x, y, t) = z_2 - z_1$, where z_2 and z_1, are the heights of the upper and lower boundaries, respectively. For such a fluid the continuity equation (2.31) simplifies to $\nabla \cdot \mathbf{U} = 0$, or in Cartesian coordinates

$$\left(\frac{\partial u}{\partial x} + \frac{\partial v}{\partial y}\right) = -\frac{\partial w}{\partial z}$$

so that the vorticity equation (4.22) may be written

$$\frac{D_h(\zeta + f)}{Dt} = (\zeta + f)\left(\frac{\partial w}{\partial z}\right) \tag{4.23}$$

(Here we have retained the relative vorticity in the divergence term for reasons that will become apparent below.)

We showed in Section 3.4 that in a barotropic fluid the thermal wind vanishes so that the geostrophic wind is independent of height. Letting the vorticity in (4.23) be approximated by the geostrophic vorticity ζ_g and the wind by the geostrophic wind (u_g, v_g), we can integrate vertically from z_1 to z_2 to get

$$h\frac{D_h(\zeta_g + f)}{Dt} = (\zeta_g + f)[w(z_2) - w(z_1)] \tag{4.24}$$

But since $w \equiv Dz/Dt$ and $h \equiv h(x, y, t)$,

$$w(z_2) - w(z_1) = \frac{Dz_2}{Dt} - \frac{Dz_1}{Dt} = \frac{D_h h}{Dt} \tag{4.25}$$

Substituting from (4.25) into (4.24) we get

$$\frac{1}{\zeta_g + f}\frac{D_h(\zeta_g + f)}{Dt} = \frac{1}{h}\frac{D_h h}{Dt}$$

or

$$\frac{D_h \ln(\zeta_g + f)}{Dt} = \frac{D_h \ln h}{Dt}$$

which implies that

$$\frac{D_h}{Dt}\left(\frac{\zeta_g + f}{h}\right) = 0 \tag{4.26}$$

This is just the potential vorticity conservation theorem for a barotropic fluid, which was first obtained by C. G. Rossby [see Eq. (4.13)]. The quantity conserved following the motion in (4.26) is the barotropic potential vorticity.

If the flow is purely horizontal ($w = 0$) the divergence term vanishes in (4.23) and we obtain the *barotropic vorticity equation*

$$\frac{D_h(\zeta_g + f)}{Dt} = 0 \tag{4.27}$$

which states that absolute vorticity is conserved following the horizontal motion. More generally, (4.27) is valid for any fluid layer in which the divergence of the horizontal wind vanishes. Since the flow in the mid-troposphere is often nearly nondivergent on the synoptic scale, (4.27) provides a surprisingly good model for short-term forecasts of the synoptic-scale 500-mb flow field (see Section 13.4).

4.6 The Baroclinic (Ertel) Potential Vorticity Equation

In Section 4.3 we used the circulation theorem and mass continuity to show that Ertel's potential vorticity, $P \equiv (\zeta_\theta + f)(-g \, \partial\theta/\partial p)$, is conserved following the motion in adiabatic flow. If diabatic heating or frictional torques are present, P is no longer conserved. An equation governing the rate of change of P following the motion in such circumstances can, however, be derived fairly simply starting from the equations of motion in their isentropic coordinate form.

4.6.1 EQUATIONS OF MOTION IN ISENTROPIC COORDINATES

If the atmosphere is stably stratified, so that potential temperature θ is a monotonically increasing function of height, θ may be used as an independent vertical coordinate. The vertical "velocity" in this coordinate system is just $\dot{\theta} \equiv D\theta/Dt$. Thus, adiabatic motions are two-dimensional when viewed in an isentropic coordinate frame. An infinitesimal control volume in isentropic coordinates with cross-sectional area δA and vertical extent $\delta\theta$ has a mass

$$\delta M = \rho \, \delta A \, \delta z = \delta A \left(-\frac{\delta p}{g} \right) = \frac{\delta A}{g} \left(-\frac{\partial p}{\partial \theta} \right) \delta\theta = \sigma \, \delta A \, \delta\theta \tag{4.28}$$

Here the "density" in (x, y, θ) space (i.e., as shown in Fig. 4.7 the quantity that when multiplied by the "volume" element $\delta A \, \delta\theta$ yields the mass element δM) is defined as

$$\sigma \equiv -g^{-1} \partial p/\partial\theta \tag{4.29}$$

The horizontal momentum equation in isentropic coordinates may be obtained by transforming the isobaric form (4.19) to yield

$$\frac{\partial \mathbf{V}}{\partial t} + \mathbf{\nabla}_\theta \left(\frac{\mathbf{V} \cdot \mathbf{V}}{2} + \Psi \right) + (\zeta_\theta + f)\mathbf{k} \times \mathbf{V} = -\dot{\theta}\frac{\partial \mathbf{V}}{\partial \theta} + \mathbf{F}_r \tag{4.30}$$

where $\mathbf{\nabla}_\theta$ is the gradient on an isentropic surface, $\zeta_\theta \equiv \mathbf{k} \cdot \mathbf{\nabla}_\theta \times \mathbf{V}$ is the isentropic relative vorticity originally introduced in (4.11), and $\Psi \equiv c_p T + \Phi$ is the Montgomery streamfunction (see Problem 2.11). Here we have included a frictional term $\mathbf{F}_r$ on the right-hand side, along with the diabatic vertical advection term. The continuity equation can be derived with the aid of (4.28) in a manner analogous to that used for the isobaric system in Section 3.1.2. The result is

$$\frac{\partial \sigma}{\partial t} + \mathbf{\nabla}_\theta \cdot (\sigma \mathbf{V}) = -\frac{\partial}{\partial \theta}(\sigma \dot{\theta}) \tag{4.31}$$

The Ψ and σ fields are linked through the pressure field by the hydrostatic equation, which in the isentropic system takes the form

$$\frac{\partial \Psi}{\partial \theta} = \Pi(p) \equiv c_p \left(\frac{p}{p_s} \right)^{R/c_p} = c_p \frac{T}{\theta} \tag{4.32}$$

where Π is called the Exner function. Equations (4.29)–(4.32) form a closed set for prediction of $\mathbf{V}$, σ, Ψ, and p, provided that $\dot{\theta}$ and $\mathbf{F}_r$ are known.

4.6.2   THE POTENTIAL VORTICITY EQUATION

If we take $\mathbf{k} \cdot \mathbf{\nabla}_\theta \times (4.30)$ and rearrange the resulting terms we obtain the isentropic vorticity equation:

$$\frac{\tilde{D}}{Dt}(\zeta_\theta + f) + (\zeta_\theta + f)\mathbf{\nabla}_\theta \cdot \mathbf{V} = \mathbf{k} \cdot \mathbf{\nabla}_\theta \times \left(\mathbf{F}_r - \dot{\theta}\frac{\partial \mathbf{V}}{\partial \theta} \right) \tag{4.33}$$

where

$$\frac{\tilde{D}}{Dt} = \frac{\partial}{\partial t} + \mathbf{V} \cdot \mathbf{\nabla}_\theta \tag{4.34}$$

is the total derivative following the horizontal motion on an isentropic surface.

Noting that $\sigma^{-2}\,\partial\sigma/\partial t = -\partial\sigma^{-1}/\partial t$, we can rewrite (4.31) in the form

$$\frac{\tilde{D}}{Dt}(\sigma^{-1}) - (\sigma^{-1})\mathbf{\nabla}_\theta \cdot \mathbf{V} = \sigma^{-2}\frac{\partial}{\partial \theta}(\sigma \dot{\theta}) \tag{4.35}$$

Multiplying each term in (4.33) by σ^{-1} and in (4.35) by $(\zeta_\theta + f)$ and adding, we obtain the desired conservation law:

$$\frac{\tilde{D}P}{Dt} = \frac{\partial P}{\partial t} + \mathbf{V} \cdot \nabla_\theta P = \frac{P}{\sigma} \frac{\partial}{\partial \theta} (\sigma \dot\theta) + \sigma^{-1} \mathbf{k} \cdot \nabla_\theta \times \left(\mathbf{F}_r - \dot\theta \frac{\partial \mathbf{V}}{\partial \theta} \right) \quad (4.36)$$

where $P \equiv (\zeta_\theta + f)/\sigma$ is the Ertel potential vorticity defined in (4.12). If the diabatic and frictional terms on the right-hand side of (4.36) can be evaluated, it is possible to determine the evolution of P following the *horizontal* motion on an isentropic surface. When the diabatic and frictional terms are small, potential vorticity is approximately conserved following the motion on isentropic surfaces.

Weather disturbances that have sharp gradients in dynamical fields, such as jets and fronts, are associated with large anomalies in the Ertel potential vorticity. In the upper troposphere such anomalies tend to be rapidly advected under nearly adiabatic conditions. Thus, the potential vorticity anomaly patterns are materially conserved on isentropic surfaces. This material conservation property makes potential vorticity anomalies particularly useful in identifying and tracing the evolution of meteorological disturbances.

4.6.3   INTEGRAL CONSTRAINTS ON ISENTROPIC VORTICITY

The isentropic vorticity equation (4.33) can be written in the form

$$\frac{\partial \zeta_\theta}{\partial t} = -\nabla_\theta \cdot [(\zeta_\theta + f)\mathbf{V}] + \mathbf{k} \cdot \nabla_\theta \times \left(\mathbf{F}_r - \dot\theta \frac{\partial \mathbf{V}}{\partial \theta} \right) \quad (4.37)$$

Using the fact that any vector $\mathbf{A}$ satisfies the relationship

$$\mathbf{k} \cdot (\nabla_\theta \times \mathbf{A}) = \nabla_\theta \cdot (\mathbf{A} \times \mathbf{k})$$

we can rewrite (4.37) in the form

$$\frac{\partial \zeta_\theta}{\partial t} = -\nabla_\theta \cdot \left[(\zeta_\theta + f)\mathbf{V} - \left(\mathbf{F}_r - \dot\theta \frac{\partial \mathbf{V}}{\partial \theta} \right) \times \mathbf{k} \right] \quad (4.38)$$

Equation (4.38) expresses the remarkable fact that isentropic vorticity can be changed only by the divergence or convergence of the horizontal flux vector in brackets on the right-hand side. The vorticity cannot be changed by vertical transfer across the isentropes. Furthermore, integration of (4.38) over the area of an isentropic surface and application of the divergence theorem (Appendix C.2) shows that for an isentrope that does not intersect the surface of the earth the global average of ζ_θ is constant. Furthermore, integration of ζ_θ over the sphere shows that the global average ζ_θ is exactly zero. Vorticity on such an isentrope is neither created nor destroyed, it is merely concentrated or diluted by horizontal fluxes along the isentropes.

Problems

4.1. What is the circulation about a square of 1000 km on a side for an easterly (that is, westward-flowing) wind that decreases in magnitude toward the north at a rate of 10 m s^{-1} per 500 km? What is the mean relative vorticity in the square?

4.2. A cylindrical column of air at 30°N with radius 100 km expands to twice its original radius. If the air is initially at rest, what is the mean tangential velocity at the perimeter after expansion?

4.3. An air parcel at 30°N moves northward conserving absolute vorticity. If its initial relative vorticity is $5 \times 10^{-5} \text{ s}^{-1}$, what is its relative vorticity upon reaching 90°N?

4.4. An air column at 60°N with $\zeta = 0$ initially stretches from the surface to a fixed tropopause at 10 km height. If the air column moves until it is over a mountain barrier 2.5 km high at 45°N, what are its absolute vorticity and relative vorticity as it passes the mountaintop, assuming that the flow satisfies the barotropic potential vorticity equation?

4.5. Find the average vorticity within a cylindrical annulus of inner radius 200 km and outer radius 400 km if the tangential velocity distribution is given by $V = A/r$, where $A = 10^6 \text{ m}^2 \text{ s}^{-1}$ and r is in meters. What is the average vorticity within the inner circle of radius 200 km? How is the vorticity distributed within the inner circle?

4.6. Show that the anomalous gradient wind cases discussed in Section 3.2.5 have negative absolute circulation in the Northern Hemisphere and hence have negative average absolute vorticity.

4.7. Compute the rate of change of circulation about a square in the x, y plane with corners at $(0, 0)$, $(0, L)$, (L, L), and $(L, 0)$ if temperature increases eastward at a rate of 1°C per 200 km and pressure increases northward at a rate of 1 mb per 200 km. Let $L = 1000$ km and the pressure at the point $(0, 0)$ be 1000 mb.

4.8. Verify the identity (4.18) by expanding the vectors in Cartesian components.

4.9. Derive a formula for the dependence of depth on radius for an incompressible fluid in solid-body rotation in a cylindrical tank with a flat bottom and free surface at the upper boundary. Let H be the depth at the center of the tank Ω the angular velocity of rotation of the tank, and a the radius of the tank.

4.10. By how much does the relative vorticity change for a column of fluid in a rotating cylinder if the column is moved from the center of the tank to a distance 50 cm from the center? The tank is rotating at the rate of 20 revolutions per minute, the depth of the fluid at the center is 10 cm, and the fluid is initially in solid-body rotation.

4.11. A cyclonic vortex is in cyclostrophic balance with a tangential velocity profile given by the expression $V = V_0(r/r_0)^n$ where V_0 is the tangential velocity component at the distance r_0 from the vortex center. Compute the circulation about a streamline at radius r, the vorticity at radius r, and the pressure at radius r. (Let p_0 be the pressure at r_0 and assume that density is a constant.)

4.12. A westerly zonal flow at 45° is forced to rise adiabatically over a north–south-oriented mountain barrier. Before striking the mountain the westerly wind increases linearly toward the south at a rate of 10 m s^{-1} per 1000 km. The crest of the mountain range is at the 80-kPa (800-mb) level and the tropopause is located at 30 kPa (300 mb). What is the initial relative vorticity of the air? What is its relative vorticity when it reaches the crest if it is deflected 5° latitude toward the south during the forced ascent? If the current assumes a uniform speed of 20 m s^{-1} during its ascent to the crest, what is the radius of curvature of the streamlines at the crest?

4.13. A cylindrical vessel of radius a and constant depth H rotating at an angular velocity Ω about its vertical axis of symmetry is filled with a homogeneous, incompressible fluid which is initially at rest with respect to the vessel. A volume of fluid V is then withdrawn through a point sink at the center of the cylinder, thus creating a vortex. Neglecting friction, derive an expression for the resulting relative azimuthal velocity as a function of radius (i.e., the velocity in a coordinate system rotating with the tank). Assume that the motion is independent of depth and that $V \ll \pi a^2 H$. Also compute the relative vorticity and the relative circulation.

4.14. (a) How far must a zonal ring of air initially at rest with respect to the earth's surface at 60° latitude and 100-km height be displaced latitudinally in order to acquire an easterly (east to west) component of 10 m s^{-1} with respect to the earth's surface?

(b) To what height must it be displaced vertically in order to acquire the same velocity? Assume a frictionless atmosphere.

4.15. The horizontal motion within a cylindrical annulus with permeable walls of inner radius 10 cm, outer radius 20 cm, and 10-cm depth is independent of height and azimuth and is represented by the expressions

$$u = 7 - 0.2r, \qquad v = 40 + 2r$$

where u and v are the radial and tangential velocity components in cm s^{-1}, positive outward and counterclockwise, respectively, and r is distance from the center of the annulus in cm. Assuming an incompressible fluid, find

(a) the circulation about the annular ring
(b) the average vorticity within the annular ring
(c) the average divergence within the annular ring
(d) the average vertical velocity at the top of the annulus if it is zero at the base

4.16. Prove that, as stated below Eq. (4.38), the globally averaged isentropic vorticity on an isentropic surface that does not intersect the ground must be zero. Show that the same result holds for the isobaric vorticity on an isobaric surface.

Suggested References

Williams and Elder, *Fluid Physics for Oceanographers and Physicists*, provides an introduction to vorticity dynamics at an elementary level. This book also provides a good general introduction to fluid dynamics.

Brown, *Fluid Mechanics of the Atmosphere*, provides a good introduction to vorticity at the graduate level.

Hoskins *et al.* (1985) provide an advanced discussion of Ertel potential vorticity and its uses in diagnosis and prediction of synoptic-scale disturbances.

Chapter

5 | The Planetary Boundary Layer

The *planetary boundary layer* is that portion of the atmosphere in which the flow field is strongly influenced directly by interaction with the surface of the earth. Ultimately this interaction depends on molecular viscosity. It is, however, only within a few millimeters of the surface, where vertical shears are very intense, that molecular diffusion is comparable to other terms in the momentum equation. Outside this *viscous sublayer* molecular diffusion is not important in the boundary layer equations for the mean wind, although it is still important for small-scale turbulent eddies. However, viscosity still has an important indirect role; it causes the velocity to vanish at the surface. As a consequence of this *no-slip boundary condition* even a fairly weak wind will cause a large velocity shear near the surface, which continually leads to the development of turbulent eddies. These turbulent motions have spatial and temporal variations at scales much smaller than those resolved by the meteorological observing network. Such shear-induced eddies, together with convective eddies caused by surface heating, are very effective in transferring momentum to the surface and transferring heat (latent and sensible) away from the surface many orders of magnitude faster than can be done by molecular processes. The depth of the planetary boundary layer produced by this turbulent transport may range from as little as 30 m in conditions of large static stability to more than 3 km in highly convective conditions. For average midlatitude conditions the planetary boundary layer extends through the lowest kilometer of the atmosphere and thus contains about 10% of the mass of the atmosphere.

The dynamical structure of the flow in the planetary boundary layer is not directly produced by viscosity. Rather, it is largely determined by the fact that the atmospheric flow is turbulent. In the *free atmosphere* (i.e., the region above the planetary boundary layer) this turbulence can be ignored in an approximate treatment of synoptic-scale motions, except perhaps in the vicinity of jet streams, fronts, and convective clouds. But in the boundary layer the dynamical equations of the previous chapters must be modified to properly represent the effects of turbulence.

5.1 Atmospheric Turbulence

Turbulent flow contains irregular quasi-random motions spanning a continuous spectrum of spatial and temporal scales. Such eddies cause nearby air parcels to drift apart and thus mix properties such as momentum and potential temperature across the boundary layer. Unlike the large-scale rotational flows discussed in earlier chapters, which have depth scales that are small compared to their horizontal scales, the turbulent eddies of concern in the planetary boundary layer tend to have similar scales in the horizontal and vertical. The maximum eddy length scale is thus limited by the boundary layer depth to be about 10^3 m. The minimum length scale (10^{-3} m) is that of the smallest eddies that can exist in the presence of diffusion by molecular friction.

Even when observations are taken with very short temporal and spatial separations, a turbulent flow will always have scales that are unresolvable because they have frequencies greater than the observation frequency and spatial scales smaller than the scale of separation of the observations. Outside the boundary layer, in the free atmosphere, the problem of unresolved scales of motion is usually not a serious one for diagnosis or forecasting of synoptic- and larger-scale circulations (although it is crucial for the mesoscale circulations to be discussed in Chapter 9). The eddies that contain the bulk of the energy in the free atmosphere are resolved by the synoptic network. However, in the boundary layer unresolved turbulent eddies are of critical importance. Through their transport of heat and moisture away from the surface they maintain the surface energy balance, and through their transport of momentum to the surface they maintain the momentum balance. The latter process dramatically alters the momentum balance of the large-scale flow in the boundary layer, so that geostrophic balance is no longer an adequate approximation to the large-scale wind field. It is this aspect of boundary layer dynamics that is of primary importance for dynamical meteorology.

5.1.1 THE BOUSSINESQ APPROXIMATION

The basic state (standard atmosphere) density varies across the lowest kilometer of the atmosphere by only about 10%, and the fluctuating component of density deviates from the basic state by only a few percent. These circumstances might suggest that boundary layer dynamics could be modeled by setting density constant and using the theory of homogeneous incompressible fluids. This is, however, generally not the case. Density fluctuations cannot be totally neglected since they are essential for representing the buoyancy force (see the discussion in Section 2.7.3).

Nevertheless, it is still possible to make some important simplifications in the dynamical equations for application in the boundary layer. The *Boussinesq approximation* is a form of the dynamical equations that is valid for this situation. In this approximation density is replaced by a constant mean value, ρ_0, everywhere except in the buoyancy term in the vertical momentum equation. The horizontal momentum equations (2.24)–(2.25) can then be expressed in Cartesian coordinates as

$$\frac{Du}{Dt} = -\frac{1}{\rho_0}\frac{\partial p}{\partial x} + fv + F_{rx} \tag{5.1}$$

and

$$\frac{Dv}{Dt} = -\frac{1}{\rho_0}\frac{\partial p}{\partial y} - fu + F_{ry} \tag{5.2}$$

while with the aid of (2.28) and (2.51) the vertical momentum equation becomes

$$\frac{Dw}{Dt} = -\frac{1}{\rho_0}\frac{\partial p}{\partial z} + g\frac{\theta}{\theta_0} + F_{rz} \tag{5.3}$$

Here, as in Section 2.7.3, θ designates the departure of potential temperature from its basic state value $\theta_0(z)$.[1]   Thus, the total potential temperature field

---

[1] The reason that we have not used the notation θ' to designate the fluctuating part of the potential temperature field will become apparent in Section 5.1.2.

is given by $\theta_{tot} = \theta(x, y, z, t) + \theta_0(z)$, and the adiabatic thermodynamic energy equation has the form of (2.54):

$$\frac{D\theta}{Dt} = -w\frac{d\theta_0}{dz} \tag{5.4}$$

Finally, the continuity equation (2.34) under the Boussinesq approximation is

$$\frac{\partial u}{\partial x} + \frac{\partial v}{\partial y} + \frac{\partial w}{\partial z} = 0 \tag{5.5}$$

5.1.2 REYNOLDS AVERAGING

In a turbulent fluid a field variable such as velocity measured at a point generally fluctuates rapidly in time as eddies of various scales pass the point. In order that measurements be truly representative of the large-scale flow it is necessary to average the flow over an interval of time long enough to average out the small-scale eddy fluctuations but still short enough to preserve the trends in the large-scale flow field. To do this we assume that the field variables can be separated into slowly varying mean fields and rapidly varying turbulent components. Following the scheme introduced by Reynolds, we then assume that for any field variables, w and θ for example, the corresponding means are indicated by overbars and the fluctuating components by primes. Thus, $w = \bar{w} + w'$ and $\theta = \bar{\theta} + \theta'$. By definition, the means of the fluctuating components vanish; the product of a deviation with a mean also vanishes when the time average is applied. Thus,

$$\overline{w'\bar{\theta}} = \overline{w'}\bar{\theta} = 0$$

where we have used the fact that a mean variable is constant over the period of averaging. The average of the product of deviation components (called the *covariance*) does not generally vanish. For example, if on average the turbulent vertical velocity is upward where the potential temperature deviation is positive and downward where it is negative, the product $\overline{w'\theta'}$ is positive and the variables are said to be positively correlated. These averaging rules imply that the average of the product of two variables will be the product of the average of the means plus the product of the average of the deviations:

$$\overline{w\theta} = \overline{(\bar{w} + w')(\bar{\theta} + \theta')} = \bar{w}\bar{\theta} + \overline{w'\theta'}$$

Before applying the Reynolds decomposition to (5.1)–(5.4) it is convenient to rewrite the total derivatives in each equation in flux form. For example, the term on the left in (5.1) may be manipulated with the aid of the continuity equation (5.5) to yield

$$\frac{Du}{Dt} = \frac{\partial u}{\partial t} + u\frac{\partial u}{\partial x} + v\frac{\partial u}{\partial y} + w\frac{\partial u}{\partial z} + u\left(\frac{\partial u}{\partial x} + \frac{\partial v}{\partial y} + \frac{\partial w}{\partial z}\right)$$

$$= \frac{\partial u}{\partial t} + \frac{\partial u^2}{\partial x} + \frac{\partial uv}{\partial y} + \frac{\partial uw}{\partial z} \tag{5.6}$$

Separating each dependent variable into mean and fluctuating parts, substituting into (5.6), and averaging then yields

$$\frac{\overline{Du}}{Dt} = \frac{\partial \bar{u}}{\partial t} + \frac{\partial}{\partial x}(\bar{u}\bar{u} + \overline{u'u'}) + \frac{\partial}{\partial y}(\bar{u}\bar{v} + \overline{u'v'}) + \frac{\partial}{\partial z}(\bar{u}\bar{w} + \overline{u'w'}) \tag{5.7}$$

Noting that the mean velocity fields satisfy the continuity equation (5.5), we can rewrite (5.7) as

$$\frac{\overline{Du}}{Dt} = \frac{\bar{D}\bar{u}}{Dt} + \frac{\partial}{\partial x}(\overline{u'u'}) + \frac{\partial}{\partial y}(\overline{u'v'}) + \frac{\partial}{\partial z}(\overline{u'w'}) \tag{5.8}$$

where

$$\frac{\bar{D}}{Dt} = \frac{\partial}{\partial t} + \bar{u}\frac{\partial}{\partial x} + \bar{v}\frac{\partial}{\partial y} + \bar{w}\frac{\partial}{\partial z}$$

is the rate of change following the mean motion.

The mean equations thus have the form,

$$\frac{\bar{D}\bar{u}}{Dt} = -\frac{1}{\rho_0}\frac{\partial \bar{p}}{\partial x} + f\bar{v} - \left[\frac{\partial \overline{u'u'}}{\partial x} + \frac{\partial \overline{u'v'}}{\partial y} + \frac{\partial \overline{u'w'}}{\partial z}\right] + \bar{F}_{rx} \tag{5.9}$$

$$\frac{\bar{D}\bar{v}}{Dt} = -\frac{1}{\rho_0}\frac{\partial \bar{p}}{\partial y} - f\bar{u} - \left[\frac{\partial \overline{u'v'}}{\partial x} + \frac{\partial \overline{v'v'}}{\partial y} + \frac{\partial \overline{v'w'}}{\partial z}\right] + \bar{F}_{ry} \tag{5.10}$$

$$\frac{\bar{D}\bar{w}}{Dt} = -\frac{1}{\rho_0}\frac{\partial \bar{p}}{\partial z} + g\frac{\bar{\theta}}{\theta_0} - \left[\frac{\partial \overline{u'w'}}{\partial x} + \frac{\partial \overline{v'w'}}{\partial y} + \frac{\partial \overline{w'w'}}{\partial z}\right] + \bar{F}_{rz} \tag{5.11}$$

$$\frac{\bar{D}\bar{\theta}}{Dt} = -\bar{w}\frac{d\theta_0}{dz} - \left[\frac{\partial \overline{u'\theta'}}{\partial x} + \frac{\partial \overline{v'\theta'}}{\partial y} + \frac{\partial \overline{w'\theta'}}{\partial z}\right] \tag{5.12}$$

$$\frac{\partial \bar{u}}{\partial x} + \frac{\partial \bar{v}}{\partial y} + \frac{\partial \bar{w}}{\partial z} = 0 \qquad (5.13)$$

The various covariance terms in square brackets in (5.9)–(5.12) represent turbulent fluxes. For example, $\overline{w'\theta'}$ is a vertical turbulent heat flux in kinematic form. Similarly, $\overline{w'u'} = \overline{u'w'}$ is a vertical turbulent flux of zonal momentum. For many boundary layers the magnitudes of the turbulent flux divergence terms are of the same order as the other terms in (5.9)–(5.12). In such cases, it is not possible to neglect the turbulent flux terms even when only the mean flow is of direct interest. Outside the boundary layer the turbulent fluxes are often sufficiently weak that the terms in square brackets in (5.9)–(5.12) can be neglected in analysis of large-scale flows. This assumption was implicitly made in Chapters 3 and 4.

The complete equations for the mean flow (5.9)–(5.12), unlike the equations for the total flow (5.1)–(5.5) and the approximate equations of Chapters 3 and 4, are not a closed set, since in addition to the five unknown mean variables $\bar{u}, \bar{v}, \bar{w}, \bar{\theta}, \bar{p}$ there are the unknown turbulent fluxes. To solve these equations *closure* assumptions must be made to approximate the unknown fluxes in terms of the five known mean state variables. Away from regions with horizontal inhomogeneities (e.g., shorelines, towns, forest edges) we can simplify by assuming that the turbulent fluxes are horizontally homogeneous so that it is possible to neglect the horizontal derivative terms in square brackets in comparison to the terms involving vertical differentiation.

5.2 Turbulent Kinetic Energy

Vortex stretching and twisting associated with turbulent eddies always tends to cause turbulent energy to flow toward the smallest scales, where it is dissipated by viscous diffusion. Thus, there must be continuing production of turbulence if the turbulent kinetic energy is to remain statistically steady. The primary source of boundary layer turbulence depends critically on the structure of the wind and temperature profiles near the surface. If the lapse rate is unstable, boundary layer turbulence is convectively generated. If it is stable, then instability associated with wind shear must be responsible for generating turbulence in the boundary layer. The comparative roles of these processes can best be understood by examining the budget for turbulent kinetic energy.

To investigate the production of turbulence we subtract the component mean momentum equations (5.9)–(5.11) from the corresponding unaveraged equations (5.1)–(5.3). We then multiply the results by u', v', w', respectively, add the resulting three equations, and average to obtain the turbulent kinetic energy equation. The complete statement of this equation is quite complicated, but its essence can be expressed symbolically as follows:

$$\frac{\bar{D}(\text{TKE})}{Dt} = \text{MP} + \text{BPL} + \text{TR} - \varepsilon \qquad (5.14)$$

where $\text{TKE} \equiv (\overline{u'^2} + \overline{v'^2} + \overline{w'^2})/2$ is the turbulent kinetic energy per unit mass, MP is the mechanical production, BPL is the buoyant production or loss, TR designates redistribution by transport and pressure forces, and ε designates frictional dissipation. ε is always positive, reflecting the dissipation of the smallest scales of turbulence by molecular viscosity.

The buoyancy term in (5.14) represents a conversion of energy between mean flow potential energy and turbulent kinetic energy. It is positive (negative) for motions that systematically tend to lower (raise) the center of mass of the atmosphere. It has the form[2]

$$\text{BPL} \equiv \overline{w'\theta'}(g/\theta_0)$$

Positive buoyancy production occurs when there is heating at the surface so that an unstable temperature lapse rate (see Section 2.7.2) develops near the ground and spontaneous convective overturning can occur. As shown in the schematic of Fig. 5.1, convective eddies have positively correlated vertical velocity and potential temperature fluctuations and hence provide a source of turbulent kinetic energy and positive heat flux. This is the dominant source in a convectively unstable boundary layer. For a statically stable atmosphere BPL is negative, which tends to reduce or eliminate turbulence.

For both statically stable and unstable boundary layers, turbulence can be generated mechanically by dynamical instability owing to wind shear. This process is represented by the mechanical production term in (5.14), which represents a conversion of energy between the mean flow and the turbulent fluctuations. This term is proportional to the shear in the mean flow and has the form

$$\text{MP} \equiv -\overline{u'w'}\frac{\partial \bar{u}}{\partial z} - \overline{v'w'}\frac{\partial \bar{v}}{\partial z} \qquad (5.15)$$

[2] In practice, buoyancy in the boundary layer is modified by the presence of water vapor, which has a density significantly lower than that of dry air. The potential temperature should be replaced by virtual potential temperature in (5.14) in order to include this effect. (See, for example, Wallace and Hobbs, 1977, p. 52)

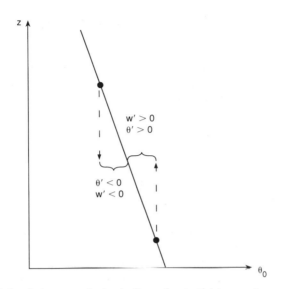

Fig. 5.1 Correlation between vertical velocity and potential temperature perturbations for upward or downward parcel displacements when the mean potential temperature $\theta_0(z)$ decreases with height.

Thus, MP is positive when the momentum flux is directed down the gradient of the mean momentum.

In a statically stable layer turbulence can exist only if mechanical production is large enough to overcome the damping effects of stability and viscous dissipation. This condition is measured by a quantity called the *flux Richardson number*. It is defined as

$$Rf \equiv -BPL/MP$$

If the boundary layer is statically unstable, then $Rf < 0$ and turbulence is sustained by convection. For stable conditions Rf will be greater than zero. Observations suggest that only when Rf is less than about 0.25 (i.e., mechanical production exceeds buoyancy damping by a factor of 4) is the mechanical production intense enough to sustain turbulence in a stable layer. Since MP depends on the shear it always becomes large close enough to the surface. However, as the static stability increases, the depth of the layer in which there is net production of turbulence shrinks. Thus, when there is a strong temperature inversion, such as produced by nocturnal radiative cooling, the boundary layer depth may be only a few decameters and vertical mixing is strongly suppressed.

5.3 Planetary Boundary Layer Momentum Equations

For the special case of horizontally homogeneous turbulence above the viscous sublayer, molecular viscosity and horizontal turbulent momentum flux divergence terms can be neglected. The mean flow horizontal momentum equations (5.9) and (5.10) then become

$$\frac{\bar{D}\bar{u}}{Dt} = -\frac{1}{\rho_0}\frac{\partial \bar{p}}{\partial x} + f\bar{v} - \frac{\partial \overline{u'w'}}{\partial z} \tag{5.16}$$

$$\frac{\bar{D}\bar{v}}{Dt} = -\frac{1}{\rho_0}\frac{\partial \bar{p}}{\partial y} - f\bar{u} - \frac{\partial \overline{v'w'}}{\partial z} \tag{5.17}$$

In general, (5.16) and (5.17) can be solved for $\bar{u}$ and $\bar{v}$ only if the vertical distribution of the turbulent momentum flux is known. Since this depends on the structure of the turbulence, no general solution is possible. Rather, a number of approximate semi-empirical methods are used.

For midlatitude synoptic-scale motions, we showed in Section 2.4 that to a first approximation the inertial acceleration terms [the terms on the left in (5.16) and (5.17)] could be neglected compared to the Coriolis force and pressure gradient force terms. Outside the boundary layer, the resulting approximation was then simply geostrophic balance. In the boundary layer the inertial terms are still small compared to the Coriolis force and pressure gradient force terms, but the turbulent flux terms must be included. Thus, to a first approximation the planetary boundary layer equations express a three-way balance among the Coriolis force, the pressure gradient force, and the turbulent momentum flux divergence:

$$f(\bar{v} - \bar{v}_{g}) - \frac{\partial \overline{u'w'}}{\partial z} = 0 \tag{5.18}$$

$$-f(\bar{u} - \bar{u}_{g}) - \frac{\partial \overline{v'w'}}{\partial z} = 0 \tag{5.19}$$

where (2.23) is used to express the pressure gradient force in terms of the geostrophic velocity.

5.3.1 WELL-MIXED BOUNDARY LAYER

If a convective boundary layer is topped by a stable layer, turbulent mixing can lead to formation of a well-mixed layer. Such boundary layers occur commonly over land during the day when surface heating is strong and over the oceans when the air near the sea surface is colder than the surface water temperature. The tropical oceans typically have boundary layers of this type.

In a well-mixed boundary layer the wind speed and potential temperature are nearly independent of height, as shown schematically in Fig. 5.2, and to a first approximation it is possible to treat the layer as a slab in which the velocity and potential temperature profiles are constant with height and turbulent fluxes vary linearly with height. For simplicity we assume that the turbulence vanishes at the top of the boundary layer. Observations indicate that the surface momentum flux can be represented by a *bulk aerodynamic formula*[3]

$$\overline{(u'w')}_s = -C_d|\bar{V}|\bar{u} \quad \text{and} \quad \overline{(v'w')}_s = -C_d|\bar{V}|\bar{v}$$

where C_d is a nondimensional *drag coefficient*, $|\bar{V}| = (\bar{u}^2 + \bar{v}^2)^{1/2}$, and the subscript s denotes surface values (referred to the standard anemometer height). Observations show that C_d is of order 1.5×10^{-3} over the oceans but may be several times as large over rough ground.

The approximate planetary boundary layer equations (5.18) and (5.19) can then be integrated from the surface to the top of the boundary layer at $z = h$ to give

$$f(\bar{v} - \bar{v}_g) = -\overline{(u'w')}_s/h = C_d|\bar{V}|\bar{u}/h \tag{5.20}$$

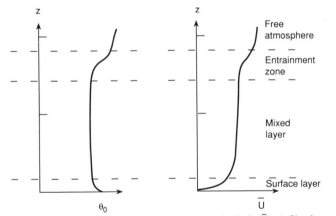

Fig. 5.2 Mean potential temperature, θ_0, and mean zonal wind, $\bar{U}$, profiles in a well-mixed boundary layer. Adapted from Stull (1988).

---

[3] The turbulent momentum flux is often represented in terms of an "eddy stress" by defining, for example, $\tau_{ex} = \rho_0 \overline{u'w'}$. We prefer to avoid this terminology to eliminate possible confusion with molecular friction.

$$-f(\bar{u} - \bar{u}_g) = -(\overline{v'w'})_s / h = C_d |\bar{V}| \bar{v} / h \tag{5.21}$$

Without loss of generality we can choose axes such that $\bar{v}_g = 0$. Then (5.20) and (5.21) can be rewritten as

$$\bar{u} = \bar{u}_g - \kappa_s |\bar{V}| \bar{v}, \qquad \bar{v} = \kappa_s |\bar{V}| \bar{u} \tag{5.22}$$

where $\kappa_s \equiv C_d / (fh)$. Thus, in the mixed layer the wind speed is less than the geostrophic speed and there is a component of motion directed toward lower pressure (i.e., to the left of the geostrophic wind in the Northern Hemisphere and to the right in the Southern Hemisphere) whose magnitude depends on κ_s. For example, if $\bar{u}_g = 10$ m s^{-1} and $\kappa_s = 0.05$ m^{-1} s, then $\bar{u} = 8.28$ m s^{-1}, $\bar{v} = 3.77$ m s^{-1}, and $|\bar{V}| = 9.10$ m s^{-1} at all heights within this idealized slab mixed layer.

It is the work done by the flow toward lower pressure that balances the frictional dissipation at the surface. Since boundary layer turbulence tends to reduce wind speeds, the turbulent momentum flux terms are often referred to as boundary layer *friction*. It should be kept in mind, however, that the forces involved are due to turbulence, not molecular viscosity.

Qualitatively, the cross-isobar flow in the boundary layer can be understood as a direct result of the three-way balance among the pressure gradient force, the Coriolis force, and turbulent drag:

$$f\mathbf{k} \times \bar{\mathbf{V}} = -\frac{1}{\rho_0} \nabla \bar{p} - \frac{C_d}{h} |\bar{V}| \bar{\mathbf{V}} \tag{5.23}$$

This balance is illustrated in Fig. 5.3. Since the Coriolis force is always normal to the velocity and the turbulent drag is a retarding force, their sum can exactly balance the pressure gradient force only if the wind is directed toward lower pressure. Furthermore, it is easy to see that as the turbulent drag becomes increasingly dominant the cross-isobar angle must increase.

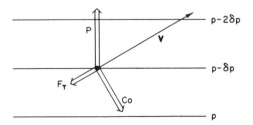

Fig. 5.3 Balance of forces in the well-mixed planetary boundary layer; P designates the pressure gradient force, Co the Coriolis force, F_T the turbulent drag.

5.3.2 THE FLUX-GRADIENT THEORY

In neutrally or stably stratified boundary layers the wind speed and direction vary significantly with height. The simple slab model is no longer appropriate; some means is needed to determine the vertical dependence of the turbulent momentum flux divergence in terms of mean variables in order to obtain closed equations for the boundary layer variables. The traditional approach to this closure problem is to assume that turbulent eddies act in a manner analogous to molecular diffusion so that the flux of a given field is proportional to the local gradient of the mean. In this case the turbulent flux terms in (5.18) and (5.19) are written as

$$\overline{u'w'} = -K_{\mathrm{m}}\left(\frac{\partial \bar{u}}{\partial z}\right), \qquad \overline{v'w'} = -K_{\mathrm{m}}\left(\frac{\partial \bar{v}}{\partial z}\right)$$

and the potential temperature flux can be written as

$$\overline{\theta'w'} = -K_{\mathrm{h}}\left(\frac{\partial \bar{\theta}}{\partial z}\right)$$

where $K_{\mathrm{m}}(\mathrm{m}^2\,\mathrm{s}^{-1})$ is the *eddy viscosity* coefficient and K_{h} is the eddy diffusivity of heat. This closure scheme is often referred to as K *theory*.

K theory has many limitations. Unlike the molecular viscosity coefficient, eddy viscosities depend on the flow rather than the physical properties of the fluid and must be determined empirically for each situation. The simplest models have assumed that the eddy exchange coefficient is constant throughout the flow. This approximation may be adequate for estimating small-scale diffusion of passive tracers in the free atmosphere. But it is a very poor approximation in the boundary layer, where the scales and intensities of typical turbulent eddies are strongly dependent on the distance to the surface as well as the static stability. Furthermore, in many cases the most energetic eddies have dimensions comparable to the boundary layer depth, and neither the momentum flux nor the heat flux is proportional to the local gradient of the mean. For example, in much of the mixed layer heat fluxes are positive even though the mean stratification may be very close to neutral.

5.3.3 THE MIXING LENGTH HYPOTHESIS

The simplest approach to determining a suitable model for the eddy diffusion coefficient in the boundary layer is based on the mixing length hypothesis introduced by the famous fluid dynamicist L. Prandtl. This hypothesis assumes that a parcel of fluid that is displaced vertically will

carry the mean properties of its original level for a characteristic distance
ξ' and then will mix with its surroundings just as an average molecule
travels a mean free path before colliding and exchanging momentum with
another molecule. By further analogy to the molecular mechanism, this
displacement is postulated to create a turbulent fluctuation whose magnitude
depends upon ξ' and the gradient of the mean property. For example,

$$\theta' = -\xi' \frac{\partial \bar{\theta}}{\partial z}, \qquad u' = -\xi' \frac{\partial \bar{u}}{\partial z}, \qquad v' = -\xi' \frac{\partial \bar{v}}{\partial z}$$

where it must be understood that $\xi' > 0$ for upward parcel displacement
and $\xi' < 0$ for downward parcel displacement. For a conservative property
like potential temperature this hypothesis is reasonable provided that the
eddy scales are small compared to the mean flow scale, or provided that
the mean gradient is constant with height. However, the hypothesis is less
justified in the case of velocity, since pressure gradient forces may cause
substantial changes in the velocity during an eddy displacement.

Nevertheless, if we use the mixing length hypothesis the vertical turbulent
flux of zonal momentum can be written as

$$-\overline{u'w'} = \overline{w'\xi'} \frac{\partial \bar{u}}{\partial z} \tag{5.24}$$

with analogous expressions for the momentum flux in the meridional direction and the potential temperature flux. In order to estimate w' in terms of
the mean fields we assume that the vertical stability of the atmosphere is
nearly neutral so that buoyancy effects are small. The horizontal scale of
the eddies should then be comparable to the vertical scale so that $|w'| \sim |\mathbf{V}'|$
and we can set

$$w' \approx \xi' \left| \frac{\partial \bar{\mathbf{V}}}{\partial z} \right|$$

where $\mathbf{V}'$ and $\bar{\mathbf{V}}$ designate the turbulent and mean parts of the horizontal
velocity field, respectively. Here the absolute value of the mean velocity
gradient is needed because if $\xi' > 0$, then $w' > 0$ (that is, upward parcel
displacements are associated with upward eddy velocities). Thus the
momentum flux can be written

$$-\overline{u'w'} = \overline{\xi'^2} \left| \frac{\partial \bar{\mathbf{V}}}{\partial z} \right| \frac{\partial \bar{u}}{\partial z} = K_{\mathrm{m}} \frac{\partial \bar{u}}{\partial z} \tag{5.25}$$

where the eddy viscosity is now defined as $K_m = \overline{\xi'^2}|\partial\overline{\mathbf{V}}/\partial z| = \overline{l^2}|\partial\overline{\mathbf{V}}/\partial z|$ and the mixing length

$$l \equiv (\overline{\xi'^2})^{1/2}$$

is the root-mean-square parcel displacement, which is a measure of average eddy size. This result suggests that larger eddies and greater shears induce greater turbulent mixing.

5.3.4 THE EKMAN LAYER

If the flux–gradient approximation is used to represent the turbulent momentum flux divergence terms in (5.18) and (5.19) and the value of K_m is taken to be constant, we obtain the equations of the classical Ekman layer:

$$K_m \frac{\partial^2 u}{\partial z^2} + f(v - v_g) = 0 \tag{5.26}$$

$$K_m \frac{\partial^2 v}{\partial z^2} - f(u - u_g) = 0 \tag{5.27}$$

where we have omitted the overbars because all fields are Reynolds averaged.

The Ekman layer equations (5.26) and (5.27) can be solved to determine the height dependence of the departure of the wind field in the boundary layer from geostrophic balance. In order to keep the analysis as simple as possible we assume that these equations apply throughout the depth of the boundary layer. The boundary conditions on u and v then require that both horizontal velocity components vanish at the ground and approach their geostrophic values far from the ground:

$$u = 0, \qquad v = 0 \qquad \text{at } z = 0$$
$$u \to u_g, \qquad v \to v_g \qquad \text{as } z \to \infty \tag{5.28}$$

To solve (5.26) and (5.27) we combine them into a single equation by first multiplying (5.27) by $i = (-1)^{1/2}$ and then adding the result to (5.26) to obtain a second-order equation in the complex velocity, $(u + iv)$:

$$K_m \frac{\partial^2(u + iv)}{\partial z^2} - if(u + iv) = -if(u_g + iv_g) \tag{5.29}$$

For simplicity, we assume that the geostrophic wind is independent of height and that the flow is oriented so that the geostrophic wind is in the zonal direction ($v_g = 0$). Then the general solution of (5.29) is

$$(u + iv) = A \exp[(if/K_m)^{1/2}z] + B \exp[-(if/K_m)^{1/2}z] + u_g \quad (5.30)$$

It can be shown that $\sqrt{i} = (1 + i)/\sqrt{2}$. Using this relationship and applying the boundary conditions (5.28), we find that for the Northern Hemisphere ($f > 0$), $A = 0$ and $B = -u_g$. Thus

$$u + iv = -u_g \exp[-\gamma(1 + i)z] + u_g$$

where $\gamma = (f/2K_m)^{1/2}$.

Applying the Euler formula $\exp(-i\theta) = \cos\theta - i \sin\theta$ and separating the real from the imaginary part, we obtain for the Northern Hemisphere

$$u = u_g(1 - e^{-\gamma z} \cos \gamma z), \qquad v = u_g e^{-\gamma z} \sin \gamma z \quad (5.31)$$

This solution is the famous Ekman spiral named for the Swedish oceanographer V. W. Ekman, who first derived an analogous solution for the surface wind drift current in the ocean. The structure of the solution (5.31) is best illustrated by a hodograph as shown in Fig. 5.4. In this figure the zonal and meridional components of the wind are plotted as functions of height. The heavy solid curve traced out on the figure connects all the points corresponding to values of u and v in (5.31) for values of γz increasing as one moves away from the origin along the spiral. Arrows show the velocities for various values of γz marked at the arrow points. When $z = \pi/\gamma$, the wind is parallel to and nearly equal to the geostrophic value. It is conventional to designate this level as the top of the Ekman layer and to define the layer depth as $De \equiv \pi/\gamma$.

Observations indicate that the wind approaches its geostrophic value at about 1 km above the surface. Letting $De = 1$ km and $f = 10^{-4}$ s^{-1}, we find

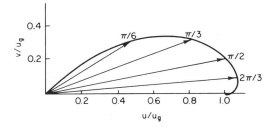

Fig. 5.4 Hodograph of the wind components in the Ekman spiral solution. The arrows show the velocity vectors for several levels in the Ekman layer, while the spiral curve traces out the velocity variation as a function of height. Points labeled on the spiral show the values of γz, which is a nondimensional measure of height.

with the aid of the definition of γ that $K_m \approx 5 \ \mathrm{m^2 \ s^{-1}}$. Referring back to (5.25) we see that for a characteristic boundary layer velocity shear of $|\delta \mathbf{V}/\delta z| \sim 5 \times 10^{-3} \ \mathrm{s^{-1}}$ this value of K_m implies a mixing length of about 30 m, which is small compared to the depth of the boundary layer, as it should be if the mixing length concept is to be useful.

Qualitatively the most striking feature of the Ekman layer solution is that, like the mixed layer solution of Section 5.3.1, it has a boundary layer wind component directed toward lower pressure. As in the mixed layer case, this is a direct result of the three-way balance between the pressure gradient force, the Coriolis force, and the turbulent drag.

The ideal Ekman layer discussed here is rarely, if ever, observed in the atmospheric boundary layer. Partly this is because turbulent momentum fluxes are usually not simply proportional to the gradient of the mean momentum. However, even if the flux–gradient model were correct it still would not be proper to assume a constant eddy viscosity coefficient, since in reality K_m must vary rapidly with height near the ground. Thus, the Ekman layer solution should not be carried all the way to the surface.

5.3.5   THE SURFACE LAYER

Some of the inadequacies of the Ekman layer model can be overcome if we distinguish a *surface layer* from the remainder of the planetary boundary layer. The surface layer, whose depth depends on stability but is usually less than 10% of the total boundary layer depth, is maintained entirely by vertical momentum transfer by the turbulent eddies; it is not directly dependent on the Coriolis or pressure gradient forces. Analysis is facilitated by supposing that the wind close to the surface is directed parallel to the x axis. The kinematic turbulent momentum flux can then be expressed in terms of a *friction velocity*, u_*, which is defined as[4]

$$u_*^2 \equiv |(\overline{u'w'})_s|$$

Measurements indicate that the magnitude of the surface momentum flux is of order $0.1 \ \mathrm{m^2 \ s^{-2}}$. Thus the friction velocity is typically of order $0.3 \ \mathrm{m \ s^{-1}}$.

According to the scale analysis in Section 2.4, the Coriolis and pressure gradient force terms in (5.16) have magnitudes of about $10^{-3} \ \mathrm{m \ s^{-2}}$ in midlatitudes. If these terms are to balance the momentum flux divergence, it is necessary that

$$\frac{\delta(u_*^2)}{\delta z} \leq 10^{-3} \ \mathrm{m \ s^{-2}}$$

---

[4] Thus the surface eddy stress (see footnote 3) is equal to $\rho_0 u_*^2$.

Thus, for $\delta z = 10$ m, $\delta(u_*^2) \le 10^{-2}$ m^2 s^{-2}, and the change in the vertical momentum flux in the lowest 10 m of the atmosphere is less than 10% of the surface flux. To a first approximation it is then permissible to assume that in the lowest several meters of the atmosphere the turbulent flux remains constant at its surface value, so that with the aid of (5.25)

$$K_m \frac{\partial \bar{u}}{\partial z} = u_*^2 \tag{5.32}$$

where we have parameterized the surface momentum flux in terms of the eddy viscosity coefficient. In applying K_m in the Ekman layer solution we assumed a constant value throughout the boundary layer. Near the surface, however, the vertical eddy scale is limited by the distance to the surface. Thus, a logical choice for the mixing length is $l = kz$ where k is a constant. In that case $K_m = (kz)^2 |\partial \bar{u}/\partial z|$. Substituting this expression into (5.32) and taking the square root of the result gives

$$\partial \bar{u}/\partial z = u_*/(kz)$$

Integrating with respect to z yields the *logarithmic wind profile*

$$\bar{u} = (u_*/k) \ln(z/z_0) \tag{5.33}$$

where z_0, the *roughness length*, is a constant of integration chosen so that $\bar{u} = 0$ at $z = z_0$. The constant k in (5.33) is a universal constant called *von Kármán's constant*, which has an experimentally determined value of $k \approx 0.4$. The roughness length z_0 varies widely depending on the physical characteristics of the surface. For grassy fields typical values are in the range of 1–4 cm. Although a number of assumptions are required in the derivation of (5.33), many experimental programs have shown that the logarithmic profile provides a generally satisfactory fit to observed wind profiles in the surface layer.

5.3.6 THE MODIFIED EKMAN LAYER

As pointed out earlier, the Ekman layer solution is not applicable in the surface layer. A more satisfactory representation for the planetary boundary layer can be obtained by combining the logarithmic surface layer profile with the Ekman spiral. In this approach the eddy viscosity coefficient is again treated as a constant, but (5.29) is applied only to the region above the surface layer and the velocity and shear at the bottom of the Ekman layer are matched to those at the top of the surface layer. The resulting

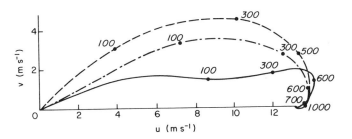

Fig. 5.5 Mean wind hodograph for Jacksonville, Florida ($\cong 30°$N), 4 April 1968 (solid line) compared with the Ekman spiral (dashed line) and the modified Ekman spiral (dash-dot line) computed with $De \cong 1200$ m. Heights shown in meters. (Adapted from Brown, 1970. Reproduced with permission of the American Meteorological Society.)

modified Ekman spiral provides a somewhat better fit to observations than the classical Ekman spiral. However, observed winds in the planetary boundary layer generally deviate substantially from the spiral pattern. Both transience and baroclinic effects (i.e., vertical shear of the geostrophic wind in the boundary layer) may cause deviations from the Ekman solution. But even in steady-state barotropic situations with near-neutral static stability the Ekman spiral is seldom observed.

It turns out that the Ekman layer wind profile is generally unstable for a neutrally buoyant atmosphere. The circulations that develop as a result of this instability have horizontal and vertical scales comparable to the depth of the boundary layer. Thus, it is not possible to parameterize them by a simple flux–gradient relationship. However, these circulations do in general transport considerable momentum vertically. The net result is usually to decrease the angle between the boundary layer wind and the geostrophic wind from that characteristic of the Ekman spiral. A typical observed wind hodograph is shown in Fig. 5.5. Although the detailed structure is rather different from the Ekman spiral, the vertically integrated horizontal mass transport in the boundary layer is still directed toward lower pressure. As we shall see in the next section it is this fact that is of primary importance for synoptic and larger-scale motions.

5.4 Secondary Circulations and Spin-Down

Both the mixed-layer solution (5.22) and the Ekman spiral solution (5.31) indicate that in the planetary boundary layer the horizontal wind has a component directed toward lower pressure. As suggested by Fig. 5.6, this implies mass convergence in a cyclonic circulation and mass divergence in

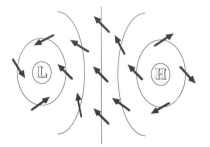

an anticyclonic circulation, which by mass continuity requires vertical motion out of and into the boundary layer, respectively. In order to estimate the magnitude of this induced vertical motion, we note that if $v_g = 0$ the cross isobaric mass transport per unit area at any level in the boundary layer is given by $\rho_0 v$. The net mass transport for a column of unit width extending vertically through the entire layer is simply the vertical integral of $\rho_0 v$. For the mixed layer this integral is simply $\rho_0 v h$ (kg m^{-1} s^{-1}), where h is the layer depth. For the Ekman spiral it is given by

$$M = \int_0^{De} \rho_0 v \, dz = \int_0^{De} \rho_0 u_g \exp(-\pi z/De) \sin(\pi z/De) \, dz \quad (5.35)$$

where $De = \pi/\gamma$ is the Ekman layer depth defined in Section 5.3.4.

Integrating the mean continuity equation (5.13) through the depth of the boundary layer gives

$$w(De) = -\int_0^{De} \left(\frac{\partial u}{\partial x} + \frac{\partial v}{\partial y} \right) dz \quad (5.36)$$

where we have assumed that $w(0) = 0$. Assuming again that $v_g = 0$ so that u_g is independent of x, we find after substituting from (5.31) into (5.36) and comparing with (5.35) that the mass transport at the top of the Ekman layer is given by

$$\rho_0 w(De) = -\frac{\partial M}{\partial y} \quad (5.37)$$

Thus, the mass flux out of the boundary layer is equal to the convergence of the cross-isobar mass transport in the layer. Noting that $-\partial u_g/\partial y = \zeta_g$ is just the geostrophic vorticity in this case, we find after integrating (5.35) and substituting into (5.37) that[5]

$$w(De) = \zeta_g \left(\frac{1}{2\gamma} \right) = \zeta_g \left| \frac{K_m}{2f} \right|^{1/2} \left(\frac{f}{|f|} \right) \tag{5.38}$$

where we have neglected the variation of density with height in the boundary layer and have assumed that $1 + e^{-\pi} \approx 1$. Hence, we obtain the important result that the vertical velocity at the top of the boundary layer is proportional to the geostrophic vorticity. In this way the effect of boundary layer fluxes is communicated directly to the free atmosphere through a forced *secondary circulation* that usually dominates over turbulent mixing. This process is often referred to as *boundary layer pumping*. It occurs only in rotating fluids and is one of the fundamental distinctions between rotating and nonrotating flow. For a typical synoptic-scale system with $\zeta_g \sim 10^{-5} \, \text{s}^{-1}, f \sim 10^{-4} \, \text{s}^{-1}$, and $De \sim 1$ km, the vertical velocity given by (5.38) is of the order of a few millimeters per second.

An analogous boundary layer pumping is responsible for the decay of the circulation created when a cup of tea is stirred. Away from the bottom and sides of the cup there is an approximate balance between the radial pressure gradient and the centrifugal force of the spinning fluid. However, near the bottom viscosity slows the motion and the centrifugal force is not sufficient to balance the radial pressure gradient. (Note that the radial pressure gradient is independent of depth since water is an incompressible fluid.) Therefore, radial inflow takes place near the bottom of the cup. Because of this inflow the tea leaves are always observed to cluster near the center at the bottom of the cup if the tea has been stirred. By continuity of mass the radial inflow in the bottom boundary layer requires upward motion and a slow compensating outward radial flow throughout the remaining depth of the tea. This slow outward radial flow approximately conserves angular momentum, and by replacing high angular momentum fluid by low angular momentum fluid serves to spin down the vorticity in the cup far more rapidly than could mere diffusion.

The characteristic time for the secondary circulation to spin down an atmospheric vortex is most easily illustrated in the case of a barotropic atmosphere. We showed previously in Section 4.5 that for synoptic-scale motions the vorticity equation could be written approximately as

---

[5] The ratio of the Coriolis parameter to its absolute value is included so that the formula will be valid in both hemispheres.

$$\frac{D\zeta_g}{Dt} = -f\left(\frac{\partial u}{\partial x} + \frac{\partial v}{\partial y}\right) = f\frac{\partial w}{\partial z} \tag{5.39}$$

where we have neglected ζ_g compared to f in the divergence term and have also neglected the latitudinal variation of f. Recalling that the geostrophic vorticity is independent of height in a barotropic atmosphere, (5.39) can easily be integrated from the top of the Ekman layer ($z = De$) to the tropopause ($z = H$) to give

$$\frac{D\zeta_g}{Dt} = +f\left[\frac{w(H) - w(De)}{(H - De)}\right] \tag{5.40}$$

Substituting for $w(De)$ from (5.38) and assuming that $w(H) = 0$ and that $H \gg De$, (5.40) may be written as

$$\frac{D\zeta_g}{Dt} = -\left|\frac{fK_m}{2H^2}\right|^{1/2} \zeta_g \tag{5.41}$$

This equation may be integrated in time to give

$$\zeta_g(t) = \zeta_g(0) \exp(-t/\tau_e) \tag{5.42}$$

where $\zeta_g(0)$ is the value of the geostrophic vorticity at time $t = 0$ and $\tau_e \equiv H|2/(fK_m)|^{1/2}$ is the time that it takes the vorticity to decrease to e^{-1} of its original value.

This *e-folding* time scale is referred to as the barotropic *spin-down time*. Taking typical values of the parameters $H \equiv 10$ km, $f = 10^{-4}\,\text{s}^{-1}$, and $K_m = 10\,\text{m}^2\,\text{s}^{-1}$, we find that $\tau_e \approx 4$ days. Thus, for midlatitude synoptic-scale disturbances in a barotropic atmosphere the characteristic spin-down time is a few days. This decay time scale should be compared to the time scale for ordinary viscous diffusion. For viscous diffusion the time scale can be estimated from scale analysis of the diffusion equation

$$\frac{\partial u}{\partial t} = K_m \frac{\partial^2 u}{\partial z^2}$$

If τ_d is the diffusive time scale and H is a characteristic vertical scale for diffusion, then from the diffusion equation

$$U/\tau_d \sim K_m U/H^2$$

so that $\tau_d \sim H^2/K_m$. For the above values of H and K_m, the diffusion time scale is thus about 100 days. Hence, in the absence of convective clouds the spin-down process is a far more effective mechanism for destroying vorticity in a rotating atmosphere than is eddy diffusion. Cumulonimbus convection can produce rapid turbulent transports of heat and momentum through the entire depth of the troposphere. These must be considered together with boundary layer pumping for intense systems such as hurricanes.

Physically, the spin-down process in the atmospheric case is similar to that described for the teacup, except that in synoptic-scale systems it is primarily the Coriolis force that balances the pressure gradient force away from the boundary, not the centrifugal force. Again the role of the secondary circulation driven by forces resulting from boundary layer drag is to provide a slow radial flow in the interior that is superposed on the azimuthal circulation of the vortex above the boundary layer. This secondary circulation is directed outward in a cyclone so that the horizontal area enclosed by any chain of fluid particles gradually increases. Since the circulation is conserved, the azimuthal velocity at any distance from the vortex center must decrease in time. Or, from another point of view, the Coriolis force for the outward-flowing fluid is directed clockwise, and this force thus exerts a torque opposite to the direction of the circulation of the vortex. In Fig. 5.7 a qualitative sketch of the streamlines of this secondary flow is shown.

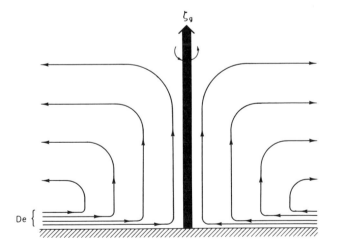

Fig. 5.7   Streamlines of the secondary circulation forced by frictional convergence in the planetary boundary layer for a cyclonic vortex in a barotropic atmosphere. The circulation extends throughout the full depth of the vortex.

It should now be obvious exactly what is meant by the term secondary circulation. It is simply a circulation superposed on the primary circulation (in this case the azimuthal circulation of the vortex) by the physical constraints of the system. In the case of the boundary layer it is viscosity that is responsible for the presence of the secondary circulation. However, other processes such as temperature advection and diabatic heating may also lead to secondary circulations as we shall see later.

The above discussion has concerned only the neutrally stratified barotropic atmosphere. An analysis for the more realistic case of a stably stratified baroclinic atmosphere would be much more complicated. However, qualitatively the effects of stratification may be easily understood. The buoyancy force (see Section 2.7.3) will act to suppress vertical motion since air lifted vertically in a stable environment will be denser than the environmental air. As a result the interior secondary circulation will be restricted in vertical extent as shown in Fig. 5.8. Most of the return flow will take place just above the boundary layer. This secondary flow will rather quickly spin-down the vorticity at the top of the Ekman layer without appreciably affecting the higher levels. When the geostrophic vorticity at the top of the boundary layer is reduced to zero, the pumping action of the Ekman layer is eliminated. The result is a baroclinic vortex with a vertical shear of the azimuthal velocity that is just strong enough to bring ζ_g to zero at the top of the boundary layer. This vertical shear of the geostrophic wind requires a radial temperature gradient that is in fact produced during the

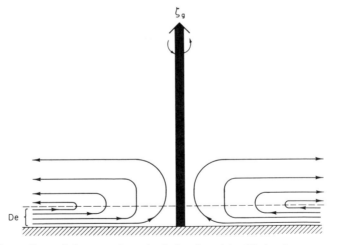

Fig. 5.8 Streamlines of the secondary circulation forced by frictional convergence in the planetary boundary layer for a cyclonic vortex in a stably stratified baroclinic atmosphere. The circulation decays with height in the interior.

spin-down phase by the adiabatic cooling of the air forced out of the Ekman layer. Thus, the secondary circulation in the baroclinic atmosphere serves two purposes: (1) it changes the azimuthal velocity field of the vortex through the action of the Coriolis force; and (2) it changes the temperature distribution so that a thermal wind balance is always maintained between the vertical shear of the azimuthal velocity and the radial temperature gradient.

Problems

5.1. Verify by direct substitution that the Ekman spiral expression (5.31) is indeed a solution of the boundary layer equations (5.26) and (5.27) for the case $v_g = 0$.

5.2. Derive the Ekman spiral solution for the more general case where the geostrophic wind has both x and y components (u_g and v_g, respectively), which are independent of height.

5.3. Letting the Coriolis parameter and density be constants, show that (5.38) is correct for the more general Ekman spiral solution obtained in Problem 5.2.

5.4. For laminar flow in a rotating cylindrical vessel filled with water (molecular kinematic viscosity $\nu = 0.01 \text{ cm}^2 \text{ s}^{-1}$), compute the depth of the Ekman layer and the spin-down time if the depth of the fluid is 30 cm and the rotation rate of the tank is ten revolutions per minute. How small would the radius of the tank have to be in order that the time scale for viscous diffusion from the side walls be comparable to the spin-down time?

5.5. Suppose that in a certain region the geostrophic wind is westerly at 15 m s^{-1}. Compute the net cross-isobaric transport in the planetary boundary layer using both the mixed-layer solution (5.22) and the Ekman layer solution (5.31). You may let $|V| = u_g$ in (5.22), $h = De = 1 \text{ km}$, $\kappa_s = 0.015 \text{ m}^{-1} \text{ s}$, and $\rho = 1 \text{ kg m}^{-3}$.

5.6. Derive an expression for the wind-driven surface Ekman layer in the ocean. Assume that the wind stress τ_w is constant and directed along the x axis. Continuity of turbulent momentum flux at the air–sea interface ($z = 0$) requires the wind stress divided by air density must equal the oceanic turbulent momentum flux at $z = 0$. Thus, if the flux–gradient theory is used the boundary condition at the surface becomes

$$\rho_0 K_m \frac{\partial u}{\partial z} = \tau_w, \qquad \rho_0 K_m \frac{\partial v}{\partial z} = 0, \qquad \text{at } z = 0$$

where K_m is the eddy viscosity in the ocean (assumed constant). As a lower boundary condition assume that u, $v \to 0$ as $z \to -\infty$. If $K_m = 10^{-3} \, m^2 \, s^{-1}$ what is the depth of the surface Ekman layer at 45°N latitude?

5.7. Show that the vertically integrated mass transport in the wind-driven oceanic surface Ekman layer is directed 90° to the right of the surface wind stress in the Northern Hemisphere. Explain this result physically.

5.8. A homogeneous barotropic ocean of depth $H = 3 \, km$ has a zonally symmetric geostrophic jet whose profile is given by the expression

$$u_g = U \exp[-(y/L)^2]$$

where $U = 1 \, m \, s^{-1}$ and $L = 200 \, km$ are constants. Compute the vertical velocity produced by convergence in the Ekman layer at the ocean bottom and show that the meridional profile of the secondary cross-stream motion forced in the interior is the same as the meridional profile of u_g. What are the maximum values of v and w if $K_m = 10^{-3} \, m^2 \, s^{-1}$ and $f = 10^{-4} \, s^{-1}$? (Assume that w and the eddy stress vanish at the surface.)

5.9. Using the approximate zonally averaged momentum equation

$$\frac{\partial \bar{u}}{\partial t} \cong f\tilde{v}$$

compute the spin-down time for the zonal jet in Problem 5.8.

5.10. Derive a formula for the vertical velocity at the top of the planetary boundary layer using the mixed-layer expression (5.22). Assume that $|\bar{\mathbf{V}}| = 10 \, m \, s^{-1}$ is independent of x and y and that $\bar{u}_g = \bar{u}_g(y)$. If $h = 1 \, km$ what value must κ_s have if the result is to agree with the vertical velocity derived from the Ekman layer solution with $De = 1 \, km$?

5.11. Show that $K_m = kzu_*$ in the surface layer.

Suggested References

Arya, *Introduction of Micrometeorology*, gives an excellent introduction to boundary layer dynamics and the elements of turbulence at the beginning undergraduate level.

Sorbjan, *Structure of the Atmospheric Boundary Layer*, is a monograph that provides a good survey of the current state of boundary layer research at the graduate level.

Panofsky and Dutton, *Atmospheric Turbulence*, has a good graduate level treatment of the statistical properties of turbulence, as well as engineering applications.

Stull, *An Introduction to Boundary Layer Meteorology*, provides a comprehensive and very nicely illustrated treatment of all aspects of the subject at the beginning graduate level.

6 | Synoptic-Scale Motions I: Quasi-geostrophic Analysis

A primary goal of dynamic meteorology is to interpret the observed structure of large-scale atmospheric motions in terms of the physical laws governing the motions. These laws, which express the conservation of momentum, mass, and energy, completely determine the relationships among the pressure, temperature, and velocity fields. As we saw in Chapter 2, these governing laws are quite complicated even when the hydrostatic approximation (which is valid for all large-scale meteorological systems) is applied. For extratropical synoptic-scale motions, however, the horizontal velocities are approximately geostrophic (see Section 2.4). Such motions, which are usually referred to as *quasi-geostrophic*, are simpler to analyze than are tropical disturbances or planetary-scale disturbances. They are also the major systems of interest in traditional short-range weather forecasting and are thus a reasonable starting point for dynamical analysis.

In this chapter we show that for extratropical synoptic-scale systems the twin requirements of hydrostatic and geostrophic balance constrain the baroclinic motions so that to a good approximation the structure and evolution of the three-dimensional velocity field are determined by the distribution of geopotential height on isobaric surfaces. The equations that express these relationships constitute the quasi-geostrophic system. Before developing this system of equations it is useful to summarize briefly the observed structure of midlatitude synoptic systems and the mean circulations in which they are embedded. We then develop the quasi-geostrophic momentum and thermodynamic energy equations and show how these can be manipulated to form the *quasi-geostrophic potential vorticity equation* and the *omega equation*. The former equation provides a method for predicting the evolution of the geopotential field, given its initial three-dimensional distribution; the latter provides a method for diagnosing the vertical motion from the known distribution of geopotential. In both cases alternative versions of the equations are discussed to help elucidate the dynamical processes responsible for the development and evolution of synoptic-scale systems.

6.1 The Observed Structure of Extratropical Circulations

Atmospheric circulation systems depicted on a synoptic chart rarely resemble the simple circular vortices discussed in Chapter 3. Rather, they are generally highly asymmetric in form with the strongest winds and largest temperature gradients concentrated along narrow bands called *fronts*. Also, such systems generally are highly baroclinic; the amplitudes and phases of the geopotential and velocity perturbations both change substantially with height. Part of this complexity is due to the fact that these synoptic systems are not superposed on a uniform mean flow but are embedded in a slowly varying planetary-scale flow that is itself highly baroclinic. Furthermore, this planetary-scale flow is influenced by *orography* (that is, by large-scale terrain variations) and continent–ocean heating contrasts, so that it is highly longitude dependent. Therefore, it is not accurate to view synoptic systems as disturbances superposed on a zonal flow that varies only with latitude and height. As is shown in Chapter 8, however, such a point of view can be useful as a first approximation in theoretical analyses of synoptic-scale wave disturbances.

Zonally averaged cross sections do provide some useful information on the gross structure of the planetary-scale circulation, in which synoptic-scale eddies are embedded. Figure 6.1 shows meridional cross sections of the longitudinally averaged zonal wind and temperature for the solstice seasons of (a) December, January, February (DJF) and (b) June, July, August (JJA). These sections extend from approximately sea level (1000 mb) to about 16 km altitude (100 mb). Thus the entire troposphere is shown while the lower stratosphere is shown only for the extratropical regions. In the present

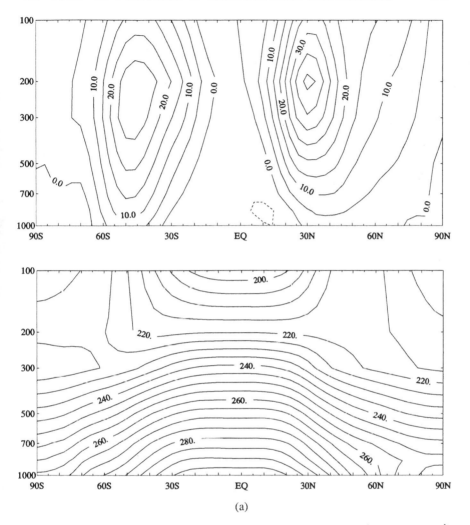

(a)

Fig. 6.1 Meridional cross sections of longitudinally averaged zonal wind (top panels, m s^{-1}) and temperature (bottom panels, contours, K) for (a) DJF and (b) JJA averaged for years 1980–1987. (Adapted from Schubert *et al.*, 1990.) (*Figure continues.*)

chapter we are concerned with the structure of the wind and temperature fields in the troposphere. The stratosphere will be discussed in Chapter 12.

The average pole-to-equator temperature gradient in the Northern Hemisphere troposphere is much larger in winter than in summer. In the Southern Hemisphere the difference between summer and winter temperature distributions is smaller, owing mainly to the large thermal inertia of the oceans

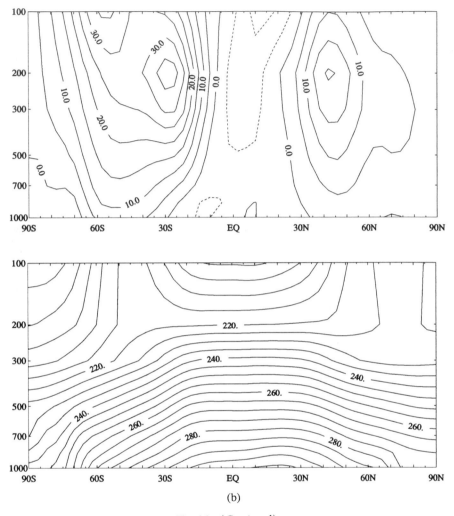

Fig. 6.1   (*Continued*)

together with the greater fraction of the surface that is covered by oceans in the Southern Hemisphere. Since the mean zonal wind and temperature fields satisfy the thermal wind relationship (3.30) to a high degree of accuracy, the maximum zonal wind speed in the Northern Hemisphere is much larger in the winter than in the summer, while there is a smaller seasonal difference in the Southern Hemisphere. Furthermore, in both seasons the core of maximum zonal wind speed (called the mean *jet stream* axis) is located just below the *tropopause* (the boundary between the troposphere and stratosphere) at the latitude where the thermal wind integrated through the troposphere is a maximum. In both hemispheres this is about 30°–35° during winter, but it moves poleward to 40°–45° during summer.

That the zonally averaged meridional cross sections of Fig. 6.1 are not representative of the mean wind structure at all longitudes can be seen in Fig. 6.2, which shows the distribution of the time-averaged zonal wind component for DJF on the 200-mb surface. It is clear from this figure that at some longitudes there are very large deviations of the time-mean zonal flow from its longitudinally averaged distribution. In particular, there are strong zonal wind maxima (jets) near 30°N just east of the Asian and Northern American continents and distinct minima in the eastern Pacific and eastern Atlantic. Synoptic-scale disturbances tend to develop preferentially in the regions of maximum time-mean zonal winds associated with the western Pacific and western Atlantic jets and to propagate downstream along *storm tracks* that approximately follow the jet axes.

The large departure of the northern winter climatological jet stream from zonal symmetry can also be readily inferred from examination of Fig. 6.3, which shows the mean 500-mb geopotential contours for January in the Northern Hemisphere. Even after averaging the height field for a month, very striking departures from zonal symmetry remain. These are clearly linked to the distributions of continents and oceans. The most prominent asymmetries are the troughs to the east of the American and Asian continents. Referring back to Fig. 6.2, we see that the intense jet at 35°N and 140°E is a result of the semipermanent trough in that region. Thus, it is

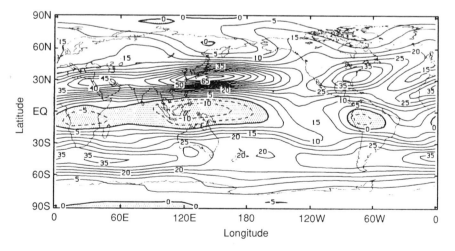

Fig. 6.2 Latitude–longitude cross section of time-averaged zonal wind speed at 200 mb for DJF averaged for years 1980–1987. (After Schubert *et al.*, 1990.)

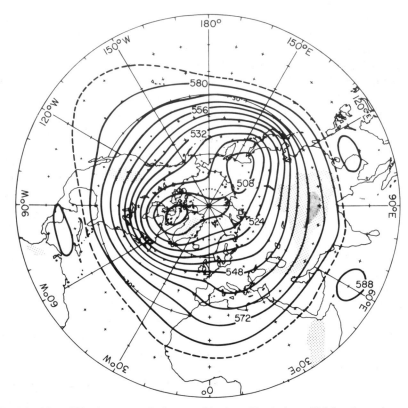

Fig. 6.3   Mean 500-mb contours in January, Northern Hemisphere. Heights shown in tens of meters. (After Palmén and Newton, 1969.)

apparent that the mean flow in which synoptic systems are embedded should really be regarded as a longitude-dependent time-averaged flow.

In addition to its longitudinal dependence, the planetary-scale flow also varies from day to day owing to its interactions with transient synoptic-scale disturbances. In fact, observations show that the transient planetary-scale flow amplitude is comparable to that of the time mean. As a result, monthly mean charts tend to smooth out the actual structure of the instantaneous jet stream since the position and intensity of the jet vary. Thus, at any time the planetary-scale flow in the region of the jet stream has much greater baroclinicity than indicated on time-averaged charts. This point is illustrated schematically in Fig. 6.4, which shows a vertical cross section through an observed jet stream over North America. Panel (a) shows the wind and temperature cross section, while panel (b) shows the wind and potential temperature. The latter provides vivid evidence of the strong static stability

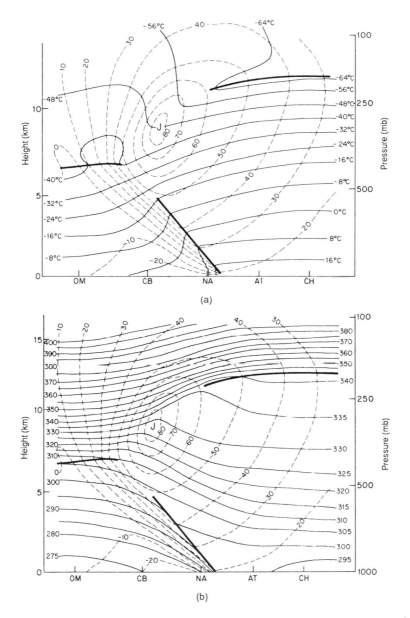

Fig. 6.4 (a) Schematic isotherms (thin solid lines, °C) and isotachs (dashed lines, m s⁻¹) in a vertical cross section through a cold front. Heavy lines mark the tropopause and frontal boundaries. The section extends approximately 1200 km in the horizontal direction. (b) Same as panel (a), but thin solid lines are potential temperature isolines in kelvins (after Wallace and Hobbs, 1977.)

in the stratosphere. It also illustrates the fact that isentropes (constant-θ surfaces) cross the tropopause in the vicinity of the jet so that adiabatic motions can exchange tropospheric and stratospheric air in that region.

At any instant the axis of the jet stream tends to be located above a narrow sloping zone of strong temperature gradients called the *polar-frontal* zone. This is the zone that in general separates the cold air of polar origin from warm tropical air. The occurrence of an intense jet core above this zone of large-magnitude temperature gradients is, of course, not mere coincidence but rather a consequence of the thermal wind balance.

It is a common observation in fluid dynamics that jets in which strong velocity shears occur may be unstable with respect to small perturbations. By this is meant that any small disturbance introduced into the jet will tend to amplify, drawing energy from the jet as it grows. Most synoptic-scale systems in midlatitudes appear to develop as the result of an instability of the jet stream flow. This instability, called *baroclinic instability*, depends on the meridional temperature gradient, particularly at the surface. Hence, through the thermal wind relationship, baroclinic instability depends on vertical shear and tends to occur in the region of the polar frontal zone.

Baroclinic instability is not, however, identical to frontal instability because most baroclinic instability models describe only geostrophically scaled motions, while disturbances in the vicinity of strong frontal zones must be highly nongeostrophic. As we shall see in Chapter 9, baroclinic disturbances may themselves act to intensify preexisting temperature gradients and hence generate frontal zones.

The stages in the development of a typical baroclinic cyclones that develops as a result of baroclinic instability are shown schematically in Fig. 6.5. In the stage of rapid development there is a cooperative interaction

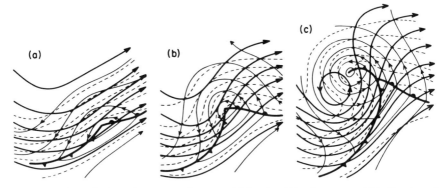

Fig. 6.5   Schematic 500-mb contours (heavy solid lines), 1000-mb contours (thin lines), and 1000–500-mb thickness (dashed) for a developing baroclinic wave at three stages of development. (After Palmén and Newton, 1969.)

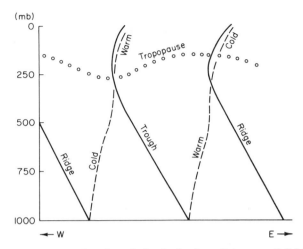

Fig. 6.6 West–east cross section through developing baroclinic wave. Solid lines are trough and ridge axes; dashed lines are axes of temperature extrema; the chain of open circles denotes the tropopause.

between the upper level and surface flows; strong cold advection is seen to occur west of the trough at the surface, with weaker warm advection to the east. This pattern of thermal advection is a direct consequence of the fact that the trough at 500 mb lags (lies to the west of) the surface trough so that the mean geostrophic wind in the 1000–500-mb layer is directed across the 1000–500-mb thickness lines toward larger thickness west of the surface trough and toward smaller thickness east of the surface trough. This dependence of the phase of the disturbance on height is better illustrated by Fig. 6.6, which shows a schematic downstream (or west–east) cross section through an idealized developing baroclinic system. Throughout the troposphere the trough and ridge axes tilt westward (or upstream) with height,[1] while the axes of warmest and coldest air are observed to have the opposite tilt. As we shall see later the westward tilt of the troughs and ridges is necessary in order that the mean flow give up potential energy to the developing wave. In the mature stage (not shown in Fig. 6.5) the troughs at 500 and 1000 mb are nearly in phase. As a consequence, the thermal advection and energy conversion are quite weak.

6.2 The Quasi-geostrophic Approximation

The main goal of this chapter is to show how the observed structure of midlatitude synoptic systems can be interpreted in terms of the constraints

[1] In reality the phase tilts tend to be concentrated below the 700-mb level.

imposed on synoptic-scale motions by the dynamical equations. Specifically we show that for motions that are hydrostatic and nearly geostrophic, the three-dimensional flow field is determined approximately by the isobaric distribution of geopotential $[\Phi(x, y, p, t)]$. For this analysis, it is convenient to use the isobaric coordinate system both because meteorological measurements are generally referred to constant-pressure surfaces and because the dynamical equations are somewhat simpler in isobaric coordinates than in height coordinates. Thus, use of the isobaric coordinate system simplifies the development of approximate prognostic and diagnostic equations.

6.2.1   SCALE ANALYSIS IN ISOBARIC COORDINATES

The dynamical equations in isobaric coordinates were developed in Section 3.1 and for reference are repeated here. The horizontal momentum equation, the hydrostatic equation, the continuity equation, and the thermodynamic energy equation may be expressed as follows:

$$\frac{D\mathbf{V}}{Dt}+f\mathbf{k}\times\mathbf{V}=-\nabla\Phi \tag{6.1}$$

$$\frac{\partial\Phi}{\partial p}=-\alpha=-\frac{RT}{p} \tag{6.2}$$

$$\nabla\cdot\mathbf{V}+\frac{\partial\omega}{\partial p}=0 \tag{6.3}$$

$$\left(\frac{\partial}{\partial t}+\mathbf{V}\cdot\nabla\right)T-S_p\omega=\frac{J}{c_p} \tag{6.4}$$

Here the total derivative in (6.1) is defined by

$$\frac{D}{Dt}\equiv\left(\frac{\partial}{\partial t}\right)_p+(\mathbf{V}\cdot\nabla)_p+\omega\frac{\partial}{\partial p} \tag{6.5}$$

where $\omega\equiv Dp/Dt$ is the individual pressure change, and in (6.4) $S_p\equiv -T\,\partial\ln\theta/\partial p$ is the static stability parameter $[S_p\approx 5\times 10^{-4}\ \mathrm{K\ Pa^{-1}}$ in the midtroposphere].

These equations, although simplified by use of the hydrostatic approximation and by neglect of some small terms that appear in the complete spherical coordinate form, still contain several terms that are of secondary significance for midlatitude synoptic-scale systems. They can be further simplified by recalling from Section 2.4 that the horizontal flow is nearly geostrophic and that the magnitude of the ratio of vertical velocity to horizontal velocity is of order 10^{-3}.

We first separate the horizontal velocity into geostrophic and ageostrophic parts by letting

$$\mathbf{V} = \mathbf{V}_g + \mathbf{V}_a \qquad (6.6)$$

where the geostrophic wind is defined as

$$\mathbf{V}_g \equiv f_0^{-1} \mathbf{k} \times \nabla \Phi \qquad (6.7)$$

and $\mathbf{V}_a$ is just the difference between the total horizontal wind and the geostrophic wind. Here we have assumed that the meridional length scale, L, is small compared to the radius of the earth so that the geostrophic wind (6.7) may be defined using a constant reference latitude value of the Coriolis parameter.[2]

For the systems of interest $|\mathbf{V}_g| \gg |\mathbf{V}_a|$. More precisely,

$$|\mathbf{V}_a|/|\mathbf{V}_g| \sim O(\mathrm{Ro})$$

That is, the ratio of the magnitudes of the ageostrophic and geostrophic winds is the same order of magnitude as the Rossby number introduced in Section 2.4.2.

The momentum can then be approximated to $O(\mathrm{Ro})$ by its geostrophic value, and the rate of change of momentum or temperature following the horizontal motion can be approximated to the same order by the rate of change following the geostrophic wind. Thus, in (6.5) $\mathbf{V}$ can be replaced by $\mathbf{V}_g$ and the vertical advection, which arises only from the ageostrophic flow, can be neglected. The rate of change of momentum following the total motion is then approximately equal to the rate of change of the geostrophic momentum following the geostrophic wind:

$$\frac{D\mathbf{V}}{Dt} \approx \frac{D_g \mathbf{V}_g}{Dt}$$

where

$$\frac{D_g}{Dt} \equiv \frac{\partial}{\partial t} + \mathbf{V}_g \cdot \nabla = \frac{\partial}{\partial t} + u_g \frac{\partial}{\partial x} + v_g \frac{\partial}{\partial y} \qquad (6.8)$$

[2] This definition of the geostrophic wind will be referred to as *constant-f* (CF) geostrophy, while the definition given in (3.4) will be called *variable-f* (VF) geostrophy. The CF geostrophic wind is nondivergent, while the VF geostrophic wind has a divergent portion (see Problem 3.19). The interpretation of the ageostrophic wind depends strongly on which type of geostrophy is used, as explained in Blackburn (1985).

Although a constant f_0 can be used in defining $\mathbf{V_g}$, it is still necessary to retain the dynamical effect of the variation of the Coriolis parameter with latitude in the Coriolis force term in the momentum equation. This variation can be approximated by expanding the latitudinal dependence of f in a Taylor series about a reference latitude ϕ_0 and retaining only the first two terms to yield

$$f = f_0 + \beta y \tag{6.9}$$

where $\beta \equiv (df/dy)_{\phi_0} = 2\Omega \cos \phi_0 / a$ and $y = 0$ at ϕ_0. This approximation is usually referred to as the *midlatitude beta-plane* approximation. For synoptic-scale motions the ratio of the first two terms in the expansion of f has order of magnitude

$$\frac{\beta L}{f_0} \sim \frac{\cos \phi_0}{\sin \phi_0} \frac{L}{a} \sim O(\text{Ro}) \ll 1$$

This justifies letting the Coriolis parameter have a constant value f_0 in the geostrophic approximation and approximating its variation in the Coriolis force term by (6.9).

From (6.1) the acceleration following the motion is equal to the difference between the Coriolis force and the pressure gradient force. This difference depends on the departure of the actual wind from the geostrophic wind. Thus, it is not permissible simply to replace the horizontal velocity by its geostrophic value in the Coriolis term. Rather, we use (6.6), (6.7), and (6.9) to write

$$f\mathbf{k} \times \mathbf{V} + \nabla \Phi = (f_0 + \beta y)\mathbf{k} \times (\mathbf{V_g} + \mathbf{V_a}) - f_0\mathbf{k} \times \mathbf{V_g}$$
$$\tag{6.10}$$
$$\approx f_0\mathbf{k} \times \mathbf{V_a} + \beta y \mathbf{k} \times \mathbf{V_g}$$

where we have used the geostrophic relation (6.7) to eliminate the pressure gradient force and neglected the ageostrophic wind compared to the geostrophic wind in the term proportional to βy. The approximate horizontal momentum equation thus has the form

$$\frac{D_g\mathbf{V_g}}{Dt} = -f_0\mathbf{k} \times \mathbf{V_a} - \beta y \mathbf{k} \times \mathbf{V_g} \tag{6.11}$$

Each term in (6.11) is thus $O(\text{Ro})$ compared to the pressure gradient force, while terms neglected are $O(\text{Ro}^2)$ or smaller.

The geostrophic wind defined in (6.7) is nondivergent. Thus,

$$\nabla \cdot \mathbf{V} = \nabla \cdot \mathbf{V_a} = \frac{\partial u_a}{\partial x} + \frac{\partial v_a}{\partial y}$$

and the continuity equation (6.3) can be rewritten as

$$\mathbf{\nabla} \cdot \mathbf{V}_a + \frac{\partial \omega}{\partial p} = 0 \tag{6.12}$$

which shows that if the geostrophic wind is defined by (6.7), ω is determined only by the ageostrophic part of the wind field.

In the thermodynamic energy equation (6.4) the horizontal advection can be approximated by its geostrophic value. However, the vertical advection is not neglected but forms part of the adiabatic heating and cooling term. This term must be retained because the static stability is usually large enough on the synoptic scale that the adiabatic heating or cooling owing to vertical motion is of the same order as the horizontal temperature advection despite the smallness of the vertical velocity. It can be somewhat simplified, though, by dividing the total temperature field, T_{tot}, into a basic state (standard atmosphere) portion that depends only on pressure, $T_0(p)$, plus a deviation from the basic state, $T(x, y, z, t)$, as was done for potential temperature in Section 2.7.4. Thus we let

$$T_{tot}(x, y, p, t) = T_0(p) + T(x, y, p, t)$$

Now since $|dT_0/dp| \gg |\partial T/\partial p|$ only the basic state portion of the temperature field need be included in the static stability term and the quasi-geostrophic thermodynamic energy equation may be expressed in the form

$$\left(\frac{\partial}{\partial t} + \mathbf{V}_g \cdot \mathbf{\nabla}\right) T - \left(\frac{\sigma p}{R}\right) \omega = \frac{J}{c_p} \tag{6.13}$$

where $\sigma \equiv -RT_0 p^{-1} \, d \ln \theta_0/dp$ and θ_0 is the potential temperature corresponding to the basic state temperature T_0 ($\sigma \approx 2 \times 10^{-6} \, m^2 \, Pa^{-2} \, s^{-2}$ in the midtroposphere).

Equations (6.2), (6.7), (6.11), (6.12), and (6.13) constitute the quasi-geostrophic equations. If J is known these form a complete set in the dependent variables Φ, T, $\mathbf{V}_g$, $\mathbf{V}_a$, and ω. This form is particularly useful when information is needed on the distribution of the ageostrophic velocity. However, when $\mathbf{V}_a$ is not required it may be eliminated to produce a somewhat simpler set.

6.2.2 THE QUASI-GEOSTROPHIC VORTICITY EQUATION

Just as the horizontal momentum can be approximated to $O(\mathrm{Ro})$ by its geostrophic value, the vertical component of vorticity can also be approximated geostrophically. If the CF geostrophic relationship (6.7) is expanded in Cartesian coordinates as

$$f_0 v_g = \frac{\partial \Phi}{\partial x}, \qquad f_0 u_g = -\frac{\partial \Phi}{\partial y} \qquad (6.14)$$

the geostrophic vorticity, $\zeta_g = \mathbf{k} \cdot \nabla \times \mathbf{V}_g$, can be expressed in terms of the horizontal Laplacian of the geopotential:

$$\zeta_g = \frac{\partial v_g}{\partial x} - \frac{\partial u_g}{\partial y} = \frac{1}{f_0}\left(\frac{\partial^2 \Phi}{\partial x^2} + \frac{\partial^2 \Phi}{\partial y^2}\right) = \frac{1}{f_0}\nabla^2 \Phi \qquad (6.15)$$

Equation (6.15) can be used to determine $\zeta_g(x, y)$ from a known field $\Phi(x, y)$. Alternatively, (6.15) can be solved by inverting the Laplacian operator to determine Φ from a known distribution of ζ_g provided that suitable conditions on Φ are specified on the boundaries of the region in question. This *invertibility* is one reason why vorticity is such a useful forecast diagnostic; if the evolution of the vorticity can be predicted, then inversion of (6.15) yields the evolution of the geopotential field, from which it is possible to determine the geostrophic wind. Since the Laplacian of a function tends to be a maximum where the function itself is a minimum, positive vorticity implies low values of geopotential and vice versa, as illustrated for a simple sinusoidal disturbance in Fig. 6.7.

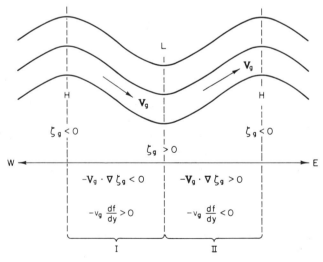

Fig. 6.7 Schematic 500-mb geopotential field showing regions of positive and negative advections of relative and planetary vorticity.

The quasi-geostrophic vorticity equation can be obtained from the x and y components of the quasi-geostrophic momentum equation (6.11), which can be expressed respectively as

$$\frac{D_g u_g}{Dt} - f_0 v_a - \beta y v_g = 0 \tag{6.16}$$

and

$$\frac{D_g v_g}{Dt} + f_0 u_a + \beta y u_g = 0 \tag{6.17}$$

Taking $\partial(6.17)/\partial x - \partial(6.16)/\partial y$ and using the fact that the divergence of the CF geostrophic wind vanishes immediately yields the vorticity equation

$$\frac{D_g \zeta_g}{Dt} = -f_0 \left(\frac{\partial u_a}{\partial x} + \frac{\partial v_a}{\partial y} \right) - \beta v_g \tag{6.18}$$

which should be compared with (4.22).

Noting that since f depends only on y so that $D_g f/Dt = \mathbf{V}_g \cdot \nabla f = \beta v_g$ and that the divergence of the ageostrophic wind can be eliminated in favor of ω using (6.12), we can rewrite (6.18) as

$$\frac{\partial \zeta_g}{\partial t} = -\mathbf{V}_g \cdot \nabla(\zeta_g + f) + f_0 \frac{\partial \omega}{\partial p} \tag{6.19}$$

which states that the local rate of change of geostrophic vorticity is given by the sum of the advection of the absolute vorticity by the geostrophic wind plus the concentration or dilution of vorticity by stretching or shrinking of fluid columns (the divergence effect).

The vorticity tendency owing to vorticity advection [the first term on the right in (6.19)] may be rewritten as

$$-\mathbf{V}_g \cdot \nabla(\zeta_g + f) = -\mathbf{V}_g \cdot \nabla \zeta_g - \beta v_g$$

The two terms on the right represent the geostrophic advections of relative vorticity and planetary vorticity, respectively. For disturbances in the westerlies, these two effects tend to have opposite signs as illustrated schematically in Fig. 6.7 for an idealized 500-mb flow.

In region I upstream of the 500-mb trough, the geostrophic wind is directed from the negative vorticity maximum at the ridge toward the positive vorticity maximum at the trough so that $-\mathbf{V}_g \cdot \nabla \zeta_g < 0$. But at the same time, since $v_g < 0$ in region I, the geostrophic wind has its y component directed down the gradient of planetary vorticity so that $-\beta v_g > 0$. Hence, in region I the advection of relative vorticity tends to decrease the local vorticity, whereas advection of planetary vorticity tends to increase the local vorticity. Similar arguments (but with reversed signs) apply to region II. Therefore, advection of relative vorticity tends to move the vorticity pattern and hence the troughs and ridges eastward (downstream). But advection of planetary vorticity tends to move the troughs and ridges westward against the advecting wind field. The latter motion is called *retrograde* motion or *retrogression*.

The net effect of advection on the evolution of the vorticity pattern depends upon which type of vorticity advection dominates. In order to compare the magnitudes of the relative and planetary vorticity advections, we assume that Φ on the midlatitude β plane can be represented as the sum of a time and zonally averaged part, which depends on y and p, and a fluctuating part that has a sinusoidal dependence in x, and y:

$$\Phi(x, y, p, t) = \bar{\Phi}(y, p) + \Phi'(p, t) \sin kx \cos ly$$

where Φ' is the amplitude of the fluctuating component and the *wave numbers* k and l are defined as $k = 2\pi/L_x$ and $l = 2\pi/L_y$ with L_x and L_y the wavelengths in the x and y directions, respectively. We further assume that

$$\bar{\Phi}(y, p) = \Phi_0(p) - f_0 Uy$$

where Φ_0 is a standard atmosphere geopotential distribution and U is a constant mean zonal wind. The geostrophic vorticity is then simply

$$\zeta_g = f_0^{-1} \nabla^2 \Phi = -f_0^{-1}(k^2 + l^2)\Phi' \sin kx \cos ly$$
$$= -f_0^{-1}(k^2 + l^2)(\Phi - \bar{\Phi})$$

(6.20)

The patterns of geopotential and relative vorticity for this case are shown in Fig. 6.8. For a disturbance with a given amplitude of geopotential disturbance, Φ', the amplitude of the vorticity increases as the square of the wave number or inversely as the square of the horizontal scale. As a consequence, the advection of relative vorticity dominates over planetary vorticity advection for short waves ($L_x < 3000$ km), while for long waves

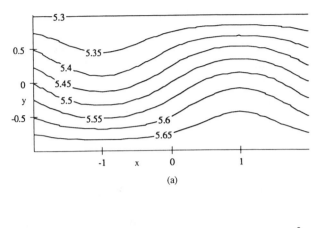

(a)

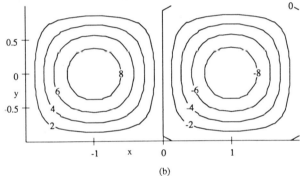

(b)

Fig. 6.8 Geopotential in units of 10^4 m^2 s^{-2} (a) and relative vorticity in units of 10^{-6} s^{-1} (b) for the sinusoidal disturbance of Eq. (6.20). Here $\Phi_0 = 5 \times 10^4$ m^2 s^{-2}, $f_0 = 10^{-4}$ s^{-1}, $\Phi' = 800$ m^2 s^{-2}, $U = 10$ m s^{-1}, and $k = l = (\pi/2) \times 10^{-6}$ m^{-1}. Axes are labeled in units of 10^3 km.

($L_x > 10,000$ km) the planetary vorticity advection tends to dominate. Therefore, as a general rule short-wavelength synoptic-scale systems should move eastward with the advecting zonal flow while long planetary waves should tend to retrogress (move westward against the mean flow).[3] Waves of intermediate wavelength may be quasi-stationary or move eastward much slower than the mean geostrophic wind speed. Since positive maxima in relative vorticity are associated with cyclonic disturbances, regions of positive vorticity advection, which can easily be estimated from upper-level

---

[3] Observed long waves tend to remain stationary rather than to retrogress. This is believed to be a result of processes such as nonlinear interactions with transient short waves, forcing owing to topographic influences, and diabatic heating contrasts associated with land–sea differences, as was mentioned in Section 6.1.

maps, are commonly used as aids in forecasting synoptic-scale weather disturbances.

Vorticity advection does not alone determine the evolution of meteorological systems. A change in the vertical shear of the horizontal wind associated with differential (i.e., height-dependent) vorticity advection will drive an ageostrophic vertical circulation, which adiabatically adjusts the horizontal temperature gradient to maintain thermal wind balance. The convergence and divergence fields associated with this vertical circulation will not only modify the effects of vorticity advection at upper levels but also force changes in the vorticity distribution in the lower troposphere, where advection may be very weak.

In an analogous manner, thermal advection, which is often strong near the surface, does not merely force changes in the temperature in the lower troposphere. Rather, it induces a vertical circulation, which through its associated divergence and convergence patterns alters the vorticity fields both near the surface and aloft so that thermal wind balance is maintained.

The vertical circulation induced by quasi-geostrophic differential vorticity advection and thermal advection is generally an order of magnitude larger than that induced by boundary layer pumping (5.38). Thus, it is reasonable to neglect boundary layer effects to a first approximation in quasi-geostrophic theory.

6.3 Quasi-geostrophic Prediction

The characteristics of the geostrophic circulation forced by the vertical motions associated with vorticity and thermal advection can be determined without explicitly determining the distribution of ω. Since T, ζ_g, and $\mathbf{V}_g$ are all functions of Φ, the quasi-geostrophic vorticity equation (6.19) and the thermodynamic energy equation (6.13) each can be written so that they contain only the two dependent variables Φ and ω. (For simplicity we ignore the diabatic heating term in the thermodynamic energy equation even though it may be important in some synoptic disturbances.) It is thus possible to eliminate ω between these two equations and obtain an equation relating Φ to $\partial\Phi/\partial t$. Defining the geopotential tendency $\chi \equiv \partial\Phi/\partial t$, using the hydrostatic equation (6.2) to eliminate T in favor of $\partial\Phi/\partial p$, and recalling that the order of partial differentiation may be reversed without changing the result, the thermodynamic energy equation (6.13) and the geostrophic vorticity equation (6.19) can be expressed respectively as

$$\frac{\partial\chi}{\partial p} = -\mathbf{V}_g \cdot \nabla\left(\frac{\partial\Phi}{\partial p}\right) - \sigma\omega \tag{6.21}$$

and

$$\nabla^2 \chi = -f_0 \mathbf{V}_g \cdot \nabla \left(\frac{1}{f_0} \nabla^2 \Phi + f \right) + f_0^2 \frac{\partial \omega}{\partial p} \tag{6.22}$$

where σ was defined below (6.13) and we have used (6.15) to write the geostrophic vorticity and its tendency in terms of the Laplacian of geopotential.

Equations (6.21) and (6.22) are one form of the *quasi-geostrophic system.* The first of these indicates that the vertical derivative of the geopotential tendency is equal to the sum of thickness advection and adiabatic thickness change owing to vertical motion. The second indicates that the horizontal Laplacian of the geopotential tendency is equal to the sum of vorticity advection plus vorticity generation by the divergence effect. Purely geostrophic motion ($\omega = 0$) is a solution to (6.21)-(6.22) only in very special situations such as barotropic flow (no pressure dependence) or zonally symmetric flow (no x dependence). More general purely geostrophic flows cannot satisfy both these equations simultaneously, since there are then two independent equations in a single unknown, so that the system is overdetermined. Thus, it should be clear that the role of the vertical motion distribution must be to maintain consistency between the geopotential tendencies required by thermal advection in (6.21) and vorticity advection in (6.22), respectively.

6.3.1 GEOPOTENTIAL TENDENCY

If we multiply (6.21) by f_0^2/σ and then differentiate with respect to pressure and add the result to (6.22), ω is eliminated and we obtain an equation that determines the local rate of change of geopotential in terms of the three-dimensional distribution of the Φ field:

$$\underbrace{\left[\nabla^2 + \frac{\partial}{\partial p} \left(\frac{f_0^2}{\sigma} \frac{\partial}{\partial p} \right) \right] \chi}_{A} = \underbrace{-f_0 \mathbf{V}_g \cdot \nabla \left(\frac{1}{f_0} \nabla^2 \Phi + f \right)}_{B}$$

$$\underbrace{- \frac{\partial}{\partial p} \left[-\frac{f_0^2}{\sigma} \mathbf{V}_g \cdot \nabla \left(-\frac{\partial \Phi}{\partial p} \right) \right]}_{C} \tag{6.23}$$

This equation is often referred to as the *geopotential tendency equation.* It provides a relationship between the local change of geopotential and the distributions of vorticity and thickness advection as can be seen by analyzing the three terms labeled A, B, and C, respectively. If the distribution of Φ is known at a given time, terms B and C may be regarded as known forcing functions, and (6.23) is a linear partial differential equation in the unknown χ.

Although (6.23) appears to be quite complicated, a qualitative notion of its implications can be gained by examining the solution for a simple wave pattern. Term A in (6.23) involves only second derivatives in space of the χ field. We showed in (6.20) that for wavelike disturbances the horizontal Laplacian of Φ is proportional to the negative of the deviation of Φ from its mean. A similar relationship holds for χ. Thus, letting

$$\chi(x, y, p, t) \approx X(p, t) \sin kx \cos ly$$

we have $\nabla^2 \chi \approx -(k^2 + l^2)\chi$. The forcing terms B and C in (6.23) are also assumed to have sinusoidal behavior in x and y:

$$-f_0 \mathbf{V}_g \cdot \nabla\left(\frac{1}{f_0}\nabla^2\Phi + f\right) \approx F_v(p) \sin kx \cos ly$$

$$-\frac{f_0^2}{\sigma}\mathbf{V}_g \cdot \nabla\left(-\frac{\partial\Phi}{\partial p}\right) \approx F_T(p) \sin kx \cos ly$$

where $F_v(p)$ and $F_T(p)$ represent the vertical dependences of the vorticity advection and thermal advection, respectively. Substituting these expressions into (6.23) and eliminating the $\sin kx \cos ly$ dependence, which is common to each term, we obtain an ordinary differential equation for the vertical dependence of the geopotential tendency:

$$\frac{d^2X}{dp^2} - \lambda^2 X = \frac{\sigma}{f_0^2}\left(F_v - \frac{dF_T}{dp}\right) \tag{6.24}$$

where $\lambda^2 \equiv (k^2 + l^2)\sigma f_0^{-2}$ and we have neglected the pressure dependence of static stability.[4] Equation (6.24) shows that forcing at a given altitude will

---

[4] Actually, σ varies substantially with pressure even in the troposphere. However, the qualitative discussion in this section would not be changed if we were to include this additional complication.

generate a response whose vertical scale (measured in pressure units) is λ^{-1}. Thus, for example, upper-level vorticity advection associated with disturbances of large horizontal scale (small k and l) will produce geopotential tendencies that extend down to the surface with little loss of amplitude, while for disturbances of small horizontal scale the response is confined close to the levels of forcing (see Fig. 6.9). In mathematical terms, the differential operator in (6.24) spreads the response in the vertical so that forcing at one altitude influences other altitudes.

The role of thermal advection in changing upper-level geopotential heights can be simply illustrated by considering the special case of $\beta = 0$ and very large horizontal scale so that $\lambda^2 \to 0$ and $F_v \approx 0$. Eq. (6.24) can then be approximated as

$$d^2 X / dp^2 \approx -\sigma f_0^{-2} (dF_T / dp)$$

Using the definitions of X and F_T and integrating twice with respect to pressure yields

$$X(p) - X(p_0) \approx -\frac{\sigma}{f_0^2} \int_{p_0}^{p} F_I \, dp$$

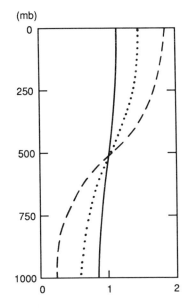

(mb)

Fig. 6.9 Vertical structure of the geopotential tendency (units of 10^{-2} m^2 s^{-3}) owing to forcing by vorticity advection. Here, $l = 0$, $k = 2\pi/L_x$, $F_v = 2 \times 10^{-2} k^2$ s^{-3}, for $p < 500$ mb and $F_v = 0$ for $p > 500$ mb, $\lambda^2 = 200 \, k^2$ Pa^{-2} m^{-2}, and $L_x = 2{,}000$ km (dashed line), $L_x = 4{,}000$ km (dotted line), and $L_x = 8{,}000$ km (solid line).

After multiplying both sides by sin kx cos ly and using the definitions of X and F_T, this may be rewritten as

$$\frac{\partial}{\partial t}[\Phi(p) - \Phi(p_0)] = \frac{\partial}{\partial t}\,\delta\Phi \approx -R \int_p^{p_0} \mathbf{V_g} \cdot \nabla T\, d\ln p$$

which states that the thickness tendency for the column between pressure levels p_0 and p is proportional to the vertically integrated temperature advection.

Term B in (6.23), the vorticity advection term, generally is the main forcing term in the upper troposphere. As the discussion in the previous subsection indicated, for short waves term B is negative in region I (upstream of the 500-mb trough.) Thus, since the sign of the geopotential tendency is opposite to that of the forcing in this case, χ will be positive and a ridge will tend to develop. This ridging is, of course, necessary for the development of a negative geostrophic vorticity. Similar arguments, but with the signs reversed, apply to region II downstream from the 500-mb trough, where falling geopotential heights are associated with a positive relative vorticity advection. It is also important to note that the vorticity advection term is zero along both the trough and ridge axes since both $\nabla \zeta_g$ and v_g are zero at the axes. Thus, vorticity advection cannot change the strength of this type of disturbance at the levels where the advection is occurring but only acts to propagate the disturbance horizontally and spread it vertically.

The major mechanism for amplification or decay of midlatitude synoptic systems is contained in term C of (6.23). This term involves the rate of change with pressure of the horizontal thickness advection. (If we had retained the diabatic heating term it would have contributed in a similar fashion.) The thickness advection tends to be largest in magnitude in the lower troposphere beneath the 500-mb trough and ridge lines in a developing baroclinic wave. Now since $-\partial\Phi/\partial p$ is proportional to temperature, the thickness advection is proportional to the temperature advection. Thus, term C in (6.23) is proportional to minus the rate of change of temperature advection with respect to pressure (i.e., plus the rate of change with respect to height). This term is sometimes referred to as the *differential temperature advection.*

To examine the influence of differential temperature advection on the geopotential tendency we consider the idealized developing wave shown in Fig. 6.5. Below the 500-mb ridge there is strong warm advection associated with the warm front, while below the 500-mb trough there is strong cold advection associated with the cold front. Above the 500-mb level the temperature gradient is usually weaker, and the isotherms often become nearly parallel to the height lines, so that thermal advection tends to be small.

Thus, in contrast to term B in (6.23), the forcing term C is concentrated in the lower troposphere. But again, the geopotential tendency response is not limited to the levels of forcing but is spread in the vertical, so that for developing waves it will deepen upper-level troughs and build upper-level ridges.

In the region of warm advection $-\mathbf{V}_g \cdot \mathbf{\nabla}(-\partial\Phi/\partial p) > 0$ since $\mathbf{V}_g$ has a component down the temperature gradient. But, as explained above, the warm advection decreases with height (increases with pressure) so that $\partial[-\mathbf{V}_g \cdot \mathbf{\nabla}(-\partial\Phi/\partial p)]/\partial p > 0$. Conversely, beneath the 500-mb trough where there is cold advection decreasing with height, the opposite signs obtain. Therefore, along the 500-mb trough and ridge axes where the vorticity advection is zero the tendency equation states that for a developing wave

$$\chi \sim \left\{ \frac{\partial}{\partial p} \left[-\mathbf{V}_g \cdot \mathbf{\nabla}\left(-\frac{\partial\Phi}{\partial p} \right) \right] \right\} \begin{array}{l} >0 \quad \text{at the ridge} \\ <0 \quad \text{at the trough} \end{array}$$

Therefore, as indicated in Fig. 6.10, the effect of cold advection below the 500-mb trough is to *deepen* the trough in the upper troposphere, and the effect of warm advection below the 500-mb ridge is to *build* the ridge in

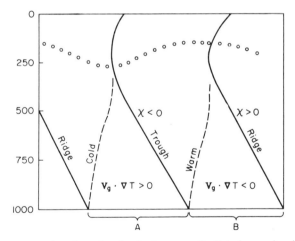

Fig. 6.10 East–west section through a developing synoptic disturbance showing the relationship of temperature advection to the upper-level height tendencies. A and B designate, respectively, regions of cold advection and warm advection in the lower troposphere.

the upper troposphere. Hence, differential temperature or thickness advection, even if limited to the lower troposphere, intensifies the upper-level troughs and ridges in a developing system.

Qualitatively, the effects of differential temperature advection are easily understood since the advection of cold air into the air column below the 500-mb trough reduces the thickness of that column and hence lowers the height of the 500-mb surface unless there is a compensating rise in the surface pressure. Obviously, warm advection into the air column below the ridge has the opposite effect.

In summary, we have shown that in the absence of diabatic heating the horizontal temperature advection must be nonzero in order that a mid-latitude synoptic system intensify through baroclinic processes. As we shall see in Chapter 8, the temperature advection pattern described above indirectly implies conversion of potential energy to kinetic energy.

6.3.2 The Quasi-geostrophic Potential Vorticity Equation

The geopotential tendency equation stated in the form (6.23) is useful for physical motivation of processes leading to geopotential changes (and hence upper-level troughing and ridging) since the tendency χ is related to the easily comprehended processes of vorticity and temperature advection. However, this form of the equation actually conceals its true character as a conservation equation for a field commonly referred to as *quasi-geostrophic potential vorticity*. To put (6.23) in conservation form, we again neglect the diabatic heating term and simplify the right-hand side by using the chain rule of differentiation to write term C as

$$-\left(\mathbf{V}_g \cdot \nabla \frac{\partial}{\partial p}\left(\frac{f_0^2}{\sigma}\frac{\partial \Phi}{\partial p}\right) + \frac{f_0^2}{\sigma}\frac{\partial \mathbf{V}_g}{\partial p}\cdot\nabla\frac{\partial \Phi}{\partial p}\right)$$

But $f_0\,\partial\mathbf{V}_g/\partial p = \mathbf{k}\times\nabla(\partial\Phi/\partial p)$, which is perpendicular to $\nabla(\partial\Phi/\partial p)$. Thus, the second part of the above expression vanishes and the first part can be combined with term B in (6.23) to yield

$$\left(\frac{\partial}{\partial t}+\mathbf{V}_g\cdot\nabla\right)\left[\frac{1}{f_0}\nabla^2\Phi+f+\frac{\partial}{\partial p}\left(\frac{f_0}{\sigma}\frac{\partial\Phi}{\partial p}\right)\right]=\frac{D_g q}{Dt}=0 \qquad (6.25)$$

where q is the quasi-geostrophic potential vorticity defined by

$$q \equiv \left[\frac{1}{f_0}\nabla^2\Phi+f+\frac{\partial}{\partial p}\left(\frac{f_0}{\sigma}\frac{\partial\Phi}{\partial p}\right)\right] \qquad (6.26)$$

The three parts of (6.26), reading from left to right, are the *relative* vorticity, the *planetary* vorticity, and the *stretching* vorticity. As a parcel moves about in the atmosphere, the relative vorticity, planetary vorticity, and stretching vorticity terms may each change. But according to (6.25) their sum is conserved following the geostrophic motion. The scalar q, commonly called the quasi-geostrophic potential vorticity, is proportional to a linearized form of the Ertel potential vorticity discussed in Section 4.3. However, unlike Ertel's potential vorticity, q has units of vorticity and is determined solely by the isobaric geopotential distribution. Note that whereas Ertel's potential vorticity is conserved following the total motion on an isentropic surface, q is conserved following the geostrophic motion on an isobaric surface. For this reason some authors prefer the term *pseudo-potential vorticity*. However, we will generally refer to q as the quasi-geostrophic potential vorticity.

The stretching vorticity portion of q can be interpreted by substituting from the hydrostatic equation (6.2) to give

$$\frac{\partial}{\partial p}\left(\frac{f_0}{\sigma}\frac{\partial \Phi}{\partial p}\right) = -Rf_0\frac{\partial}{\partial p}\left(\frac{T}{\sigma p}\right) = -f_0\frac{\partial}{\partial p}\left(\frac{T}{S_p}\right) \approx -\frac{f_0}{S_p}\frac{\partial T}{\partial p}$$

where we have used the fact that S_p [as defined below (6.5)] varies only slowly with height in the troposphere. Referring back to Fig. 4.7, it is clear that as an air column moves adiabatically from left to right in the figure it is stretched vertically through upward motion in the upper portion and downward motion in the lower portion of the column. Thus, the upper portion must cool and the lower portion warm adiabatically so that $\partial T/\partial p$ must increase, and the stretching vorticity term becomes increasingly negative. If the planetary vorticity changes are small, the relative vorticity must then become increasingly positive in order that q remain constant following the geostrophic motion.

Comparing (6.23) and (6.25), we see that term A in (6.23) is actually just the tendency of q. Thus, the tendency of quasi-geostrophic potential vorticity is proportional to minus the geopotential tendency. A local increase (decrease) in q is associated with trough (ridge) development. Since q is a conserved quantity following the geostrophic motion we can diagnose the tendency purely from the geostrophic advection of q. Furthermore, (6.25) shows that the tendency will be zero (i.e., the flow will be steady) provided that the geostrophic wind is everywhere parallel to lines of constant q. Given the distribution of Φ, (6.25) can be integrated in time to provide a forecast of the evolution of the Φ field. However, because $\mathbf{V}_g$ depends on the distribution of Φ, the equation is highly nonlinear and numerical methods must be used for obtaining solutions.

6.4   Diagnosis of Vertical Motion

Since ζ_g and $\mathbf{V}_g$ are both defined in terms of $\Phi(x, y, p, t)$, (6.19) can be used to diagnose the ω field provided that the fields of both Φ and $\partial\Phi/\partial t$ are known. The former is a primary product of operational weather analysis. However, since upper-level analyses are generally available only twice per day, the latter can only be crudely approximated from observations by taking differences over 12 hours. Despite this limitation, the vorticity equation method of estimating ω is usually more accurate than the continuity equation method discussed in Section 3.5.1. However, neither of these methods of estimating ω uses the information available in the thermodynamic energy equation. An alternative method of estimating the vertical motion that utilizes both the vorticity equation and the thermodynamic equation is developed in this section.

6.4.1   OMEGA EQUATION

If we eliminate χ instead of ω between equations (6.21) and (6.22) we obtain a diagnostic equation that relates the field of ω at any instant to the Φ field at the same time. This equation is called the vertical motion or *omega equation*. Unlike the vorticity equation or thermodynamic energy equation methods discussed earlier, the omega equation method of diagnosing the vertical velocity requires information on the geopotential distribution at only a single time. Furthermore, since the omega equation arises from a combination of the vorticity and thermodynamic equations, the values of ω determined are consistent with both equations.

To obtain the omega equation we take the horizontal Laplacian of (6.21) to yield

$$\nabla^2 \frac{\partial \chi}{\partial p} = -\nabla^2 \left[\mathbf{V}_g \cdot \nabla \left(\frac{\partial \Phi}{\partial p} \right) \right] - \sigma \nabla^2 \omega \tag{6.27}$$

We next differentiate (6.22) with respect to pressure, yielding

$$\frac{\partial}{\partial p} (\nabla^2 \chi) = -f_0 \frac{\partial}{\partial p} \left[\mathbf{V}_g \cdot \nabla \left(\frac{1}{f_0} \nabla^2 \Phi + f \right) \right] + f_0^2 \frac{\partial^2 \omega}{\partial p^2} \tag{6.28}$$

Since the order of the operators on the left-hand side in (6.27) and (6.28) may be reversed, the result of subtracting (6.27) from (6.28) is to eliminate χ. After some rearrangement of terms, we obtain the *omega equation*

$$\underbrace{\left(\nabla^2 + \frac{f_0^2}{\sigma}\frac{\partial^2}{\partial p^2}\right)}_{A}\omega = \underbrace{\frac{f_0}{\sigma}\frac{\partial}{\partial p}\left[\mathbf{V}_g \cdot \nabla\left(\frac{1}{f_0}\nabla^2\Phi + f\right)\right]}_{B}$$

$$\underbrace{+\frac{1}{\sigma}\nabla^2\left[\mathbf{V}_g \cdot \nabla\left(-\frac{\partial\Phi}{\partial p}\right)\right]}_{C} \tag{6.29}$$

Equation (6.29) involves only derivatives in space. It is, therefore, a diagnostic equation for the field of ω in terms of the instantaneous Φ field. The omega equation, unlike the continuity equation, provides a method of estimating ω that does not depend on observations of the ageostrophic wind. In fact, direct wind observations are not required at all, nor does the omega equation require information on the vorticity tendency, as required in the vorticity equation method, or on the temperature tendency, as required in the adiabatic method discussed in Section 3.5.2. Only observations of Φ at a single time are needed to determine the ω field using (6.29). The terms in (6.29), however, employ higher-order derivatives than are involved in the other methods of estimating ω. Accurately estimating such terms from noisy observational data can be quite difficult.

The terms in (6.29) can be physically interpreted in a manner analogous to the corresponding terms in the tendency equation. The differential operator in A is very similar to the operator in term A of the tendency equation (6.23). Thus, term A acts to spread the response to a localized forcing. Because the forcing in (6.29) tends to be a maximum in the midtroposphere and ω is required to vanish at the upper and lower boundaries, for qualitative discussion it is permissible to assume that ω has sinusoidal behavior not only in the horizontal but also in the vertical:

$$\omega = W_0 \sin(\pi p/p_0) \sin kx \sin ly$$

we can then write

$$\left(\nabla^2 + \frac{f_0^2}{\sigma}\frac{\partial^2}{\partial p^2}\right)\omega \approx -\left[k^2 + l^2 + \frac{1}{\sigma}\left(\frac{f_0\pi}{p_0}\right)^2\right]\omega$$

which shows that term A in (6.29) is proportional to $-\omega$. Recalling that ω is proportional to $-w$ so that $\omega < 0$ implies *upward* vertical motion, we see that term A is proportional to the vertical velocity. Thus, there will be upward (downward) motion where the sum of terms B and C is positive (negative). The physical significance of each of these forcing terms is discussed in turn below.

Term B is called the *differential vorticity advection*. Clearly this term is proportional to the rate of increase with height of the advection of absolute vorticity. To understand the role of differential vorticity advection we again consider an idealized developing baroclinic system. Figure 6.11 indicates schematically the geopotential contours at 500 and 1000 mb for such a system. At the centers of the surface high and surface low, designated H and L, respectively, the vorticity advection at 1000 mb must be very small. However, at 500 mb the positive relative vorticity advection is a *maximum* above the surface low, while negative relative vorticity advection is strongest above the surface high. Thus, for a short-wave system where relative vorticity advection is larger than the planetary vorticity advection the pattern of vertical motion owing to the influence of term B alone is

$$w \propto \left\{ \frac{\partial}{\partial z} [-\mathbf{V}_\mathrm{g} \cdot \mathbf{\nabla}(\zeta_\mathrm{g}+f)] \begin{array}{ll} <0 & \text{above point H} \\ >0 & \text{above point L} \end{array} \right.$$

Thus, differential vorticity advection is associated with rising motion above the surface low and subsidence above the surface high. This pattern of vertical motion is in fact just what is required to produce the thickness tendencies in the 500–1000-mb layer above the surface highs and lows. For example, above the surface low there is positive vorticity associated with negative geopotential deviations since vorticity is proportional to the Laplacian of geopotential. Increasing vorticity thus implies a falling geopotential ($\chi < 0$). Hence the 500–1000-mb thickness is decreasing in that region. Since horizontal temperature advection is small above the center of the surface low, the only way to cool the atmosphere as required by the thickness tendency is by adiabatic cooling through the vertical motion field. Thus, the vertical motion maintains a hydrostatic temperature field (that is, a field in which temperature and thickness are proportional) in the presence of differential vorticity advection. Without this compensating vertical motion

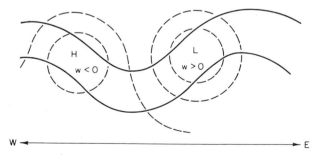

Fig. 6.11 Schematic 500-mb contours (solid lines) and 1000-mb contours (dashed lines) indicating regions of strong vertical motion owing to differential vorticity advection.

either the vorticity changes at 500 mb could not remain geostrophic or the temperature changes in the 500–1000-mb layer could not remain hydrostatic.

Term C of (6.29), which is merely the negative of the horizontal Laplacian of the thickness advection, is proportional to the thickness advection. The vertical velocity forced by term C acting alone is given by

$$w \propto \nabla^2 \left[\mathbf{V}_g \cdot \mathbf{\nabla} \left(-\frac{\partial \Phi}{\partial p} \right) \right] \propto -\mathbf{V}_g \cdot \mathbf{\nabla} \left(-\frac{\partial \Phi}{\partial p} \right)$$

If there is warm (cold) advection, term C will be positive (negative) so that in the absence of differential vorticity advection w would be positive (negative). Thus, as indicated in Fig. 6.12, rising motion will occur to the east of the surface low in the warm front zone and sinking motion will occur west of the surface low behind the cold front. Physically, this vertical motion pattern is required to keep the upper-level vorticity field geostrophic in the presence of the height changes caused by the thermal advection. For example, warm advection increases the 500–1000-mb thickness in the region of the 500-mb ridge. Thus, the geopotential height rises at the ridge and the anticyclonic vorticity must increase if geostrophic balance is to be maintained. Since vorticity advection cannot produce additional anticyclonic vorticity at the ridge, horizontal divergence is required to account for the negative vorticity tendency. Continuity of mass then requires that there be upward motion to replace the diverging air at the upper levels. By analogous arguments it can be shown that subsidence is required in the cold advection region beneath the 500-mb trough.

To summarize, we have shown as a result of scaling arguments that for synoptic-scale motions where vorticity is constrained to be geostrophic and

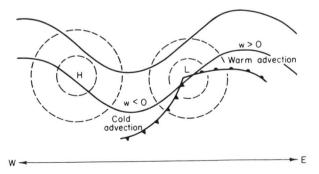

Fig. 6.12 Schematic 500-mb contours (thin solid lines), 1000-mb contours (dashed lines), and surface fronts (heavy lines) indicating regions of strong vertical motion owing to temperature advection.

temperature is constrained to be hydrostatic, the vertical motion field is determined uniquely by the geopotential field. Further, we have shown that this vertical motion field is just what is required to ensure that changes in vorticity will be geostrophic and changes in temperature will be hydrostatic. These constraints, whose importance can hardly be overemphasized, will be elaborated in the next subsection.

6.4.2 THE Q VECTOR

Although terms B and C in the omega equation (6.29) apparently have clear interpretations as separate physical processes, in practice there is often a significant amount of cancellation between them. They also are not invariant under a Galilean transformation of the zonal coordinate. That is, adding a constant mean zonal velocity will change the magnitude of each of these terms without changing the net forcing of the vertical motion. For these reasons an alternative form of the omega equation, the Q-vector form, has been developed in which the forcing of the vertical motion is expressed in terms of the divergence of a horizontal vector forcing field.

In order to keep the mathematical development as simple as possible we will consider the Q-vector formulation of the omega equation only for the case in which β is neglected. This is usually referred to as an f plane because it is equivalent to approximating the geometry by a Cartesian planar geometry with constant rotation.

On the f plane the quasi-geostrophic prediction equations may be expressed simply as follows:

$$\frac{D_g u_g}{Dt} - f_0 v_a = 0 \tag{6.30}$$

$$\frac{D_g v_g}{Dt} + f_0 u_a = 0 \tag{6.31}$$

$$\frac{D_g T}{Dt} - S_p \omega = 0 \tag{6.32}$$

These are coupled by the thermal wind relationship

$$p\frac{\partial u_g}{\partial p} = \frac{R}{f_0}\frac{\partial T}{\partial y}, \qquad p\frac{\partial v_g}{\partial p} = -\frac{R}{f_0}\frac{\partial T}{\partial x} \tag{6.33a, b}$$

We now eliminate the time derivatives between (6.30) and (6.32) by first taking

$$p\frac{\partial}{\partial p}(6.30) - \frac{R}{f_0}\frac{\partial}{\partial y}(6.32)$$

to obtain

$$p\frac{\partial}{\partial p}\left[\frac{\partial u_g}{\partial t} + u_g\frac{\partial u_g}{\partial x} + v_g\frac{\partial u_g}{\partial y} - f_0 v_a\right] - \frac{R}{f_0}\frac{\partial}{\partial y}\left[\frac{\partial T}{\partial t} + u_g\frac{\partial T}{\partial x} + v_g\frac{\partial T}{\partial y} - S_p\omega\right] = 0$$

Using the chain rule of differentiation, this may be rewritten as

$$\frac{RS_p}{f_0}\frac{\partial \omega}{\partial y} - f_0 p\frac{\partial v_a}{\partial p} = -\left(\frac{\partial}{\partial t} + u_g\frac{\partial}{\partial x} + v_g\frac{\partial}{\partial y}\right)\left(p\frac{\partial u_g}{\partial p} - \frac{R}{f_0}\frac{\partial T}{\partial y}\right)$$

$$-p\left[\frac{\partial u_g}{\partial p}\frac{\partial u_g}{\partial x} + \frac{\partial v_g}{\partial p}\frac{\partial u_g}{\partial y}\right] + \frac{R}{f_0}\left[\frac{\partial u_g}{\partial y}\frac{\partial T}{\partial x} + \frac{\partial v_g}{\partial y}\frac{\partial T}{\partial y}\right]$$

But, by the thermal wind relation (6.33) the term in parenthesis on the right-hand side vanishes and

$$-p\left[\frac{\partial u_g}{\partial p}\frac{\partial u_g}{\partial x} + \frac{\partial v_g}{\partial p}\frac{\partial u_g}{\partial y}\right] = -\frac{R}{f_0}\left[\frac{\partial T}{\partial y}\frac{\partial u_g}{\partial x} - \frac{\partial T}{\partial x}\frac{\partial u_g}{\partial y}\right]$$

Using these facts, plus the fact that

$$\partial u_g/\partial x + \partial v_g/\partial y = 0$$

we finally obtain the simplified form

$$\sigma\frac{\partial \omega}{\partial y} - f_0^2\frac{\partial v_a}{\partial p} = -2Q_2 \tag{6.34a}$$

where

$$Q_2 \equiv -\frac{R}{p}\left[\frac{\partial u_g}{\partial y}\frac{\partial T}{\partial x} + \frac{\partial v_g}{\partial y}\frac{\partial T}{\partial y}\right] = -\frac{R}{p}\frac{\partial \mathbf{V}_g}{\partial y}\cdot\mathbf{\nabla}T$$

Similarly, if we take

$$p\frac{\partial}{\partial p}(6.31) + \frac{R}{f_0}\frac{\partial}{\partial x}(6.32)$$

followed by application of (6.33b) we obtain

$$\sigma \frac{\partial \omega}{\partial x} - f_0^2 \frac{\partial u_a}{\partial p} = -2Q_1 \tag{6.34b}$$

where

$$Q_1 \equiv -\frac{R}{p}\left[\frac{\partial u_g}{\partial x}\frac{\partial T}{\partial x} + \frac{\partial v_g}{\partial x}\frac{\partial T}{\partial y}\right] = -\frac{R}{p}\frac{\partial \mathbf{V}_g}{\partial x}\cdot \mathbf{\nabla} T$$

If we now take $\partial(6.34b)/\partial x + \partial(6.34a)/\partial y$ and use (6.12) to eliminate the ageostrophic wind, we obtain the **Q**-vector form of the omega equation:

$$\sigma \nabla^2 \omega + f_0^2 \frac{\partial^2 \omega}{\partial p^2} = -2\mathbf{\nabla}\cdot\mathbf{Q} \tag{6.35}$$

where

$$\mathbf{Q} \equiv (Q_1, Q_2) = \left(-\frac{R}{p}\frac{\partial \mathbf{V}_g}{\partial x}\cdot\mathbf{\nabla} T, -\frac{R}{p}\frac{\partial \mathbf{V}_g}{\partial y}\cdot\mathbf{\nabla} T\right) \tag{6.36}$$

Equation (6.35) shows that on the f plane vertical motion is forced only by the divergence of **Q**. Unlike the traditional form of the omega equation, the **Q**-vector form does not have forcing terms that partly cancel. The forcing of ω can be represented simply by the pattern of the **Q** vector. By the arguments of the last subsection the left-hand side in (6.35) is proportional to the vertical velocity (w). Hence, regions where **Q** is convergent (divergent) correspond to ascent (descent).

The **Q** vector may be interpreted physically by considering the special case of baroclinic motion that is purely geostrophic so that the vertical velocity vanishes. Then

$$\frac{D_g T}{Dt} = \left(\frac{\partial}{\partial t} + \mathbf{V}_g\cdot\mathbf{\nabla}\right) T = 0$$

Thus,

$$\frac{\partial}{\partial x}\left(\frac{\partial}{\partial t} + \mathbf{V}_g\cdot\mathbf{\nabla}\right) T = \left(\frac{\partial}{\partial t} + \mathbf{V}_g\cdot\mathbf{\nabla}\right)\frac{\partial T}{\partial x} + \frac{\partial \mathbf{V}_g}{\partial x}\cdot\mathbf{\nabla} T = 0$$

which implies that

$$\left(\frac{\partial}{\partial t}+\mathbf{V}_g \cdot \boldsymbol{\nabla}\right)\frac{\partial T}{\partial x}=\frac{Q_1 p}{R}$$

By symmetry,

$$\left(\frac{\partial}{\partial t}+\mathbf{V}_g \cdot \boldsymbol{\nabla}\right)\frac{\partial T}{\partial y}=\frac{Q_2 p}{R}$$

so that in vector form

$$\frac{D_g}{Dt}\left(\frac{R}{p}\boldsymbol{\nabla}T\right)=\mathbf{Q} \tag{6.37}$$

Thus, $\mathbf{Q}$ is proportional to the rate of change of horizontal temperature gradient forced by geostrophic motion alone. By similar means it can be shown that the change in the vertical shear of the geostrophic wind owing to advection by purely geostrophic flow is given by

$$\frac{D_g}{Dt}\left(f_0\frac{\partial u_g}{\partial p}\right)=-Q_2, \qquad \frac{D_g}{Dt}\left(f_0\frac{\partial v_g}{\partial p}\right)=+Q_1 \tag{6.38}$$

Comparing (6.38) with the components of (6.37), we see that purely geostrophic flow will tend to destroy the thermal wind relationship since the forcing of the vertical shear of the geostrophic wind and the horizontal temperature gradient are equal in magnitude but have opposite signs. Only in the presence of ageostrophic winds and their accompanying vertical motions can the thermal wind balance be maintained.

Although (6.37) provides a useful physical interpretation of the $\mathbf{Q}$ vector, it is not easy to use this expression to estimate the direction and magnitude of the $\mathbf{Q}$ vector at a given point on a weather map. Such an estimate can be made quite readily, however, by utilizing an alternative expression for the $\mathbf{Q}$ vector. If the motion is referred to a Cartesian coordinate system in which the x axis is parallel to the local isotherm with cold air on the left, then (6.36) can be simplified to give

$$\mathbf{Q}=-\frac{R}{p}\left(\frac{\partial T}{\partial y}\right)\left(\frac{\partial v_g}{\partial x}\mathbf{i}-\frac{\partial u_g}{\partial x}\mathbf{j}\right)$$

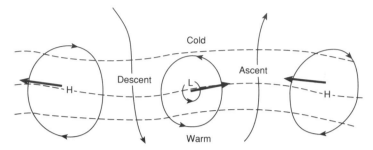

Fig. 6.13 Q vectors (bold arrows) for idealized pattern of isobars (solid) and isotherms (dashed) for a family of cyclones and anticyclones. (After Sanders and Hoskins, 1990.)

where we have again used the fact that $\partial u_g/\partial x = -\partial v_g/\partial y$. From the rules for cross multiplication of unit vectors the above expression for $\mathbf{Q}$ can be rewritten as

$$\mathbf{Q} = -\frac{R}{p}\left|\frac{\partial T}{\partial y}\right|\left(\mathbf{k}\times\frac{\partial\mathbf{V}_g}{\partial x}\right) \tag{6.39}$$

Thus, the $\mathbf{Q}$ vector can be obtained by evaluating the vectorial change of $\mathbf{V}_g$ along the isotherm (with cold air on the left), rotating the resulting change vector by 90° clockwise, and multiplying the resulting vector by $|\partial T/\partial y|$.

The $\mathbf{Q}$ vector, and hence the forcing of vertical motion, can be estimated with the aid of (6.39) from observations of Φ and T on a single isobaric surface. Examples for two simple cases, both of which have temperature decreasing toward the north, are shown in Figs. 6.13 and 6.14. Figure 6.13

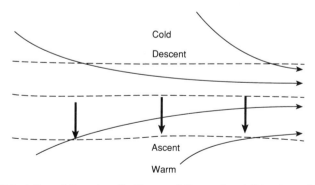

Fig. 6.14 Orientation of Q vectors (bold arrows) for confluent (jet entrance) flow. Dashed lines are isotherms. (After Sanders and Hoskins, 1990.)

shows an idealized pattern of cyclones and anticyclones in a slightly perturbed westerly thermal wind. Near the center of the low the geostrophic
wind change moving eastward along the isotherm (i.e., with cold air to the
left) is from northerly to southerly. Thus the geostrophic wind change vector
points northward, and a 90° clockwise rotation produces a **Q** vector parallel
to the thermal wind. In the highs, by the same reasoning the **Q** vectors are
antiparallel to the thermal wind. The pattern of $\nabla \cdot \mathbf{Q}$ thus yields descent
in the region of cold air advection west of the trough and ascent in the
warm air advection region east of the trough.

In the situation shown in Fig. 6.14 the geostrophic flow is confluent so
that the geostrophic wind increases eastward along the isotherms. In this
case the vectorial change in $\mathbf{V}_g$ is parallel to the isotherms so that the **Q**
vectors are normal to the isotherms and are directed up the temperature
gradient. Again, rising motion occurs where the **Q** vectors are convergent.
Since such rising must imply vorticity stretching in the column below,
cyclonic vorticity will tend to increase below a region of upper level
convergent **Q** vectors.

6.4.3 THE AGEOSTROPHIC CIRCULATION

In the traditional form of quasi-geostrophic theory given in Section 6.3
the ageostrophic velocity component is not explicitly determined. Rather,
its role in the secondary vertical circulation is implicitly included through
diagnostic determination of the ω vertical motion field. There are some
dynamical aspects of the ageostrophic motion that are not, however, obvious
from analysis of vertical motion alone. In particular, in some synoptic
situations advection by the ageostrophic wind may be important in the
evolution of the temperature and vorticity fields.

Since the ageostrophic wind generally has both irrotational and nondivergent components, the total ageostrophic flow field cannot be obtained from
knowledge of the divergence alone. Rather, it is necessary to use the
quasi-geostrophic momentum equation (6.11). If for simplicity we neglect
the β effect and solve (6.11) for the ageostrophic wind we obtain

$$\mathbf{V}_a = \frac{1}{f_0}\mathbf{k} \times \frac{D_g \mathbf{V}_g}{Dt} = \frac{1}{f_0}\left[\mathbf{k} \times \frac{\partial \mathbf{V}_g}{\partial t} + \mathbf{k} \times (\mathbf{V}_g \cdot \nabla)\mathbf{V}_g\right] \qquad (6.40)$$

which shows that in the Northern Hemisphere the ageostrophic wind vector
is directed to the left of the geostrophic acceleration following the geostrophic motion.

The forcing of the ageostrophic wind can conveniently be divided into
the two terms shown in brackets on the right in (6.40). The first term is

referred to as the *isallobaric* wind. It can easily be shown to be proportional to the gradient of the geopotential tendency (see Problem 6.13), and is directed down the gradient of the tendency field. This contribution to $\mathbf{V}_a$ is shown schematically for a baroclinic wave disturbance by the black arrows in Fig. 6.15. The second term in brackets in (6.40) may be called the advective part of the ageostrophic wind. For baroclinic waves in the jet stream the advective term is dominated by zonal advection so that

$$\mathbf{k} \times (\mathbf{V}_g \cdot \nabla)\mathbf{V}_g \approx \bar{u}\frac{\partial}{\partial x}(\mathbf{k} \times \mathbf{V}_g) = -f_0^{-1}\,\bar{u}\frac{\partial}{\partial x}(\nabla\Phi)$$

where $\bar{u}$ is the mean zonal flow, and we have used the definition of the

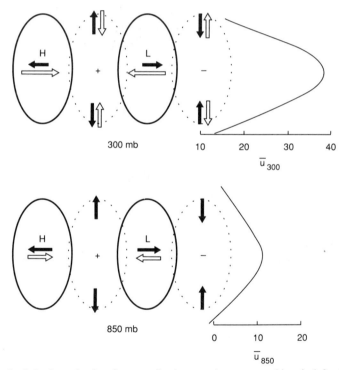

Fig. 6.15 Isallobaric and advective contributions to the ageostrophic wind for baroclinic waves in the westerlies. Solid ellipses indicate perturbation geopotential patterns at 300 and 850 mb. Dashed ellipses show geopotential tendency pattern with positive and negative tendencies indicated by + and − signs, respectively. The mean zonal flow distribution in which the waves are embedded is indicated on the right for each level. Solid arrows show the isallobaric part of the ageostrophic wind and open arrows show the advective part. (Adapted from Lim *et al.*, 1991.)

geostrophic wind (6.7). The advective contribution to the ageostrophic flow is shown by the open arrows in Fig. 6.15. Note that owing to the strong jet stream at 300 mb at the center of the waves the advective contribution dominates over the isallobaric contribution. On the flanks of the waves at 300 mb the two contributions are of comparable amplitude, so the net ageostrophic wind is small. At the 850-mb level, on the other hand, the two contributions nearly cancel at the center of the perturbations, while the advective contribution is nearly zero on the flanks. The net result is that the ageostrophic motion for baroclinic waves is primarily zonal in the upper troposphere and primarily meridional in the lower troposphere.

6.5   Idealized Model of a Baroclinic Disturbance

In Section 6.2 we showed that for synoptic-scale systems the fields of vertical motion and geopotential tendency are determined to a first approximation by the three-dimensional distribution of geopotential. The results of our diagnostic analyses using the geopotential tendency and omega equations can now be combined to illustrate the essential structural characteristics of a developing baroclinic wave. For reference, we restate here the qualitative content of the tendency and omega equations:

Geopotential Tendency Equation

$$\text{Geopotential}\begin{pmatrix}\text{fall}\\\text{rise}\end{pmatrix} \propto \begin{pmatrix}+\\-\end{pmatrix}\text{vorticity advection}$$

$$+\begin{pmatrix}\text{cold}\\\text{warm}\end{pmatrix}\text{advection decreasing with height}$$

Omega Equation

$$\begin{pmatrix}\text{Rising}\\\text{Sinking}\end{pmatrix}\text{motion} \propto \text{rate of increase with height of}\begin{pmatrix}+\\-\end{pmatrix}\text{vorticity advection}$$

$$+\begin{pmatrix}\text{warm}\\\text{cold}\end{pmatrix}\text{advection}$$

In Fig. 6.16 the relationship of the vertical motion field to the 500- and 1000-mb geopotential fields is illustrated schematically for a developing baroclinic wave. Also indicated are the physical processes that give rise to the vertical circulation in various regions.

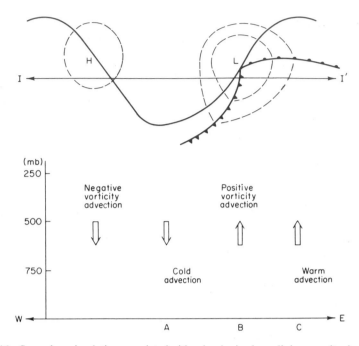

Fig. 6.16   Secondary circulation associated with a developing baroclinic wave: (top) schematic 500-mb contour (solid line), 1000-mb contours (dashed lines), and surface fronts; (bottom) vertical profile through the line II' indicating the vertical motion field.

Additional structural features, including those that can be diagnosed with the tendency equation, are summarized in Table 6.1. In this table the signs of various physical parameters are indicated for vertical columns located at the position of (A) the 500-mb trough, (B) the surface low, and (C) the 500-mb ridge. It can be seen from this table that in all cases the vertical motion and divergence fields act to keep the temperature changes hydrostatic and vorticity changes geostrophic in order to preserve thermal wind balance.

Following the nomenclature of Chapter 5, we may regard the vertical and divergent ageostrophic motions as constituting a secondary circulation imposed by the simultaneous constraints of geostrophic and hydrostatic balance. The secondary circulation described in this chapter is, however, completely independent of the circulation driven by boundary layer pumping. In fact, it is observed that in midlatitude synoptic-scale systems, the vertical velocity forced by frictional convergence in the boundary layer is generally much smaller than the vertical velocity owing to differential vorticity advection. For this reason we have neglected boundary layer friction in developing the equations of the quasi-geostrophic system.

Table 6.1 *Characteristics of a Developing Baroclinic Disturbance*

Physical parameter	A 500-mb trough	B Surface low	C 500-mb ridge
$\partial(\delta\Phi)/\partial t$ (500–1000 mb)	Negative (thickness advection partly canceled by adiabatic warming)	Negative (adiabatic cooling)	Positive (thickness advection partly canceled by adiabatic cooling)
w (500 mb)	Negative	Positive	Positive
$\partial\Phi/\partial t$ (500 mb)	Negative (differential thickness advection)	Negative (vorticity advection)	Positive (differential thickness advection)
$\partial\zeta_g/\partial t$ (1000 mb)	Negative (divergence)	Positive (convergence)	Positive (convergence)
$\partial\zeta_g/\partial t$ (500 mb)	Positive (convergence)	Positive (advection partly canceled by divergence)	Negative (divergence)

It is also of interest to note that the secondary circulation in a developing baroclinic system always acts to oppose the horizontal advection fields. Thus, the divergent motions tend partly to cancel the vorticity advection and the adiabatic temperature changes owing to vertical motion tend to cancel partly the thermal advection. This tendency of the secondary flow to cancel partly the advective changes has important implications for the flow evolution. These will be discussed in Chapter 8.

It should now be clear that a secondary divergent circulation is necessary to satisfy the twin constraints of geostrophic and hydrostatic balance for a baroclinic system. Without such a circulation geostrophic advection tends to destroy the thermal wind balance. The secondary circulation is itself forced, however, by slight departures from geostrophy. Referring again to Fig. 6.16, we see that in the region of the upper-level trough (column A) cold advection causes the geopotential height to fall and thus intensifies the horizontal pressure gradient. The wind, therefore, becomes slightly subgeostrophic and experiences an acceleration across the isobars toward lower pressure. It is this cross-isobaric ageostrophic wind component that is responsible for the convergence that spins up the vorticity in the upper troposphere so that it adjusts geostrophically to the new geopotential distribution. In terms of the momentum balance, the cross-isobaric flow is accelerated by the pressure gradient force so that the wind speed adjusts back toward geostrophic balance. In the region of the upper-level ridge analogous arguments apply, but in this case the subgeostrophic flow leads to a divergent secondary circulation. In both cases, as we will see in Chapter 8, the ageostrophic flow toward lower pressure is associated with conversion of energy from potential energy to kinetic energy.

Problems

6.1. Show that the static stability parameter $\sigma = -\alpha \partial \ln \theta / p$ may be written in terms of Φ as

$$\sigma = \frac{\partial^2 \Phi}{\partial p^2} - \frac{1}{p}\frac{\partial \Phi}{\partial p}\left(\frac{R}{c_p}-1\right) = \frac{1}{p^2}\left(\frac{\partial}{\partial \ln p}-\frac{R}{c_p}\right)\frac{\partial \Phi}{\partial \ln p}$$

6.2. Show that for an isothermal atmosphere σ, as defined in Problem 6.1, varies inversely as the square of the pressure.

6.3. Suppose that on the 50-kPa (500-mb) surface the relative vorticity at a certain location at 45°N latitude is increasing at a rate of $3 \times 10^{-6}\,\text{s}^{-1}$ per 3 h. The wind is from the southwest at $20\,\text{m s}^{-1}$ and the relative vorticity decreases toward the northeast at a rate of $4 \times 10^{-6}\,\text{s}^{-1}$ per 100 km. Use the quasi-geostrophic vorticity equation to estimate the horizontal divergence at this location on a β plane.

6.4. Given the following expression for the geopotential field:

$$\Phi = \Phi_0(p) + cf_0\{-y[\cos(\pi p/p_0)+1]+k^{-1}\sin k(x-ct)\}$$

where Φ_0 is a function of p alone, c is a constant speed, k a zonal wave number, and $p_0 = 100\,\text{kPa}$ (1000 mb):

(a) Use the quasi-geostrophic vorticity equation to obtain the horizontal divergence field consistent with this Φ field. (Assume that $df/dy = 0$.)

(b) Assuming that $\omega(p_0) = 0$, obtain an expression for $\omega(x, y, p, t)$ by integrating the continuity equation with respect to pressure.

(c) Sketch the geopotential fields at 75 kPa (750 mb) and 25 kPa (250 mb). Indicate regions of maximum divergence and convergence and positive and negative vorticity advection.

6.5. For the geopotential distribution of Problem 6.4 obtain an alternative expression for ω by using the adiabatic form of the thermodynamic energy equation (6.13). Assume that σ is a constant. For what value of k does this expression for ω agree with that obtained in Problem 6.4?

6.6. As an additional check on the results of Problems 6.4 and 6.5 use the omega equation (6.29) to obtain an expression for ω. Note that the three expressions for ω agree only for one value of k. Thus, the geopotential field $\Phi(x, y, p, t)$ of Problem 6.4 is consistent with quasi-geostrophic dynamics only for one value of the zonal wave number.

6.7. Suppose that the geopotential distribution at a certain time has the form

$$\Phi(x, y, p) = \Phi_0(p) - f_0 U_0 y \cos(\pi p/p_0) + f_0 c k^{-1} \sin kx$$

where U_0 is a constant zonal speed and all other constants are as in Problem 6.4. Assuming that f and σ are constants, show by evaluating the terms on the right-hand side of the tendency equation (6.23) that $\chi = 0$ provided that $k^2 = \sigma^{-1}(f_0 \pi/p_0)^2$. Make qualitative sketches of the geopotential fields at 750 mb and 250 mb for this case. Indicate regions of maximum positive and negative vorticity advection at each level. (Note: the wavelength corresponding to this value of k is called the *radius of deformation*.)

6.8. For the geopotential field of Problem 6.7 use the omega equation (6.29) to find an expression for ω for the conditions in which $\chi = 0$. *Hint*: let $\omega = W_0 \cos kx \sin(\pi p/p_0)$ where W_0 is a constant to be determined. Sketch a cross section in the x, p plane indicating trough and ridge lines, vorticity maxima and minima, vertical motion and divergence patterns, and locations of maximum cold and warm temperature advection.

6.9. Given the following expression for the geopotential field:

$$\Phi(x, y, p) = \Phi_0(p) + f_0[-Uy + k^{-1} V \cos(\pi p/p_0) \sin k(x - ct)]$$

where U, V, and c are constant speeds, use the quasi-geostrophic vorticity equation (6.19) to obtain an estimate of ω. Assume that $df/dy = \beta$ is a constant (*not zero*) and that ω vanishes for $p = p_0$.

6.10. For the conditions given in Problem 6.9, use the adiabatic thermodynamic energy equation to obtain an alternative estimate for ω. Determine the value of c for which this estimate of ω agrees with that found in Problem 6.9.

6.11. For the conditions given in Problem 6.9, use the omega equation (6.29) to obtain an expression for ω. Verify that this result agrees with the results of Problems 6.9 and 6.10. Sketch the phase relationship between Φ and ω at 250 mb and 750 mb. What is the amplitude of ω

if $\beta = 2 \times 10^{-11} \, \text{m}^{-1} \, \text{s}^{-1}$, $U = 25 \, \text{m s}^{-1}$, $V = 8 \, \text{m s}^{-1}$, $k = 2\pi/(10^4 \, \text{km})$, $f_0 = 10^{-4} \, \text{s}^{-1}$, $\sigma = 2 \times 10^{-6} \, \text{Pa}^{-2} \, \text{m}^2 \, \text{s}^{-2}$, and $p_0 = 10^2 \, \text{kPa}$?

6.12. Compute the **Q**-vector distributions corresponding to the geopotential fields given in Problems 6.4 and 6.7.

6.13. Show that the isallobaric wind may be expressed in the form

$$\mathbf{V}_{\text{isall}} = -f_0^{-2} \, \boldsymbol{\nabla}\chi$$

where $\chi = \partial\Phi/\partial t$.

Suggested References

Wallace and Hobbs, *Atmospheric Science: An Introductory Survey,* has an excellent description of the observed structure and evolution of midlatitude synoptic-scale disturbances.

Blackburn (1985) *Interpretation of ageostrophic winds and implications for jetstream maintenance,* discusses the differences between variable f(VF) and constant f(CF) ageostrophic motion.

Durran and Snellman (1987) illustrate the application of both the traditional and **Q**-vector forms of the omega equation in diagnosing the vertical motion of an observed system.

Sanders and Hoskins (1990) show how the distribution of **Q** vectors can easily be visualized on realistic weather maps.

Pedlosky, *Geophysical Fluid Dynamics, 2nd Edition,* presents a detailed formal derivation of the quasi-geostrophic system with applications to both the atmosphere and the oceans.

Chapter

7 | Atmospheric Oscillations: Linear Perturbation Theory

In Chapter 13 we will discuss numerical techniques for solving the equations governing large-scale atmospheric motions. If the objective is to produce an accurate forecast of the circulation at some future time, a detailed numerical model based on the primitive equations and including processes such as latent heating, radiative transfer, and boundary layer drag should produce the best results. However, the inherent complexity of such a model generally precludes any simple interpretation of the physical processes that produce the predicted circulation. If we wish to gain physical insight into the fundamental nature of atmospheric motions, it is helpful to employ simplified models in which certain processes are omitted and compare the results with those of more complete models. This is, of course, just what was done in deriving the quasi-geostrophic model. However, the quasi-geostrophic potential vorticity equation is still a complicated non-linear equation that must be solved numerically. It is difficult to gain an appreciation for the processes that produce the wavelike character observed in many meterological disturbances through study of numerical integrations alone.

In this chapter we discuss the *perturbation method*, a simple technique that is useful for qualitative analysis of atmospheric waves. We then use this method to examine several types of waves in the atmosphere. In Chapter 8 the perturbation theory will be used to study the development of synoptic-wave disturbances.

7.1 The Perturbation Method

In the perturbation method all field variables are divided into two parts, a *basic state* portion, which is usually assumed to be independent of time and longitude, and a perturbation portion, which is the local deviation of the field from the basic state. Thus, for example, if $\bar{u}$ designates a time- and longitude-averaged zonal velocity and u' is the deviation from that average, then the complete zonal velocity field is $u(x, t) = \bar{u} + u'(x, t)$. In that case for example, the inertial acceleration $u \, \partial u/\partial x$ can be written

$$u\frac{\partial u}{\partial x} = (\bar{u} + u')\frac{\partial}{\partial x}(\bar{u} + u') = \bar{u}\frac{\partial u'}{\partial x} + u'\frac{\partial u'}{\partial x}$$

The basic assumptions of perturbation theory are that the basic state variables must themselves satisfy the governing equations when the perturbations are set to zero and the perturbation fields must be small enough so that all terms in the governing equations that involve products of the perturbations can be neglected. The latter requirement would be met in the above example if $|u'/\bar{u}| \ll 1$ so that

$$|\bar{u} \, \partial u'/\partial x| \gg |u' \, \partial u'/\partial x|$$

If terms that are products of the perturbation variables are neglected, the nonlinear governing equations are reduced to linear differential equations in the perturbation variables in which the basic state variables are specified coefficients. These equations can then be solved by standard methods to determine the character and structure of the perturbations in terms of the known basic state. For equations with constant coefficients the solutions are sinusoidal or exponential in character. Solution of the perturbation equations then determines such characteristics as the propagation speed, vertical structure, and conditions for growth or decay of the waves. The perturbation technique is especially useful in studying the stability of a given basic state flow with respect to small superposed perturbations. This application is the subject of Chapter 8.

7.2   Properties of Waves

Wave motions are oscillations in field variables (such as velocity and pressure) that propagate in space. In this chapter we are concerned with linear sinusoidal wave motions. Many of the mechanical properties of such waves are also features of a familiar system, the linear harmonic oscillator. An important property of the harmonic oscillator is that the period, or time required to execute a single oscillation, is independent of the amplitude of the oscillation. For most natural vibratory systems this condition holds only for oscillations of sufficiently small amplitude. The classical example of such a system is the simple pendulum (Fig. 7.1) consisting of a mass M suspended by a massless string of length l, free to perform small oscillations about the equilibrium position $\theta = 0$. The component of the gravity force parallel to the direction of motion is $-Mg \sin \theta$. Thus, the equation of motion is

$$Ml\frac{d^2\theta}{dt^2} = -Mg \sin \theta$$

Now for small displacements, $\sin \theta \approx \theta$ so that the governing equation becomes

$$\frac{d^2\theta}{dt^2} + \nu^2\theta = 0 \qquad (7.1)$$

where $\nu^2 \equiv g/l$. The harmonic oscillator equation (7.1) has the general solution

$$\theta = \theta_1 \cos \nu t + \theta_2 \sin \nu t = \theta_0 \cos(\nu t - \alpha)$$

Fig. 7.1   A simple pendulum.

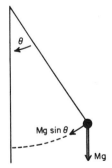

where θ_1, θ_2, θ_0 and α are constants determined by the initial conditions (see Problem 7.1 at the end of the chapter), and ν is the frequency of oscillation. The complete solution can thus be expressed in terms of an amplitue θ_0 and a phase $\phi(t) = \nu t - \alpha$. The phase varies linearly in time by a factor of 2π radians per wave period.

Propagating waves can also be characterized by their amplitudes and phases. In a propagating wave, however, phase depends not only on time but on one or more space variables as well. Thus, for a one-dimensional wave propagating in the x direction, $\phi(x, t) = kx - \nu t - \alpha$. Here the *wave number*, k, is defined as 2π divided by the wavelength. For propagating waves the phase is constant for an observer moving at the *phase speed* $c \equiv \nu/k$. This may be verified by observing that if phase is to remain constant following the motion,

$$\frac{D\phi}{Dt} = \frac{D}{Dt}(kx - \nu t - \alpha) = k\frac{Dx}{Dt} - \nu = 0$$

Thus, $Dx/Dt = c = \nu/k$ for phase to be constant. For $\nu > 0$ and $k > 0$ we have $c > 0$, so that phase propagates in the positive direction as illustrated for a sinusoidal wave in Fig. 7.2.

7.2.1. FOURIER SERIES

The representation of a perturbation as a simple sinusoidal wave might seem an oversimplification since disturbances in the atmosphere are never purely sinusoidal. It can be shown, however, that any reasonably well-behaved function of longitude can be represented in terms of a zonal mean plus a *Fourier* series of sinusoidal components:

$$f(x) = \sum_{s=1}^{\infty} (A_s \sin k_s x + B_s \cos k_s x) \tag{7.2}$$

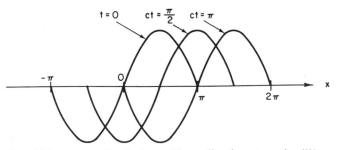

Fig. 7.2   A sinusoidal wave traveling in the positive x direction at speed c. (Wave number is assumed to be unity.)

where $k_s = 2\pi s/L$ is the *zonal* wave number (units m^{-1}), L is the distance around a latitude circle, and s, the *planetary* wave number, is an integer designating the number of waves around a latitude circle. The coefficients A_s are calculated by multiplying both sides of (7.2) by $\sin(2\pi nx/L)$, where n is an integer, and integrating around a latitude circle. Applying the orthogonality relationships

$$\int_0^L \sin\frac{2\pi sx}{L} \sin\frac{2\pi nx}{L} \, dx = \left\{ \begin{matrix} 0, & s \neq n \\ L/2, & s = n \end{matrix} \right\}$$

we obtain

$$A_s = \frac{2}{L} \int_0^L f(x) \sin\frac{2\pi sx}{L} \, dx$$

In a similar fashion, multiplying both sides in (7.2) by $\cos(2\pi nx/L)$ and integrating gives

$$B_s = \frac{2}{L} \int_0^L f(x) \cos\frac{2\pi sx}{L} \, dx$$

A_s and B_s are called the *Fourier coefficients*, and

$$f_s(x) = A_s \sin k_s x + B_s \cos k_s x \tag{7.3}$$

is called the sth *Fourier component* or sth harmonic of the function $f(x)$. If the Fourier coefficients are computed for, say, the longitudinal dependence of the (observed) geopotential perturbation, the largest-amplitude Fourier components will be those for which s is close to the observed number of troughs or ridges around a latitude circle. When only qualitative information is desired, it is usually sufficient to limit the analysis to a single typical Fourier component and assume that the behavior of the actual field will be similar to that of the component. The expression for a Fourier component may be written more compactly by using complex exponential notation. According to the Euler formula

$$\exp(i\phi) = \cos\phi + i\sin\phi$$

where $i \equiv (-1)^{1/2}$ is the imaginary unit. Thus, we can write

$$\begin{aligned} f_s(x) &= \text{Re}[C_s \exp(ik_s x)] \\ &= \text{Re}[C_s \cos k_s x + iC_s \sin k_s x] \end{aligned} \tag{7.4}$$

where Re[ ] denotes "real part of" and C_s is a complex coefficient. Comparing (7.3) and (7.4), we see that the two representations of $f_s(x)$ are identical provided that

$$B_s = \text{Re}[C_s] \quad \text{and} \quad A_s = -\text{Im}[C_s]$$

where Im[ ] stands for "imaginary part of." This exponential notation will generally be used for applications of the perturbation theory below and also in Chapter 8.

7.2.2 DISPERSION AND GROUP VELOCITY

A fundamental property of linear oscillators is that the frequency of oscillation ν depends only on the physical characteristics of the oscillator, not on the motion itself. For propagating waves, however, ν generally depends on the wave number of the perturbation as well as the physical properties of the medium. Thus, since $c = \nu / k$, the phase speed also depends on the wave number except in the special case where $\nu \propto k$. For waves in which the phase speed varies with k the various sinusoidal components of a disturbance originating at a given location are at a later time found in different places; i.e., they are dispersed. Such waves are referred to as *dispersive*, and the formula that relates ν and k is called a *dispersion relationship*. Some types of waves, such as acoustic waves, have phase speeds that are independent of the wave number. In such *nondispersive* waves a spatially localized disturbance consisting of a number of Fourier wave components (a *wave group*) will preserve its shape as it propagates in space at the phase speed of the wave.

For dispersive waves, however, the shape of a wave group will not remain constant as the group propagates. Since the individual Fourier components of a wave group may either reinforce or cancel each other, depending on the relative phases of the components, the energy of the group will be concentrated in limited regions as illustrated in Fig. 7.3.

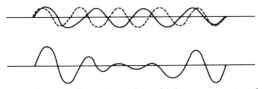

Fig. 7.3 Wave groups formed from two sinusoidal components of slightly different wavelengths. For nondispersive waves the pattern in the lower part of the diagram propagates without change of shape. For dispersive waves the shape of the pattern changes in time.

When waves are dispersive, the speed of the wave group is generally different from the average phase speed of the individual Fourier components. Hence, as shown in Fig. 7.4, individual wave components may move through the group as the group propagates along. Furthermore, the group generally broadens in the course of time; that is, the energy is *dispersed*. A familiar example is the wake of a ship, in which individual wave crests are observed to move twice as fast as the wave group.

An expression for the *group velocity*, which is the velocity at which the observable disturbance (and hence the energy) propagates, can be derived as follows: We consider the superposition of two horizontally propagating waves of equal amplitude but slightly different wavelengths with wave numbers and frequencies differing by $2\delta k$ and $2\delta\nu$, respectively. The total disturbance is thus

$$\psi(x, t) = \exp\{i[(k + \delta k)x - (\nu + \delta\nu)t]\} + \exp\{i[(k - \delta k)x - (\nu - \delta\nu)t]\}$$

where for brevity the Re[ ] notation in (7.4) is omitted, and it is understood that only the real part of the right-hand side has physical meaning. Rearranging terms we get

$$\psi = [e^{i(\delta kx - \delta\nu t)} + e^{-i(\delta kx - \delta\nu t)}]e^{i(kx - \nu t)}$$
$$= 2\cos(\delta kx - \delta\nu t)e^{i(kx - \nu t)} \tag{7.5}$$

The disturbance (7.5) is the product of a high-frequency *carrier wave* of wavelength $2\pi/k$ whose phase speed, ν/k, is the average for the two Fourier components, and a low-frequency *envelope* of wavelength $2\pi/\delta k$ that travels

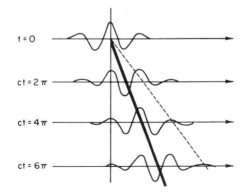

Fig. 7.4 Propagation of a wave group. Heavy line shows group speed; dashed line shows phase speed.

at the speed $\delta\nu/\delta k$. Thus, in the limit as $\delta k \to 0$, the horizontal velocity of the envelope, or *group velocity*, is just

$$c_{gx} = \partial\nu/\partial k$$

Thus, the wave energy propagates at the group velocity. This result applies generally to arbitrary wave envelopes provided that the wavelength of the wave group, $2\pi/\delta k$, is large compared to the wavelength of the dominant component, $2\pi/k$.

7.3 Simple Wave Types

Waves in fluids result from the action of restoring forces on fluid parcels that have been displaced from their equilibrium positions. The restoring forces may be due to compressibility, gravity, rotation, or electromagnetic effects. In the present section we consider the two simplest examples of linear waves in fluids: acoustic waves and shallow-water gravity waves.

7.3.1 ACOUSTIC OR SOUND WAVES

Sound waves, or acoustic waves, are *longitudinal waves.* That is, they are waves in which the particle oscillations are parallel to the direction of propagation. Sound is propagated by the alternating adiabatic compression and expansion of the medium. As an example, in Fig. 7.5 we show a schematic section along a tube that has a diaphragm at its left end. If the diaphragm is set into vibration the air adjacent to it will be alternately compressed and expanded as the diaphragm moves inward and outward.

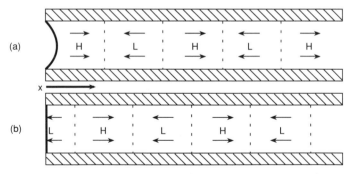

Fig. 7.5   Schematic diagram illustrating the propagation of a sound wave in a tube with a flexible diaphragm at the left end. Labels H and L designate centers of high and low perturbation pressure. Arrows show velocity perturbations. Panel (b) shows the situation 1/4 period later than panel (a) for propagation in the positive x direction.

The resulting oscillating pressure gradient force will be balanced by an oscillating acceleration of the air in the adjoining region, which will cause a pressure oscillation farther into the tube, etc. The result of this continual adiabatic increase and decrease of pressure through alternating compression and rarefaction is, as shown in Fig. 7.5, a sinusoidal pattern of pressure and velocity perturbations, which propagates to the right down the tube. Individual air parcels do not, however, have a net rightward motion; they only oscillate back and forth while the pressure pattern moves rightward at the speed of sound.

To introduce the perturbation method we consider the problem illustrated by Fig. 7.5, that is, one-dimensional sound waves propagating in a straight pipe parallel to the x axis. To exclude the possibility of *transverse* oscillations (that is, oscillations in which the particle motion is at right angles to the direction of phase propagation), we assume at the outset that $v = w = 0$. In addition we eliminate all dependence on y and z by assuming that $u = u(x, t)$. With these restrictions the momentum equation, continuity equation, and thermodynamic energy equation for adiabatic motion are, respectively,

$$\frac{Du}{Dt} + \frac{1}{\rho}\frac{\partial p}{\partial x} = 0 \tag{7.6}$$

$$\frac{D\rho}{Dt} + \rho\frac{\partial u}{\partial x} = 0 \tag{7.7}$$

$$\frac{D\ln\theta}{Dt} = 0 \tag{7.8}$$

where for this case $D/Dt = \partial/\partial t + u\,\partial/\partial x$. Recalling from (2.44) and the ideal gas law that potential temperature may be expressed as

$$\theta = (p/\rho R)(p_s/p)^{R/c_p} \quad \text{where } p_s = 1000 \text{ mb}$$

we may eliminate θ in (7.8) to give

$$\frac{1}{\gamma}\frac{D\ln p}{Dt} - \frac{D\ln\rho}{Dt} = 0 \tag{7.9}$$

where $\gamma = c_p/c_v$. Eliminating ρ between (7.7) and (7.9) gives

$$\frac{1}{\gamma}\frac{D\ln p}{Dt} + \frac{\partial u}{\partial x} = 0 \tag{7.10}$$

The dependent variables are now divided into constant basic state portions (denoted by overbars) and perturbation portions (denoted by primes):

$$u(x, t) = \bar{u} + u'(x, t)$$

$$p(x, t) = \bar{p} + p'(x, t) \tag{7.11}$$

$$\rho(x, t) = \bar{\rho} + \rho'(x, t)$$

Substituting (7.11) into (7.6) and (7.10) we obtain

$$\frac{\partial}{\partial t}(\bar{u} + u') + (\bar{u} + u')\frac{\partial}{\partial x}(\bar{u} + u') + \frac{1}{\bar{\rho} + \rho'}\frac{\partial}{\partial x}(\bar{p} + p') = 0$$

$$\frac{\partial}{\partial t}(\bar{p} + p') + (\bar{u} + u')\frac{\partial}{\partial x}(\bar{p} + p') + \gamma(\bar{p} + p')\frac{\partial}{\partial x}(\bar{u} + u') = 0$$

We next observe that provided $|\rho'/\bar{\rho}| \ll 1$ we can use the binomial expansion to approximate the density term as

$$\frac{1}{\bar{\rho} + \rho'} = \frac{1}{\bar{\rho}}\left(1 + \frac{\rho'}{\bar{\rho}}\right)^{-1} \approx \frac{1}{\bar{\rho}}\left(1 - \frac{\rho'}{\bar{\rho}}\right)$$

Neglecting products of the perturbation quantities and noting that the basic state fields are constants, we obtain the linear perturbation equations[1]

$$\left(\frac{\partial}{\partial t} + \bar{u}\frac{\partial}{\partial x}\right)u' + \frac{1}{\bar{\rho}}\frac{\partial p'}{\partial x} = 0 \tag{7.12}$$

$$\left(\frac{\partial}{\partial t} + \bar{u}\frac{\partial}{\partial x}\right)p' + \gamma\bar{p}\frac{\partial u'}{\partial x} = 0 \tag{7.13}$$

Eliminating u' by operating on (7.13) with $(\partial/\partial t + \bar{u}\,\partial/\partial x)$ and substituting from (7.12) we get[2]

$$\left(\frac{\partial}{\partial t} + \bar{u}\frac{\partial}{\partial x}\right)^2 p' - \frac{\gamma\bar{p}}{\bar{\rho}}\frac{\partial^2 p'}{\partial x^2} = 0 \tag{7.14}$$

---

[1] It is not necessary that the perturbation velocity be small compared to the mean velocity for linearization to be valid. It is only required that quadratic terms in the perturbation variables be small compared to the dominant linear terms in (7.12) and (7.13).

[2] Note that the squared differential operator in the first term expands in the usual way as

$$\left(\frac{\partial}{\partial t} + \bar{u}\frac{\partial}{\partial x}\right)^2 = \frac{\partial^2}{\partial t^2} + 2\bar{u}\frac{\partial^2}{\partial t\,\partial x} + \bar{u}^2\frac{\partial^2}{\partial x^2}$$

which is a form of the standard *wave equation* familiar from electromagnetic theory. A simple solution representing a plane sinusoidal wave propagating in x is

$$p' = A \exp[ik(x - ct)] \tag{7.15}$$

where for brevity we omit the Re{ } notation, but it is to be understood that only the real part of (7.15) has physical significance. Substituting the assumed solution (7.15) into (7.14) we find that the phase speed c must satisfy

$$(-ikc + ik\bar{u})^2 - (\gamma \bar{p} / \bar{\rho})(ik)^2 = 0$$

where we have canceled out the factor $A \exp[ik(x - ct)]$, which is common to both terms. Solving for c gives

$$c = \bar{u} + (\gamma \bar{p} / \bar{\rho})^{1/2} = \bar{u} \pm (\gamma R \bar{T})^{1/2} \tag{7.16}$$

Therefore (7.15) is a solution of (7.14) provided that the phase speed satisfies (7.16). According to (7.16) the speed of wave propagation relative to the zonal current is

$$c_s \equiv \pm (\gamma R \bar{T})^{1/2}$$

This quantity is called the *adiabatic speed of sound*. The mean zonal velocity here plays only a role of *Doppler shifting* the sound wave so that the frequency corresponding to a given wave number k,

$$\nu = kc = k(\bar{u} \pm c_s)$$

appears higher to an observer downstream from the source than to an upstream observer.

7.3.2 SHALLOW-WATER GRAVITY WAVES

As a second example of pure wave motion we consider the horizontally propagating oscillations known as shallow-water waves. Shallow-water gravity waves can exist only if the fluid has a free surface or an internal density discontinuity. As shown in the previous subsection, in acoustic waves the restoring force is parallel to the direction of propagation of the wave. In shallow-water gravity waves, however, the restoring force is in the vertical, so it is transverse to the direction of propagation.

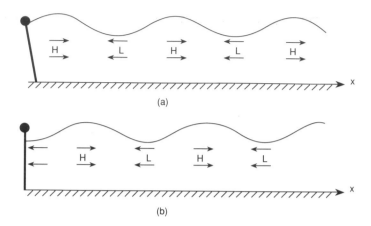

(a)

(b)

Fig. 7.6 The propagation of a surface gravity wave in a water channel generated by a paddle at the left end. Symbols as in Fig. 7.5.

The mechanism for propagation of gravity waves can be understood by considering water in a channel extending in the x direction with an oscillating paddle at the origin. The back-and-forth oscillations of the paddle generate alternating upward and downward perturbations in the free surface height, which produce alternating positive and negative accelerations. These, in turn, lead to alternating patterns of fluid convergence and divergence. The net result is a sinusoidal disturbance of the free surface height, which moves toward the right, and has perturbation velocity and free surface height exactly in phase as shown in Fig. 7.6. A similar sort of disturbance could be set up moving toward the left, but in that case the velocity and free surface height perturbations would be exactly 180° out of phase.

As a specific example we consider a fluid system consisting of two homogeneous incompressible layers of differing density as shown in Fig.

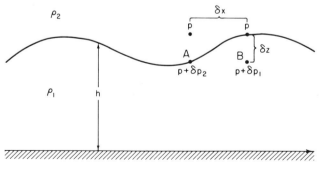

Fig. 7.7 A two-layer fluid system.

7.7. Waves may propagate along the interface between the two layers. The assumption of incompressibility is sufficient to exclude sound waves, and we can thus isolate the gravity waves. If the density of the lower layer ρ_1 is greater than the density of the upper layer ρ_2, the system is stably stratified. Since both ρ_1 and ρ_2 are constants, the horizontal pressure gradient in each layer is independent of height if the pressure is hydrostatic. This may be verified by differentiating the hydrostatic approximation with respect to x:

$$\frac{\partial}{\partial z}\left(\frac{\partial p}{\partial x}\right) = -\frac{\partial \rho}{\partial x}\, g = 0$$

For simplicity, we assume that there is no horizontal pressure gradient in the upper layer. The pressure gradient in the lower layer can be obtained by vertical integration of the hydrostatic equation. For the points A and B shown in Fig. 7.7 we find, respectively,

$$p + \delta p_1 = p + \rho_1 g\, \delta z = p + \rho_1 g(\partial h/\partial x)\, \delta x$$

$$p + \delta p_2 = p + \rho_2 g\, \delta z = p + \rho_2 g(\partial h/\partial x)\, \delta x$$

where $\partial h/\partial x$ is the slope of the interface. Taking the limit $\delta x \to 0$, we obtain the pressure gradient in the lower layer

$$\lim_{\delta x \to 0}\left[\frac{(p+\delta p_1)-(p+\delta p_2)}{\delta x}\right] = g\, \delta\rho\, \frac{\partial h}{\partial x}$$

where $\delta\rho = \rho_1 - \rho_2$.

We assume that the motion is two-dimensional in the x, z plane. The x-momentum equation for the lower layer is then

$$\frac{\partial u}{\partial t} + u\frac{\partial u}{\partial x} + w\frac{\partial u}{\partial z} = -\frac{g\,\delta\rho}{\rho_1}\frac{\partial h}{\partial x} \qquad (7.17)$$

while the continuity equation is

$$\frac{\partial u}{\partial x} + \frac{\partial w}{\partial z} = 0 \qquad (7.18)$$

Now since the pressure gradient in (7.17) is independent of z, u will also be independent of z provided that $u \neq u(z)$ initially. Thus, (7.18) can be integrated vertically from the lower boundary $z = 0$ to the interface $z = h$ to yield

$$w(h) - w(0) = -h(\partial u/\partial x)$$

But $w(h)$ is just the rate at which the interface height is changing,

$$w(h) = \frac{Dh}{Dt} = \frac{\partial h}{\partial t} + u \frac{\partial h}{\partial x}$$

and $w(0) = 0$ for a flat lower boundary. Hence, the vertically integrated continuity equation can be written

$$\frac{\partial h}{\partial t} + \frac{\partial}{\partial x}(hu) = 0 \tag{7.19}$$

Equations (7.17) and (7.19) are a closed set in the variables u and h. We now apply the perturbation technique by letting

$$u = \bar{u} + u', \qquad h = H + h'$$

where $\bar{u}$ as before is a constant basic state zonal velocity and H is the mean depth of the lower layer. The perturbation forms of (7.17) and (7.19) are then

$$\frac{\partial u'}{\partial t} + \bar{u} \frac{\partial u'}{\partial x} + \frac{g\,\delta\rho}{\rho_1} \frac{\partial h'}{\partial x} = 0 \tag{7.20}$$

$$\frac{\partial h'}{\partial t} + \bar{u} \frac{\partial h'}{\partial x} + H \frac{\partial u'}{\partial x} = 0 \tag{7.21}$$

where we assume that $H \gg |h'|$ so that products of the perturbation variables can be neglected.

Eliminating u' between (7.20) and (7.21) yields

$$\left(\frac{\partial}{\partial t} + \bar{u} \frac{\partial}{\partial x} \right)^2 h' - \frac{gH\,\delta\rho}{\rho_1} \frac{\partial^2 h'}{\partial x^2} = 0 \tag{7.22}$$

which is a wave equation similar in form to (7.14). It is easily verified by direct substitution that (7.22) has a solution of the form

$$h' = A \exp[ik(x - ct)]$$

where the phase speed c satisfies the relationship

$$c = \bar{u} \pm (gH\,\delta\rho/\rho_1)^{1/2} \tag{7.23}$$

If the upper and lower layers are air and water, respectively, then $\delta\rho \approx \rho_1$ and the phase speed formula simplifies to

$$c = \bar{u} \pm \sqrt{gH}$$

The quantity $\sqrt{gH}$ is called the *shallow-water* wave speed. It is a valid approximation only for waves whose wavelengths are much greater than the depth of the fluid. This restriction is necessary in order that the vertical velocities be small enough so that the hydrostatic approximation is valid. If the depth of the ocean is taken as 4 km, the shallow-water gravity wave speed is $\approx 200 \, \mathrm{m \, s^{-1}}$. Thus, long waves on the ocean surface travel very rapidly. It should be emphasized again that this theory applies only to waves of wavelength much greater than H. Such long waves are not ordinarily excited by the wind stresses but may be produced by very large scale disturbances such as earthquakes.[3]

Shallow-water gravity waves may also occur at interfaces within the ocean where there is a very sharp density gradient (diffusion will always prevent formation of a true density discontinuity). In particular, the surface water is separated from the deep water by a narrow region of sharp density contrast called the *thermocline*. If the horizontal pressure gradient vanishes in the layer above the thermocline, then (7.22) governs the displacement, h', of the thermocline from its mean height H. If the density changes by an amount $\delta\rho/\rho_1 \approx 0.01$ across the thermocline, then from (7.23) the wave speed for waves traveling along the thermocline will be only one-tenth of the surface wave speed for a fluid of the same depth.[4]

7.4 Internal Gravity (Buoyancy) Waves

We now consider the nature of gravity wave propagation in the atmosphere. Atmospheric gravity waves can exist only when the atmosphere is stably stratified so that a fluid parcel displaced vertically will undergo buoyancy oscillations (see Section 2.7.3). Since the buoyancy force is the restoring force responsible for gravity waves, the term *buoyancy wave* is actually more appropriate as a name for these waves. However, in this text we will generally use the traditional name *gravity wave*.

In a fluid, such as the ocean, which is bounded both above and below, gravity waves propagate primarily in the horizontal plane since vertically traveling waves are reflected from the boundaries to form standing waves.

---

[3] Long waves excited by underwater earthquakes or volcanic eruptions are called *tsunamis*.

[4] Gravity waves propagating along an internal density discontinuity are sometimes referred to as *internal* waves. We will, however, reserve that terminology for the vertically propagating waves considered in Section 7.4.

However, in a fluid that has no upper boundary, such as the atmosphere, gravity waves may propagate vertically as well as horizontally. In vertically propagating waves the phase is a function of height. Such waves are referred to as *internal* waves. Although internal gravity waves are not generally of great importance for synoptic-scale weather forecasting (and indeed are nonexistent in the filtered quasi-geostrophic models), they can be important in mesoscale motions. For example, they are responsible for the occurrence of mountain *lee waves.* They also are believed to be an important mechanism for transporting energy and momentum to high levels and are often associated with the formation of clear air turbulene (CAT).

7.4.1 PURE INTERNAL GRAVITY WAVES

For simplicity we neglect the Coriolis force and limit our discussion to two-dimensional internal gravity waves propagating in the x, z plane. An expression for the frequency of such waves can be obtained by modifying the parcel theory developed in Section 2.7.3.

Internal gravity waves are transverse waves in which the parcel oscillations are parallel to the phase lines as indicated in Fig. 7.8. A parcel displaced a distance δs along a line tilted at an angle α to the vertical as shown in Fig. 7.8 will undergo a vertical displacement $\delta z = \delta s \cos \alpha$. For such a parcel the *vertical* buoyancy force per unit mass is just $-N^2 \delta z$, as was shown in (2.52). Thus, the component of the buoyancy force parallel to the tilted path along which the parcel oscillates is just

$$-N^2(\delta s \cos \alpha) \cos \alpha = -(N \cos \alpha)^2 \delta s$$

The momentum equation for the parcel oscillation is then

$$\frac{d^2(\delta s)}{dt^2} = -(N \cos \alpha)^2 \delta s \tag{7.24}$$

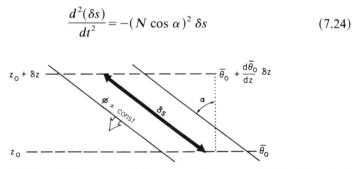

Fig. 7.8   Parcel oscillation path (heavy arrow) for pure gravity waves with phase lines tilted at an angle α to the vertical.

which has the general solution $\delta s = \exp[\pm i(N \cos \alpha)t]$. Thus, the parcels execute a simple harmonic oscillation at the frequency $\nu = N \cos \alpha$. This frequency depends only on the static stability (measured by the buoyancy frequency N) and the angle of the phase lines to the vertical.

The above heuristic derivation can be verified by considering the linearized equations for two-dimensional internal gravity waves. For simplicity we employ the *Boussinesq approximation*, in which density is treated as a constant except where it is coupled with gravity in the buoyancy term of the vertical momentum equation. Thus, in this approximation the atmosphere is considered to be incompressible and local density variations are assumed to be small perturbations of the constant basic state density field. Because the vertical variation of the basic state density is neglected except where coupled with gravity, the Boussinesq approximation is valid only for motions in which the vertical scale is less than the atmospheric scale height $H \ (\approx 8 \text{ km})$.

Neglecting effects of rotation, the basic equations for two-dimensional motion of an incompressible atmosphere may be written as follows:

$$\frac{\partial u}{\partial t} + u\frac{\partial u}{\partial x} + w\frac{\partial u}{\partial z} + \frac{1}{\rho}\frac{\partial p}{\partial x} = 0 \tag{7.25}$$

$$\frac{\partial w}{\partial t} + u\frac{\partial w}{\partial x} + w\frac{\partial w}{\partial z} + \frac{1}{\rho}\frac{\partial p}{\partial z} + g = 0 \tag{7.26}$$

$$\frac{\partial u}{\partial x} + \frac{\partial w}{\partial z} = 0 \tag{7.27}$$

$$\frac{\partial \theta}{\partial t} + u\frac{\partial \theta}{\partial x} + w\frac{\partial \theta}{\partial z} = 0 \tag{7.28}$$

where the potential temperature θ is related to pressure and density by

$$\theta = \frac{p}{\rho R}\left(\frac{p_s}{p}\right)^{\kappa} \tag{7.29}$$

We now linearize (7.25)–(7.29) by letting

$$\rho = \rho_0 + \rho', \qquad u = \bar{u} + u'$$

$$p = \bar{p}(z) + p', \qquad w = w' \tag{7.30}$$

$$\theta = \bar{\theta}(z) + \theta'$$

where the basic state zonal flow and the density ρ_0 are both assumed to be constant. The basic state pressure field must satisfy the hydrostatic equation

$$d\bar{p}/dz = -\rho_0 g \tag{7.31}$$

while the basic state potential temperature must satisfy (7.29), so that

$$\ln \bar{\theta} = \gamma^{-1} \ln \bar{p} - \ln \rho_0 + \text{const} \tag{7.32}$$

The linearized equations are obtained by substituting from (7.30) into (7.25)–(7.29) and neglecting all terms that are products of the perturbation variables. For example, the last two terms in (7.26) are approximated as follows:

$$\frac{1}{\rho} \frac{\partial p}{\partial z} + g = \frac{1}{\rho_0 + \rho'} \left(\frac{d\bar{p}}{dz} + \frac{\partial p'}{\partial z} \right) + g$$

$$\approx \frac{1}{\rho_0} \frac{d\bar{p}}{dz} \left(1 - \frac{\rho'}{\rho_0} \right) + \frac{1}{\rho_0} \frac{\partial p'}{\partial z} + g = \frac{1}{\rho_0} \frac{\partial p'}{\partial z} + \frac{\rho'}{\rho_0} g \tag{7.33}$$

where (7.31) has been used to eliminate $\bar{p}$. The perturbation form of (7.29) is obtained by noting that

$$\ln \left[\bar{\theta} \left(1 + \frac{\theta'}{\bar{\theta}} \right) \right] = \gamma^{-1} \ln \left[\bar{p} \left(1 + \frac{p'}{\bar{p}} \right) \right] - \ln \left[\rho_0 \left(1 + \frac{\rho'}{\rho_0} \right) \right] + \text{const} \tag{7.34}$$

Now, recalling that $\ln (ab) = \ln (a) + \ln (b)$ and that $\ln (1 + \varepsilon) \approx \varepsilon$ for any $\varepsilon \ll 1$ we find with the aid of (7.32) that (7.34) may be approximated by

$$\frac{\theta'}{\bar{\theta}} \approx \frac{1}{\gamma} \frac{p'}{\bar{p}} - \frac{\rho'}{\bar{\rho}}$$

Solving for ρ' yields

$$\rho' \approx -\rho_0 \frac{\theta'}{\bar{\theta}} + \frac{p'}{c_s^2} \tag{7.35}$$

where $c_s^2 \equiv \bar{p}\gamma/\rho_0$ is the square of the speed of sound. For buoyancy wave motions $|\rho_0 \theta'/\bar{\theta}| \gg |p'/c_s^2|$; i.e., density fluctuations due to pressure changes are small compared with those due to temperature changes. Therefore, to a first approximation,

$$\theta'/\bar{\theta} = -\rho'/\rho_0 \tag{7.36}$$

Using (7.33) and (7.36) the linearized version of the set (7.25)–(7.28) can be written as

$$\left(\frac{\partial}{\partial t}+\bar{u}\frac{\partial}{\partial x}\right)u'+\frac{1}{\rho_0}\frac{\partial p'}{\partial x}=0 \tag{7.37}$$

$$\left(\frac{\partial}{\partial t}+\bar{u}\frac{\partial}{\partial x}\right)w'+\frac{1}{\rho_0}\frac{\partial p'}{\partial z}-\frac{\theta'}{\bar{\theta}}g=0 \tag{7.38}$$

$$\frac{\partial u'}{\partial x}+\frac{\partial w'}{\partial z}=0 \tag{7.39}$$

$$\left(\frac{\partial}{\partial t}+\bar{u}\frac{\partial}{\partial x}\right)\theta'+w'\frac{d\bar{\theta}}{dz}=0 \tag{7.40}$$

Subtracting $\partial(7.37)/\partial z$ from $\partial(7.38)/\partial x$ we can eliminate p' to obtain

$$\left(\frac{\partial}{\partial t}+\bar{u}\frac{\partial}{\partial x}\right)\left(\frac{\partial w'}{\partial x}-\frac{\partial u'}{\partial z}\right)-\frac{g}{\bar{\theta}}\frac{\partial\theta'}{\partial x}=0 \tag{7.41}$$

which is just the y component of the vorticity equation.

With the aid of (7.39) and (7.40) u' and θ' can be eliminated from (7.41) to yield a single equation for w':

$$\left(\frac{\partial}{\partial t}+\bar{u}\frac{\partial}{\partial x}\right)^{2}\left(\frac{\partial^2 w'}{\partial x^2}+\frac{\partial^2 w'}{\partial z^2}\right)+N^{2}\frac{\partial^2 w'}{\partial x^2}=0 \tag{7.42}$$

where $N^2 \equiv g\, d\ln\bar{\theta}/dz$ is the square of the buoyancy frequency, which is assumed to be constant.[5]

Equation (7.42) has harmonic wave solutions of the form

$$w'=\mathrm{Re}[\hat{w}\exp(i\phi)]=w_\mathrm{r}\cos\phi-w_\mathrm{i}\sin\phi \tag{7.43}$$

where $\hat{w}=w_\mathrm{r}+iw_\mathrm{i}$ is a complex amplitude with real part w_r and imaginary part w_i, and $\phi=kx+mz-\nu t$ is the phase, which is assumed to depend linearly on z as well as on x and t. Here the horizontal wave number k is real since the solution is always sinusoidal in x. The vertical wave number $m=m_\mathrm{r}+m_\mathrm{i}$ may, however, be complex, in which case m_r describes

---

[5] Strictly speaking, N^2 cannot be exactly constant if ρ_0 is constant. However, for shallow disturbances the variation of N^2 with height is unimportant.

sinusoidal variation in z and m_i describes exponential decay or growth in z depending on whether m_i is positive or negative. When m is real the total wave number may be regarded as a vector $\kappa \equiv (k, m)$, directed perpendicular to lines of constant phase, and in the direction of phase increase, whose components, $k = 2\pi/L_x$ and $m = 2\pi/L_z$, are inversely proportional to the horizontal and vertical wavelengths, respectively. Substitution of the assumed solution into (7.42) yields the dispersion relationship

$$(\nu - \bar{u}k)^2(k^2 + m^2) - N^2 k^2 = 0$$

so that

$$\hat{\nu} \equiv \nu - \bar{u}k = \pm Nk/(k^2 + m^2)^{1/2} = \pm Nk/|\kappa| \qquad (7.44)$$

where $\hat{\nu}$, the *intrinsic frequency*, is the frequency relative to the mean wind, and the plus (minus) sign is to be taken for eastward (westward) phase propagation relative to the mean wind.

If we let $k > 0$ and $m < 0$ then lines of constant phase tilt eastward with respect to height as shown in Fig. 7.9 (i.e., for $\phi = kx + mz$ to remain constant as x increases, z must also increase when $k > 0$ and $m < 0$). The choice of the positive root in (7.44) then corresponds to eastward and downward phase propagation relative to the mean flow with horizontal and vertical phase speeds (relative to the mean flow) given by $c_x = \hat{\nu}/k$ and $c_z = \hat{\nu}/m$,

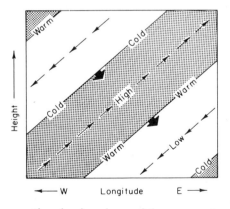

Fig. 7.9 Idealized cross section showing phases of the pressure, temperature, and velocity perturbations for an internal gravity wave. Thin arrows indicate the perturbation velocity field, blunt solid arrows the phase velocity. Shading shows regions of upward motion. (After Wallace and Kousky, 1968.)

respectively.[6] The components of the group velocity, c_{gx} and c_{gz}, on the other hand, are given by

$$c_{gx} = \frac{\partial \nu}{\partial k} = \bar{u} \pm \frac{Nm^2}{(k^2 + m^2)^{3/2}} \tag{7.45a}$$

$$c_{gz} = \frac{\partial \nu}{\partial m} = \pm \frac{-Nkm}{(k^2 + m^2)^{3/2}} \tag{7.45b}$$

where the upper or lower signs are chosen in the same way as in (7.44). Thus, the vertical component of group velocity has a sign opposite to that of the vertical phase speed relative to the mean flow (downward phase propagation implies upward energy propagation). Furthermore, it is easily shown from (7.45) that the group velocity vector is parallel to lines of constant phase. Internal gravity waves thus have the remarkable property that group velocity is perpendicular to the direction of phase propagation. Since energy propagates at the group velocity, this implies that energy propagates parallel to the wave crests and troughs, rather than perpendicular to them as in acoustic waves or shallow-water gravity waves. In the atmosphere, internal gravity waves generated in the troposphere by cumulus convection, by flow over topography, and by other processes may propagate energy upward many scale heights into the upper atmosphere, even though individual fluid parcel oscillations may be confined to vertical distances much less than a kilometer.

Referring again to Fig. 7.9 it is evident that the angle of the phase lines to the local vertical is given by

$$\cos \alpha = L_z / (L_x^2 + L_z^2)^{1/2} = \pm k / (k^2 + m^2)^{1/2} = \pm k / |\kappa|$$

Thus, $\hat{\nu} = \pm N \cos \alpha$ (i.e., gravity wave frequencies must be less than the buoyancy frequency) in agreement with the heuristic parcel oscillation model (7.24). The tilt of phase lines for internal gravity waves depends only on the ratio of the wave frequency to the buoyancy frequency and is independent of wavelength.

7.4.2 TOPOGRAPHIC WAVES

When air with mean wind speed $\bar{u}$ is forced to flow over a sinusoidal pattern of ridges under statically stable conditions, individual air parcels

---

[6] Note that phase speed is not a vector. The phase speed in the direction perpendicular to constant phase lines (i.e., the blunt arrows in Fig. 7.9) is given by $\nu / (k^2 + m^2)^{1/2}$, which is not equal to $(c_x^2 + c_z^2)^{1/2}$.

are alternately displaced upward and downward from their equilibrium levels and will thus undergo buoyancy oscillations as they move across the ridges as shown in Fig. 7.10. In this case there are solutions in the form of waves that are stationary relative to the ground [i.e. $\nu = 0$ in (7.43)]. For such stationary waves w' depends only on (x, z) and (7.42) simplifies to

$$\left(\frac{\partial^2 w'}{\partial x^2} + \frac{\partial^2 w'}{\partial z^2}\right) + \frac{N^2}{\bar{u}^2}\, w' = 0 \qquad (7.46)$$

Substituting from (7.43) into (7.46) then yields the dispersion relationship

$$m^2 = N^2/\bar{u}^2 - k^2 \qquad (7.47)$$

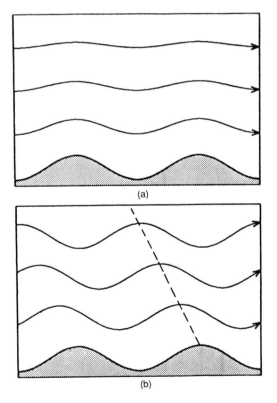

(a)

(b)

Fig. 7.10 Streamlines in steady flow over an infinite series of sinusoidal ridges for the narrow ridge case (a) and the broad ridge case (b). The dashed line in (b) shows the phase of maximum upward displacement. (After Durran, 1990.)

For given values of N, k and $\bar{u}$, (7.47) determines the vertical structure. Then if $|\bar{u}| < N/k$, (7.47) shows that $m^2 > 0$ (i.e., m must be real) and solutions of (7.46) have the form of vertically propagating waves:

$$w' = \hat{w} \exp[i(kx + mz)]$$

Here we see from (7.44) that if we set $k > 0$ then for $\bar{u} > 0$ we have $\hat{v} < 0$ so that $m > 0$, while for $\bar{u} < 0$ we have $m < 0$. In the former situation the lower signs apply on the right in (7.45), while in the latter the upper signs apply. In both cases the vertical phase propagation is downward relative to the mean flow, and vertical energy propagation is upward.

When $m^2 < 0$, $m = im_i$ is imaginary and the solution to (7.46) will have the form of vertically trapped waves:

$$w' = \hat{w} \exp(ikx) \exp(-m_i z)$$

Thus, vertical propagation is possible only when $|\bar{u}k|$, the magnitude of the frequency relative to the mean flow, is less than the buoyancy frequency. Stable stratification, wide ridges, and comparatively weak zonal flow provide favorable conditions for the formation of vertically propagating topographic waves (m real). Since the energy source for these waves is at the ground, they must transport energy upward. Hence, the phase speed relative to the mean zonal flow must have a downward component and lines of constant phase must tilt westward with height. When m is imaginary, on the other hand, the solution (7.43) has exponential behavior in the vertical with an exponential decay height of μ^{-1}, where $\mu = |m|$. Boundedness as $z \to \infty$ requires that we choose the solution with exponential decay away from the lower boundary.

In order to contrast the character of the solutions for real and imaginary m we consider a specific example in which there is westerly mean flow over topography with a height profile given by

$$h(x) = h_M \cos kx$$

where h_M is the amplitude of the topography. Then since the flow at the lower boundary must be parallel to the boundary, the vertical velocity perturbation at the boundary is given by the rate at which the boundary height changes following the motion:

$$w'(x, 0) = (Dh/Dt)_{z=0} \approx \bar{u} \, \partial h / \partial x = -\bar{u}kh_M \sin kx$$

and the solution of (7.46) that satisfies this condition can be written

$$
w(x, z) = \begin{cases} -\bar{u}h_M k e^{-\mu z} \sin kx, & \bar{u}k > N \\ -\bar{u}h_M k \sin(kx + mz), & \bar{u}k < N \end{cases} \tag{7.48}
$$

For fixed mean wind and buoyancy frequency the character of the solution depends only on the horizontal scale of the topography. The two cases of (7.48) may be regarded as narrow ridge and wide ridge cases, respectively. The streamline patterns corresponding to these cases for westerly flow are illustrated in Fig. 7.10. In the narrow ridge case (panel a) the maximum upward displacement occurs at the ridge tops and the amplitude of the disturbance decays with height. In the wide ridge case (panel b) the line of maximum upward displacement tilts back toward the west ($m > 0$) and amplitude is independent of height consistent with an internal gravity wave propagating westward relative to the mean flow.

Alternatively, for fixed zonal wave number and buoyancy frequency, the solution depends only on the speed of the mean zonal wind. As indicated in (7.48), only for mean zonal wind magnitudes less than the critical value N/k will vertical wave propagation occur.

Equation (7.46) was obtained for conditions of constant basic state flow. In reality both the zonal wind $\bar{u}$ and the stability parameter N generally vary with height, and ridges are usually isolated rather than periodic. A wide variety of responses are possible depending on the shape of the terrain and wind and stability profiles. Under certain conditions large-amplitude waves can be formed, which may generate severe downslope surface winds and zones of strong clear air turbulence. Such circulations will be discussed further in Section 9.4.

7.5   Inertio-gravity Waves

Gravity waves with horizontal scales greater than a few hundred kilometers and periods greater than a few hours are hydrostatic, but they are influenced by the Coriolis effect and are characterized by parcel oscillations that are elliptical rather than straight lines as in the pure gravity wave case. This *elliptical polarization* can be understood qualitatively by observing that the Coriolis effect resists horizontal parcel displacements in a rotating fluid, but in a manner somewhat different from that in which the buoyancy force resists vertical parcel displacements in a statically stable atmosphere. In the latter case the resistive force is opposite to the direction of parcel displacement, whereas in the former it is at right angles to the horizontal parcel velocity.

7.5.1 PURE INERTIAL OSCILLATIONS

In Section 3.2.3 it was shown that a parcel put into horizontal motion in a resting atmosphere with constant Coriolis parameter executes a circular trajectory in an anticyclonic sense. A generalization of this type of inertial motion to the case with a geostrophic mean zonal flow can be derived using a parcel argument similar to that used for the buoyancy oscillation in Section 2.7.3.

If the basic state flow is assumed to be a zonally directed geostrophic wind u_g, and it is assumed that the parcel displacement does not perturb the pressure field, the approximate equations of motion become

$$\frac{Du}{Dt} = fv = f\frac{Dy}{Dt} \tag{7.49}$$

$$\frac{Dv}{Dt} = f(u_g - u) \tag{7.50}$$

We consider a parcel that is moving with the geostrophic basic state motion at a position $y = y_0$. If the parcel is displaced across stream by a distance δy, we can obtain its new zonal velocity from the integrated form of (7.49):

$$u(y_0 + \delta y) = u_g(y_0) + f\delta y \tag{7.51}$$

The geostrophic wind at $y_0 + \delta y$ can be approximated as

$$u_g(y_0 + \delta y) = u_g(y_0) + \frac{\partial u_g}{\partial y}\delta y \tag{7.52}$$

Using (7.51) and (7.52) to evaluate (7.50) at $y_0 + \delta y$ yields

$$\frac{Dv}{Dt} = \frac{D^2\delta y}{Dt^2} = -f\left(f - \frac{\partial u_g}{\partial y}\right)\delta y = -f\frac{\partial M}{\partial y}\delta y \tag{7.53}$$

where we have defined the "absolute momentum" $M \equiv fy - u_g$.

This equation is mathematically of the same form as (2.52), the equation for the motion of a vertically displaced particle in a stratified atmosphere. Depending on the sign of the coefficient on the right-hand side in (7.53) the parcel will be forced to return to its original position or will accelerate further from that position. This coefficient thus determines the condition for *inertial instability*:

$$f\frac{\partial M}{\partial y}\begin{cases} >0 & \text{stable} \\ =0 & \text{neutral} \\ <0 & \text{unstable} \end{cases} \tag{7.54}$$

Viewed in an inertial reference frame, the instability results from an imbalance between the pressure gradient and inertial forces for a parcel displaced radially in an axisymmetric vortex. In the Northern (Southern) Hemisphere, where f is positive (negative), the flow is inertially stable provided that $f - \partial u_g/\partial y$, the absolute vorticity of the basic flow, is positive (negative). Observations show that for extratropial synoptic-scale systems the flow is always inertially stable, although near neutrality often occurs on the anticyclonic shear side of upper-level jet streaks. The occurrence of inertial instability over a large area would be expected immediately to trigger inertially unstable motions, which would mix the fluid laterally just as convection mixes it vertically and reduce the shear until the absolute vorticity times f was again positive. (This explains why anticyclonic shears cannot become arbitrarily large.) Inertial instability is further considered in a more general context in Section 9.3.

7.5.2 INERTIO-GRAVITY WAVES

When the flow is both inertially and gravitationally stable, parcel displacements are resisted by both rotation and buoyancy. The resulting oscillations are called *inertio-gravity waves*. The dispersion relation for such waves can be analyzed using a variant of the parcel method applied in Section 7.4. We consider parcel oscillations along a slantwise path in the y, z plane as shown in Fig. 7.11. For a vertical displacement δz the buoyancy force component parallel to the slope of the parcel oscillation is $-N^2 \, \delta z \cos \alpha$, and for a meridional displacement δy the Coriolis (inertial) force component parallel to the slope of the parcel path is $-f^2 \, \delta y \sin \alpha$, where we have assumed that the geostrophic basic flow is constant in latitude. Thus, the harmonic oscillator equation for the parcel (7.24) is modified to the form

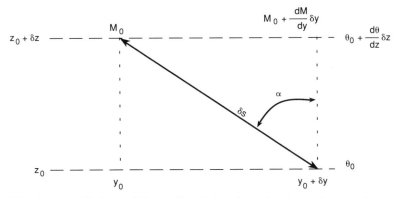

Fig. 7.11 Parcel oscillation path in meridional plane for an inertio-gravity wave. See text for definition of symbols.

$$\frac{D^2 \delta s}{Dt^2} = -(f \sin \alpha)^2 \, \delta s - (N \cos \alpha)^2 \, \delta s \tag{7.55}$$

where δs is again the perturbation parcel displacement.

The frequency now satisfies the dispersion relationship

$$\nu^2 = N^2 \cos^2 \alpha + f^2 \sin^2 \alpha \tag{7.56}$$

Since $N^2 > f^2$ (7.56) indicates that inertio-gravity wave frequencies must lie in the range $f \le |\nu| \le N$. The frequency approaches N as the trajectory slope approaches the vertical and approaches f as the trajectory slope approaches the horizontal. For typical midlatitude tropospheric conditions inertio-gravity wave periods are in the approximate range of 12 minutes to 15 hours. Rotational effects become important, however, only when the second term on the right in (7.56) is similar in magnitude to the first term. This requires that $\tan^2 \alpha \sim N^2/f^2 = 10^4$, in which case it is clear from (7.56) that $\nu \ll N$. Thus, only low-frequency gravity waves are significantly modified by the rotation of the earth, and these have very small parcel trajectory slopes.

The heuristic parcel derivation can again be verified by using the linearized dynamical equations. In this case, however, it is necessary to include rotation. The small parcel trajectory slopes of the relatively long-period waves that are significantly altered by rotation imply that the horizontal scales are much greater than the vertical scales for these waves. Therefore, we may assume that the motions are in hydrostatic balance. If in addition we assume a motionless basic state the linearized equations (7.37)-(7.40) are replaced by the set

$$\frac{\partial u'}{\partial t} - fv' + \frac{1}{\rho_0} \frac{\partial p'}{\partial x} = 0 \tag{7.57}$$

$$\frac{\partial v'}{\partial t} + fu' + \frac{1}{\rho_0} \frac{\partial p'}{\partial y} = 0 \tag{7.58}$$

$$\frac{1}{\rho_0} \frac{\partial p'}{\partial z} - \frac{\theta'}{\bar{\theta}} g = 0 \tag{7.59}$$

$$\frac{\partial u'}{\partial x} + \frac{\partial v'}{\partial y} + \frac{\partial w'}{\partial z} = 0 \tag{7.60}$$

$$\frac{\partial \theta'}{\partial t} + w' \frac{d\bar{\theta}}{dz} = 0 \tag{7.61}$$

The hydrostatic relationship (7.59) may be used to eliminate θ' in (7.61) to yield

$$\frac{\partial}{\partial t}\left(\frac{1}{\rho_0}\frac{\partial p'}{\partial z}\right) + N^2 w' = 0 \tag{7.62}$$

Letting

$$(u', v', w', p'/\rho_0) = \mathrm{Re}[(\hat{u}, \hat{v}, \hat{w}, \hat{p}) \exp i(kx + ly + mz - \nu t)]$$

and substituting into (7.57), (7.58), and (7.62) we obtain

$$\hat{u} = (\nu^2 - f^2)^{-1}(\nu k + ilf)\hat{p} \tag{7.63}$$

$$\hat{v} = (\nu^2 - f^2)^{-1}(\nu l - ikf)\hat{p} \tag{7.64}$$

$$\hat{w} = -\frac{\nu m}{N^2}\hat{p} \tag{7.65}$$

which with the aid of (7.60) yields the dispersion relation for hydrostatic waves.

$$\nu^2 = f^2 + N^2(k^2 + l^2)m^{-2} \tag{7.66}$$

Since hydrostatic waves must have $(k^2 + l^2)/m^2 \ll 1$, (7.66) indicates that for vertical propagation to be possible (m real) the frequency must satisfy the inequality $|f| < |\nu| \ll N$. Equation (7.66) is just the limit of (7.56) when we let

$$\sin^2 \alpha \to 1, \qquad \cos^2 \alpha = (k^2 + l^2)/m^2$$

which is consistent with the hydrostatic approximation.

If axes are chosen to make $l = 0$, it may be shown (see Problem 7.13) that the ratio of the vertical to horizontal components of group velocity is given by

$$|c_{\mathrm{gz}}/c_{\mathrm{gx}}| = |k/m| = (\nu^2 - f^2)^{1/2}/N \tag{7.67}$$

Thus, for fixed ν, inertio-gravity waves propagate more closely to the horizontal than do pure internal gravity waves. But as in the latter case the group velocity vector is again parallel to lines of constant phase.

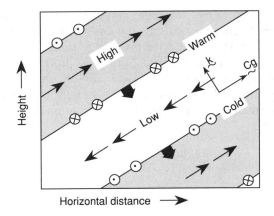

Horizontal distance ⟶

Fig. 7.12   Vertical section in a plane containing the wave vector **k** showing the phase relationships among velocity, geopotential, and temperature fluctuations in an upward-
propagating inertio-gravity wave with $m < 0$, $\nu > 0$, and $f > 0$ (Northern Hemisphere). Thin sloping lines denote the surfaces of constant phase (perpendicular to
the wave vector) and thick arrows show the direction of phase propagation. Thin
arrows show the perturbation zonal and vertical velocity fields. Meridional wind
perturbations are shown by arrows pointed into the page (northward) and out of
the page (southward). Note that the perturbation wind vector turns clockwise
(anticyclonically) with height. (After Andrews *et al.*, 1987.)

Eliminating $\hat{p}$ between (7.63) and (7.64) for the case $l = 0$ yields the
relationship $\hat{v} = -if\hat{u}/\nu$, from which it is easily verified that if $\hat{u}$ is real the
perturbation horizontal motions satisfy the relations

$$u' = \hat{u}\cos(kx + mz - \nu t), \qquad v' = \hat{u}(f/\nu)\sin(kx + mz - \nu t) \quad (7.68)$$

so that the horizontal velocity vector rotates anticyclonically (i.e., clockwise
in the Northern Hemisphere) with time. As a result, parcels follow elliptical
trajectories in a plane orthogonal to the wave number vector. Equations
(7.68) also show that the horizontal velocity vector turns anticyclonically
with height for waves with upward energy propagation (e.g., $m < 0$ and
$\nu > 0$). These characteristics are illustrated by the vertical cross section
shown in Fig. 7.12. The anticyclonic turning of the horizontal wind with
height and time is a primary method for distinguishing inertio-gravity
oscillations in meteorological data.

7.6   Adjustment to Geostrophic Balance

In Chapter 6 we showed that synoptic-scale motions in midlatitudes are
in approximate quasi-geostrophic balance. Departures from this balance

can lead to excitation of inertio-gravity waves, which act to adjust the mass and momentum distributions so that the flow tends to return toward geostrophic balance. In this section we investigate the process by which geostrophic balance is achieved, i.e., the *adjustment* process. For simplicity we utilize the prototype shallow-water system; similar considerations apply to a continuously stratified atmosphere. We also consider linearized motion about a basic state of no motion. The horizontal momentum and continuity equations are then

$$\frac{\partial u'}{\partial t} - fv' = -g\frac{\partial h'}{\partial x} \tag{7.69}$$

$$\frac{\partial v'}{\partial t} + fu' = -g\frac{\partial h'}{\partial y} \tag{7.70}$$

$$\frac{\partial h'}{\partial t} + H\left(\frac{\partial u'}{\partial x} + \frac{\partial v'}{\partial y}\right) = 0 \tag{7.71}$$

where h' is, as before, the deviation from the mean depth H. Taking $\partial(7.69)/\partial x + \partial(7.70)/\partial y$ yields

$$\frac{\partial^2 h'}{\partial t^2} - c^2\left(\frac{\partial^2 h'}{\partial x^2} + \frac{\partial^2 h'}{\partial y^2}\right) + fH\zeta' = 0 \tag{7.72}$$

where $c^2 \equiv gH$ and $\zeta' = \partial v'/\partial x - \partial u'/\partial y$.

For $f = 0$ (nonrotating system) the vorticity and height perturbations are uncoupled and (7.72) yields a two-dimensional shallow-water wave equation for h [compare with (7.22)]:

$$\frac{\partial^2 h'}{\partial t^2} - c^2\left(\frac{\partial^2 h'}{\partial x^2} + \frac{\partial^2 h'}{\partial y^2}\right) = 0 \tag{7.73}$$

which has solutions of the form

$$h' = A\exp[i(kx + ly - \nu t)] \tag{7.74}$$

with $\nu^2 = c^2(k^2 + l^2) = gH(k^2 + l^2)$. On the other hand, for $f \neq 0$ the h' and ζ' fields are coupled through (7.72). For motions with time scales longer than $1/f_0$ (which is certainly true for synoptic-scale motions) the ratio of the first two terms in (7.72) is given by

$$\frac{|\partial^2 h'/\partial t^2|}{|c^2(\partial^2 h'/\partial x^2 + \partial^2 h'/\partial y^2)|} \lesssim \frac{f_0^2 L^2}{gH}$$

which is small for $L \sim 1000$ km, provided that $H \gg 1$ km. Under such circumstances the time derivative term in (7.72) is small compared to the other two terms, and (7.72) states simply that the vorticity is in geostrophic balance.

If the flow is initially unbalanced the complete equation (7.72) can be used to describe the approach toward geostrophic balance provided that we can obtain a second relationship between h' and ζ'. Taking

$$\partial(7.70)/\partial x - \partial(7.69)/\partial y$$

yields

$$\frac{\partial \zeta'}{\partial t} + f\left(\frac{\partial u'}{\partial x} + \frac{\partial v'}{\partial y}\right) = 0 \tag{7.75}$$

which can be combined with (7.71) to give the linearized potential vorticity conservation law:

$$\frac{\partial \zeta'}{\partial t} - \frac{f}{H}\frac{\partial h'}{\partial t} = 0 \tag{7.76}$$

Thus, letting Q' designate the perturbation potential vorticity, we obtain from (7.76) the conservation relationship

$$Q'(x, y, t) = \zeta'/f - h'/H = \text{constant} \tag{7.77}$$

Hence, if we know the distribution of Q' at the initial time, we know Q' for all time:

$$Q'(x, y, t) = Q'(x, y, 0)$$

and the final adjusted state can be determined without solving the time-dependent problem.

This problem was first solved by Rossby in the 1930s and is often referred to as the Rossby adjustment problem. As a simplified, albeit somewhat unrealistic, example of the adjustment process we consider an idealized shallow-water system on a rotating plane with initial conditions

$$u', v' = 0, \qquad h' = -h_0 \, \text{sgn}(x) \tag{7.78}$$

where sgn means "sign of."

This corresponds to an initial step function in h' at $x = 0$, with the fluid motionless. Thus, from (7.77)

$$(\zeta'/f) - (h'/H) = (h_0/H) \, \text{sgn}(x) \tag{7.79}$$

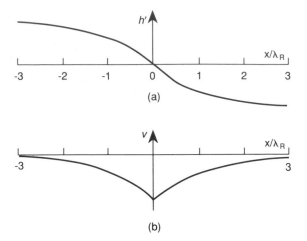

Fig. 7.13   The geostrophic equilibrium solution corresponding to adjustment from the initial
state defined in (7.78). Panel (a) shows the initial and final surface elevation profiles;
(b) shows the geostrophic velocity profile in the final state. (After Gill, 1982.)

Using (7.79) to eliminate ζ' in (7.72) yields

$$\frac{\partial^2 h'}{\partial t^2} - c^2\left(\frac{\partial^2 h'}{\partial x^2} + \frac{\partial^2 h'}{\partial y^2}\right) + f^2 h' = -f^2 h_0\, \mathrm{sgn}(x) \qquad (7.80)$$

which in the homogeneous case ($h_0 = 0$) yields the dispersion relation

$$\nu^2 = f^2 + c^2(k^2 + l^2) = f^2 + gH(k^2 + l^2) \qquad (7.81)$$

This should be compared to (7.66).

Since initially h' is independent of y it will remain so for all time. Thus,
in the final steady state (7.80) becomes

$$-c^2 \frac{d^2 h'}{dx^2} + f^2 h' = -f^2 h_0\, \mathrm{sgn}(x) \qquad (7.82)$$

which has the solution

$$\frac{h'}{h_0} = \begin{cases} -1 + \exp(-x/\lambda_R) & \text{for } x > 0 \\ +1 - \exp(+x/\lambda_R) & \text{for } x < 0 \end{cases} \qquad (7.83)$$

where $\lambda_R \equiv f^{-1}\sqrt{gH}$ is the Rossby *radius of deformation*. Hence, the radius of deformation may be interpreted as the horizontal length scale over which the height field adjusts during approach to geostrophic equilibrium. For $|x| \gg \lambda_R$ the original h' remains unchanged. Substituting from (7.83) into (7.69)–(7.71) shows that the steady velocity field is geostrophic and non-divergent:

$$u' = 0 \quad \text{and} \quad v' = \frac{g}{f}\frac{\partial h'}{\partial x} = -\frac{gh_0}{f\lambda_R}\exp\left(\frac{-|x|}{\lambda_R}\right) \qquad (7.84)$$

The steady-state solution (7.84) is shown in Fig. 7.13.

Note that the result (7.84) could not be derived merely by setting $\partial/\partial t = 0$ in (7.69)–(7.71). That would yield geostrophic balance and *any* distribution of h' would satisfy the equations

$$fu' = -g\frac{\partial h'}{\partial y}, \qquad fv' = g\frac{\partial h'}{\partial x}, \qquad \frac{\partial u'}{\partial x} + \frac{\partial v'}{\partial y} = 0$$

Only by combining (7.69)–(7.71) to obtain the potential vorticity equation and requiring the flow to satisfy potential vorticity conservation at all intermediate times can the degeneracy of the geostrophic final state be eliminated. In other words, although any height field can satisfy the steady-state versions of (7.69)–(7.71), there is only one field that is consistent with a given initial state; this field can be found readily because it can be computed from the distribution of potential vorticity, which is conserved.

Although the final state can be computed without solving the time-dependent equation, if the evolution of the adjustment process is required it is necessary to solve (7.80) subject to the initial conditions (7.78), which is beyond the scope of this discussion. We can, however, compute the amount of energy that is dispersed by gravity waves during the adjustment process. This only requires computing the energy change between the initial and final states.

The potential energy per unit horizontal area is given by

$$\int_0^{h'} \rho gz\, dz = \rho g h'^2/2$$

Thus, the potential energy released per unit length in y during adjustment is

$$\int_{-\infty}^{+\infty} \frac{\rho g h_0^2}{2}\, dx - \int_{-\infty}^{+\infty} \frac{\rho g h'^2}{2}\, dx$$

$$= 2\int_0^{+\infty} \frac{\rho g h_0^2}{2}[1 - (1 - e^{-x/\lambda_R})^2]\, dx = \frac{3}{2}\rho g h_0^2 \lambda_R \qquad (7.85)$$

In the nonrotating case ($\lambda_R \to \infty$) all potential energy available initially is released (converted to kinetic energy) so that there is an infinite energy release. (Energy is radiated away in the form of gravity waves, leaving a flat free surface extending to $|x| \to \infty$ as $t \to \infty$.)

In the rotating case only the finite amount (7.85) is converted to kinetic energy, and only a portion of this kinetic energy is radiated away. The rest remains in the steady geostrophic circulation. The kinetic energy in the steady state per unit length is

$$2 \int_0^{+\infty} \rho H \frac{v'^2}{2} \, dx = \rho H \left(\frac{gh_0}{f\lambda_R} \right)^2 \int_0^{+\infty} e^{-2x/\lambda_R} \, dx = \tfrac{1}{2} \rho g h_0^2 \lambda_R \qquad (7.86)$$

Thus, in the rotating case a finite amount of potential energy is released, but only one-third of the potential energy released goes into the steady geostrophic mode. The remaining two-thirds is radiated away in the form of inertio-gravity waves.

This simple analysis illustrates the following points:

(a) It is difficult to extract the potential energy of a rotating fluid. Although there is an infinite reservoir of potential energy in this example (since h' is finite as $|x| \to \infty$), only a finite amount is converted before geostrophic balance is achieved.

(b) Conservation of potential vorticity allows one to determine the steady-state geostrophically adjusted velocity and height fields without carrying out a time integration.

(c) The length scale for the steady solution is the Rossby radius λ_R. The dynamics of the adjustment process plays an essential role in initialization and data assimilation in numerical prediction (see Section 13.7). For example, under some conditions the adjustment process may effectively damp out new height data inserted at a grid point since the new data will generally be unbalanced and hence will tend to adjust toward geostrophic balance with the existing wind field.

7.7   Rossby Waves

The wave type that is of most importance for large-scale meteorological processes is the *Rossby wave*, or *planetary wave*. In an inviscid barotropic fluid of constant depth (where the divergence of the horizontal velocity must vanish) the Rossby wave is an absolute vorticity-conserving motion that owes its existence to the variation of the Coriolis force with latitude, the so-called β effect. More generally, in a baroclinic atmosphere the Rossby wave is a potential vorticity-conserving motion that owes its existence to the isentropic gradient of potential vorticity.

Rossby wave propagation can be understood in a qualitative fashion by considering a closed chain of fluid parcels initially aligned along a circle of latitude. Recall that the absolute vorticity η is given by $\eta = \zeta + f$, where ζ is the relative vorticity and f is the Coriolis parameter. Assume that $\zeta = 0$ at time t_0. Now suppose that at t_1, δy is the meridional displacement of a fluid parcel from the original latitude. Then at t_1 we have

$$(\zeta + f)_{t_1} = f_{t_0}$$

or

$$\zeta_{t_1} = f_{t_0} - f_{t_1} = -\beta \, \delta y \tag{7.87}$$

where $\beta \equiv df/dy$ is the planetary vorticity gradient at the original latitude.

From (7.87) it is evident that if the chain of parcels is subject to a sinusoidal meridional displacement under absolute vorticity conservation, the resulting perturbation vorticity will be positive (i.e., cyclonic) for a southward displacement and negative (anticyclonic) for a northward displacement.

This perturbation vorticity field will induce a meridional velocity field, which advects the chain of fluid parcels southward west of the vorticity maximum and northward west of the vorticity minimum, as indicated in Fig. 7.14. Thus, the fluid parcels oscillate back and forth about their equilibrium latitude, and the pattern of vorticity maxima and minima propagates to the west. This westward-propagating vorticity field constitutes a Rossby wave. Just as a positive vertical gradient of potential temperature resists vertical fluid displacements and provides the restoring force for

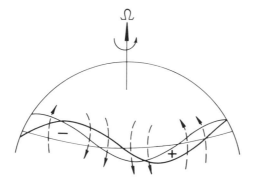

Fig. 7.14 Perturbation vorticity field and induced velocity field (dashed arrows) for a meridionally displaced chain of fluid parcels. Heavy wavy line shows original perturbation position, light line westward displacement of the pattern due to advection by the induced velocity.

gravity waves, the meridional gradient of absolute vorticity resists meridional displacements and provides the restoring mechanism for Rossby waves.

7.7.1　FREE BAROTROPIC ROSSBY WAVES

The dispersion relationship for barotropic Rossby waves may be derived formally by finding wave-type solutions of the linearized barotropic vorticity equation. The barotropic vorticity equation (4.27) states that the vertical component of absolute vorticity is conserved following the horizontal motion. For a midlatitude β plane it has the form

$$\left(\frac{\partial}{\partial t}+u\frac{\partial}{\partial x}+v\frac{\partial}{\partial y}\right)\zeta+\beta v=0 \tag{7.88}$$

We now assume that the motion consists of a basic state zonal velocity plus a small horizontal perturbation:

$$u=\bar{u}+u', \qquad v=v', \qquad \zeta=\bar{\zeta}+\zeta'$$

We define a perturbation streamfunction ψ' according to

$$u'=-\partial\psi'/\partial y, \qquad v'=\partial\psi'\partial x$$

from which $\zeta'=\nabla^2\psi'$. The perturbation form of (7.88) is then

$$\left(\frac{\partial}{\partial t}+\bar{u}\frac{\partial}{\partial x}\right)\nabla^2\psi'+\beta\frac{\partial\psi'}{\partial x}=0 \tag{7.89}$$

where as usual we have neglected terms involving the products of perturbation quantities. We seek a solution of the form

$$\psi'=\text{Re}[\Psi\exp(i\phi)] \tag{7.90}$$

where $\phi=kx+ly-\nu t$. Here k and l are wave numbers in the zonal and meridional directions, respectively. Substituting from (7.90) into (7.89) gives

$$(-\nu+k\bar{u})(-k^2-l^2)+k\beta=0$$

which may immediately be solved for ν:

$$\nu=\bar{u}k-\beta k/K^2 \tag{7.91}$$

where $K^2\equiv k^2+l^2$ is the total horizontal wave number squared.

Recalling that $c_x = \nu/k$, we find that the zonal phase speed relative to the mean wind is

$$c_x - \bar{u} = -\beta/K^2 \qquad (7.92)$$

Thus, the Rossby wave zonal phase propagation is always *westward* relative to the mean zonal flow. Furthermore, the Rossby wave phase speed depends inversely on the square of the horizontal wave number. Therefore, Rossby waves are dispersive waves whose phase speeds increase rapidly with increasing wavelength.

This result is consistent with the discussion in Section 6.2.2., in which we showed that the advection of planetary vorticity, which tends to make disturbances retrogress, increasingly dominates over relative vorticity advection as the wavelength of a disturbance increases. Equation (7.92) provides a quantitative measure of this effect in cases where the disturbance is small enough in amplitude so that perturbation theory is applicable. For a typical midlatitude synoptic-scale disturbance, with similar meridional and zonal scales $(l \approx k)$ and zonal wavelength of order 6000 km, the Rossby wave speed relative to the zonal flow calculated from (7.92) is approximately $-8\ \mathrm{m\ s^{-1}}$. Since the mean zonal wind is generally westerly and greater than $8\ \mathrm{m\ s^{-1}}$, synoptic-scale Rossby waves usually move eastward, but at a phase speed relative to the ground that is somewhat less than the mean zonal wind speed.

For longer wavelengths the westward Rossby wave phase speed may be large enough to balance the eastward advection by the mean zonal wind so that the resulting disturbance is stationary relative to the surface of the earth. From (7.92) it is clear that the free Rossby wave solution becomes stationary when

$$K^2 = \beta/\bar{u} \equiv K_s^2 \qquad (7.93)$$

The significance of this condition will be discussed in the next subsection.

Unlike the Rossby wave phase speed, which is always westward relative to the mean flow, the zonal group velocity may be either eastward or westward relative to the mean flow, depending on the ratio of the zonal and meridional wave numbers. It turns out, however, that for stationary Rossby modes (i.e., modes with $c - \bar{u} = 0$) the zonal group velocity is always eastward. Derivation of the group velocity is left as an exercise for the reader (see Problem 7.19).

It is possible to carry out a less restrictive analysis of free planetary waves using the perturbation form of the full primitive equations. In that case the structure of the free modes depends critically on the boundary conditions at the surface and the upper boundary. The results of such an analysis are mathematically complicated, but qualitatively they yield waves with horizontal dispersion properties similar to those in the shallow-water model. It turns out that the free oscillations allowed in a hydrostatic gravitationally stable atmosphere consist of eastward- and westward-moving gravity waves that are slightly modified by the rotation of the earth, and westward-moving Rossby waves that are slightly modified by gravitational stability. These free oscillations are the normal modes of oscillation of the atmosphere. As such, they are continually excited by the various forces acting on the atmosphere. The planetary-scale free oscillations, although they can be detected by careful observational studies, generally have rather weak amplitudes. Presumably this is because the forcing is quite weak at the large phase speeds characteristic of most such waves. An exception is the 16-day period zonal wave number 1 normal mode, which can be quite strong in the winter stratosphere.

7.7.2 FORCED TOPOGRAPHIC ROSSBY WAVES

Although free propagating Rossby modes are only rather weakly excited in the atmosphere, forced stationary Rossby modes are of primary importance for understanding the planetary-scale circulation pattern. Such modes may be forced by longitudinally dependent diabatic heating patterns or by flow over topography. Of particular importance for the Northern Hemisphere extratropical circulation are stationary Rossby modes forced by flow over the Rockies and the Himalayas. It is just the topographic Rossby wave that was described qualitatively in the discussion of streamline deflections in potential vorticity-conserving flows crossing mountain ranges in Section 4.3.

As the simplest possible dynamical model of topographic Rossby waves we use the barotropic vorticity equation for a homogeneous fluid of variable depth [see Eqs. (4.24) and (4.25)]. We assume that the upper boundary is at a fixed height H and the lower boundary is at the variable height $h_T(x, y)$. We also use quasi-geostrophic scaling so that $|\zeta_g| \ll f_0$. With the aid of (4.25) we can then write (4.24) in the form

$$H\left(\frac{\partial}{\partial t} + \mathbf{V} \cdot \mathbf{\nabla}\right)(\zeta_g + f) = -f_0 \frac{Dh_T}{Dt} \tag{7.94}$$

After linearizing and applying the midlatitude β-plane approximation (7.94) yields

$$\left(\frac{\partial}{\partial t} + \bar{u}\frac{\partial}{\partial x}\right)\zeta_g' + \beta v_g' = -\frac{f_0}{H}\bar{u}\frac{\partial h_T}{\partial x} \tag{7.95}$$

We now examine solutions of (7.95) for the special case of a sinusoidal lower boundary. We specify the topography to have the form

$$h_T(x, y) = \text{Re}[h_0 \exp(ikx)] \cos ly \qquad (7.96)$$

and represent the geostrophic wind and vorticity by the perturbation streamfunction

$$\psi(x, y) = \text{Re}[\psi_0 \exp(ikx)] \cos ly \qquad (7.97)$$

Then (7.95) has a steady-state solution with complex amplitude given by

$$\psi_0 = f_0 h_0 / [H(K^2 - K_s^2)] \qquad (7.98)$$

The streamfunction is either exactly in phase (ridges over the mountains) or exactly out of phase (troughs over the mountains) with the topography depending on the sign of $K^2 - K_s^2$. For long waves ($K < K_s$) the topographic vorticity source in (7.95) is primarily balanced by meridional advection of planetary vorticity (the β effect). For short waves ($K > K_s$) the source is balanced primarily by the zonal advection of relative vorticity.

The topographic wave solution (7.98) has the unrealistic characteristic that when the wave number exactly equals the critical wave number K_s the amplitude goes to infinity. From (7.93) it is clear that this singularity occurs at the zonal wind speed for which the free Rossby mode becomes stationary. Thus, it may be thought of as a resonant response of the barotropic system.

Charney and Eliassen (1947) used the topographic Rossby wave model to explain the winter mean longitudinal distribution of 500-mb heights in the Northern Hemisphere midlatitude. They removed the resonant singularity by including boundary layer drag in the form of Ekman pumping, which for the barotropic vorticity equation is simply a linear damping of the relative vorticity [see (5.41)]. The vorticity equation thus takes the form

$$\left(\frac{\partial}{\partial t} + \bar{u}\frac{\partial}{\partial x}\right)\zeta_g' + \beta v_g' + r\zeta_g' = -\frac{f_0}{H}\bar{u}\frac{\partial h_T}{\partial x} \qquad (7.99)$$

where $r \equiv \tau_e^{-1}$ is the inverse of the spin-down time defined in Section 5.4.

For steady flow (7.99) has a solution with complex amplitude

$$\psi_0 = f_0 h_0 / [H(K^2 - K_s^2 - i\varepsilon)] \qquad (7.100)$$

where $\varepsilon \equiv rK^2(k\bar{u})^{-1}$. Thus, boundary layer drag shifts the phase of the response and removes the singularity at resonance. However, the amplitude is still a maximum for $K = K_s$ and the trough in the streamfunction occurs 1/4 cycle east of the mountain crest, in approximate agreement with observations.

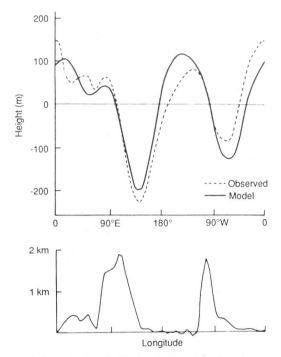

Fig. 7.15 Upper panel shows the longitudinal variation of the disturbance geopotential height ($\equiv f_0 \psi / g$) in the Charney–Eliassen model for the parameters given in the text (solid line), compared with the observed 500-mb height perturbations at 45°N in January (dashed line). The lower panel shows the smoothed profile of topography at 45°N used in the computation. (After Held, 1983.)

By use of a Fourier expansion Eq. (7.99) can be solved for realistic distributions of topography. The results for an x dependence of h_T given by a smoothed version of the earth's topography at 45°N, a meridional wave number corresponding to a latitudinal half-wavelength of 35°, $\tau_e = 5$ days, $\bar{u} = 17$ m s^{-1}, $f_0 = 10^{-4}$ s^{-1}, and $H = 8$ km are shown in Fig. 7.15. Despite its simplicity, the Charney–Eliassen model does a remarkable job of reproducing the observed 500-mb stationary wave pattern in Northern Hemisphere midlatitudes.

Problems

7.1. Show that the Fourier component $F(x) = \text{Re}[C \exp(imx)]$ can be written as

$$F(x) = |C| \cos m(x + x_0)$$

where $x_0 = m^{-1} \sin^{-1}(C_i/|C|)$ and C_i stands for the imaginary part of C.

7.2. In the study of atmospheric wave motions it is often necessary to consider the possibility of amplifying or decaying waves. In such a case we might assume that a solution has the form

$$\psi = A \cos(kx - \nu t - kx_0) \exp(\alpha t)$$

where A is the initial amplitude, α the amplification factor, and x_0 the initial phase. Show that this expression can be written more concisely as

$$\psi = \text{Re}[Be^{ik(x-ct)}]$$

where both B and c are complex constants. Determine the real and imaginary parts of B and c in terms of a, α, k, ν, and x_0.

7.3. Several of the wave types discussed in this chapter are governed by equations that are generalizations of the wave equation

$$\frac{\partial^2 \psi}{\partial t^2} = c^2 \frac{\partial^2 \psi}{\partial x^2}$$

This equation can be shown to have solutions corresponding to waves of arbitrary profile moving at the speed c in both the positive and negative x directions. We consider an arbitrary initial profile of the field ψ:

$$\psi = f(x) \quad \text{at } t = 0$$

If the profile is translated in the positive x direction at speed c without change of shape, then $\psi = f(x')$, where x' is a coordinate moving at speed c so that $x = x' + ct$. Thus, in terms of the fixed coordinate x we can write $\psi = f(x - ct)$, corresponding to a profile that moves in the positive x direction at speed c without change of shape. Verify that $\psi = f(x - ct)$ is a solution for any arbitrary continuous profile $f(x - ct)$. *Hint*: Let $x - ct = x'$ and differentiate f using the chain rule.

7.4. Assuming that the pressure perturbation for a one-dimensional acoustic wave is given by (7.15), find the corresponding solutions of the zonal wind and density perturbations. Express the amplitude and phase for u' and ρ' in terms of the amplitude and phase of p'.

7.5. Show that for isothermal motion ($DT/Dt = 0$) the acoustic wave speed is given by $(gH)^{1/2}$ where $H = RT/g$ is the scale height.

7.6. In Section 7.3.1 the linearized equations for acoustic waves were developed for the special situation of one-dimensional propagation in a horizontal tube. Although this situation does not appear to be directly applicable to the atmosphere, there is a special atmospheric mode, the *Lamb wave*, which is a horizontally propagating acoustic mode with no vertical velocity perturbation ($w' = 0$). Such oscillations have been observed following violent explosions such as volcanic eruptions and atmospheric nuclear tests. Using (7.12) and (7.13) plus the linearized forms of the hydrostatic equation and the continuity equation (7.7), derive the height dependence of the perturbation fields for the Lamb mode in an isothermal basic state atmosphere assuming that the pressure perturbation at the lower boundary ($z = 0$) has the form (7.15). Determine the vertically integrated kinetic energy density per unit horizontal area for this mode.

7.7. If the surface height perturbation in a shallow-water gravity wave is given by

$$h' = \text{Re}[Ae^{ik(x-ct)}]$$

find the corresponding velocity perturbation $u'(x, t)$. Sketch the phase relationship between h' and u' for an eastward-propagating wave.

7.8. Assuming that the vertical velocity perturbation for a two-dimensional internal gravity wave is given by (7.43), obtain the corresponding solution for the u', p', and θ' fields. Use these results to verify the approximation

$$|\rho_0 \theta'/\bar{\theta}| \gg |p'/c_s^2|$$

which was used in (7.36).

7.9. For the situation in Problem 7.8, express the vertical flux of horizontal momentum, $\rho_0 \overline{u'w'}$, in terms of the amplitude A of the vertical velocity perturbation. Hence, show that the momentum flux is positive for waves in which phase speed propagates eastward and downward.

7.10. Show that if (7.38) is replaced by the hydrostatic equation (i.e., $\partial w/\partial t$ is neglected) the resulting frequency equation for internal gravity waves is just the asymptotic limit of (7.44) for waves in which $|k| \ll |m|$.

7.11. (a) Show that the group velocity vector in two-dimensional internal gravity waves is parallel to lines of constant phase.

(b) Show that in the long-wave limit ($|k| \ll |m|$) the magnitude of the zonal component of the group velocity equals the magnitude of the zonal phase speed so that energy propagates one wavelength per wave period.

7.12. Determine the perturbation horizontal and vertical velocity fields for stationary gravity waves forced by flow over sinusoidally varying topography, given the following conditions: the height of the ground is $h = h_0 \cos kx$, where $h_0 = 50$ m is a constant; $N = 2 \times 10^{-2}$ s^{-1}; $\bar{u} = 5$ m s^{-1}; and $k = 3 \times 10^{-3}$ m^{-1}. *Hint*: For small amplitude topography ($h_0 k \ll 1$) we can approximate the lower boundary condition by

$$w' = Dh/Dt = \bar{u}\, \partial h/\partial x \quad \text{at } z = 0$$

7.13. Verify the group velocity relationship for inertio-gravity waves given in (7.67).

7.14. Show that when $u = 0$ the wave number vector κ for an internal gravity wave is perpendicular to the group velocity vector.

7.15. Using the linearized form of the vorticity equation (6.18) and the β-plane approximation, derive the Rossby wave speed for a homogeneous incompressible ocean of depth h. Assume a motionless basic state and small perturbations that depend only on x and t,

$$u = u'(x, t), \qquad v = v'(x, t), \qquad h = H + h'(x, t)$$

where H is the mean depth of the ocean. With the aid of the continuity equation for a homogeneous layer (7.21) and the geostrophic wind relationship $v' = g f_0^{-1} \partial h'/\partial x$, show that the perturbation potential vorticity equation can be written in the form

$$\frac{\partial}{\partial t}\left(\frac{\partial^2}{\partial x^2} - \frac{f_0^2}{gH} \right)h' + \beta \frac{\partial h'}{\partial x} = 0$$

and that $h' = h_0 e^{ik(x-ct)}$ is a solution provided that

$$c = -\beta(k^2 + f_0^2/gH)^{-1}$$

If the ocean is 4 km deep, what is the Rossby wave speed at latitude 45° for a wave of 10,000 km zonal wavelength?

7.16. In Section 4.3 we showed that for a homogeneous incompressible fluid a decrease in depth with latitude has the same dynamic effect as a latitudinal dependence of the Coriolis parameter. Thus, Rossby-type waves can be produced in a rotating cylindrical vessel if the depth of the fluid is dependent on the radial coordinate. To determine the Rossby wave speed formula for this *equivalent β* effect, we assume that the flow is confined between rigid lids in an annular region whose distance from the axis of rotation is large enough so that the curvature terms in the equations can be neglected. We then can refer the motion to Cartesian coordinates with x directed azimuthally and y directed toward the axis of rotation. If the system is rotating at angular velocity Ω and the depth is linearly dependent on y,

$$H(y) = H_0 - \gamma y$$

show that the perturbation continuity equation can be written as

$$H_0\left(\frac{\partial u'}{\partial x} + \frac{\partial v'}{\partial y}\right) - \gamma v' = 0$$

and that the perturbation quasi-geostrophic vorticity equation is thus

$$\frac{\partial}{\partial t}\nabla^2\psi' + \beta\frac{\partial\psi'}{\partial x} = 0$$

where ψ' is the perturbation geostrophic streamfunction and $\beta = 2\Omega\gamma/H_0$. What is the Rossby wave speed in this situation for waves of wavelength 100 cm in both the x and y directions if $\Omega = 1\,\text{s}^{-1}$, $H_0 = 20$ cm, and $\gamma = 0.05$? (*Hint*: Assume that the velocity field is geostrophic except in the divergence term.)

7.17. Show by scaling arguments that if the horizontal wavelength is much greater than the depth of the fluid, two-dimensional surface gravity waves will be hydrostatic so that the shallow-water approximation applies.

7.18. The linearized form of the quasi-geostrophic vorticity equation (6.18) can be written as follows:

$$\left(\frac{\partial}{\partial t} + \bar{u}\frac{\partial}{\partial x}\right)\nabla^2\psi' + \beta\frac{\partial\psi'}{\partial x} = -f_0\boldsymbol{\nabla}\cdot\mathbf{V}$$

Suppose that the horizontal divergence field is given by

$$\boldsymbol{\nabla}\cdot\mathbf{V} = A\cos[k(x - ct)]$$

where A is a constant. Find a solution for the corresponding relative vorticity field. What is the phase relationship between vorticity and divergence? For what value of c does the vorticity become infinite?

7.19. Derive an expression for the group velocity of a barotropic Rossby wave. Show that for stationary waves the group velocity always has an eastward zonal component relative to the earth. Hence, Rossby wave energy propagation must be downstream of topographic sources.

Suggested References

Hildebrand, *Advanced Calculus for Applications*, is one of many standard textbooks that discuss the mathematical techniques used in this chapter, including the representation of functions in Fourier series and the general properties of the wave equation.

Turner, *Buoyancy Effects in Fluids*, contains an excellent discussion of internal gravity waves.

Gill, *Atmosphere-Ocean Dynamics*, has a very complete treatment of gravity, inertio-gravity, and Rossby waves, with particular emphasis on observed oscillations in the oceans.

Smith (1979) discusses many aspects of waves generated by flow over mountains.

Chapman and Lindzen, *Atmospheric Tides: Thermal and Gravitational*, is the classic reference on both observational and theoretical aspects of tides, a class of atmospheric motions for which the linear perturbations method has proved to be particularly successful.

Scorer, *Natural Aerodynamics*, contains an excellent qualitative discussion on many aspects of waves generated by barriers such as lee waves.

Chapter

8 Synoptic-Scale Motions II: Baroclinic Instability

In Chapter 6 we showed that the quasi-geostrophic system can qualitatively account for the observed relationships among the vorticity, temperature, and vertical velocity fields in midlatitude synoptic-scale systems. The diagnostic approach used in that chapter provided useful insights into the structure of synoptic-scale systems; it also demonstrated the key role of potential vorticity in dynamical analysis. It did not, however, provide quantitative information on the origins, growth rates, and propagation speeds of such disturbances. In this chapter we show how linear perturbation analysis can be used to obtain such information.

The development of synoptic-scale weather disturbances is often referred to as *cyclogenesis*, a term that emphasizes the role of relative vorticity in developing synoptic-scale systems. In this chapter we analyze the processes that lead to cyclogenesis. Specifically, we discuss the role of dynamical instability of the mean flow in accounting for the growth of synoptic-scale disturbances. We show that the quasi-geostrophic equations can, indeed, provide a reasonable theoretical basis for understanding the development of synoptic-scale storms, although, as will be discussed in Section 9.2, ageostrophic effects must be included to model the development of fronts and subsynoptic-scale storms.

8.1 Hydrodynamic Instability

A zonal-mean flow field is said to be hydrodynamically unstable if a small disturbance introduced into the flow grows spontaneously, drawing energy from the mean flow. It is useful to divide fluid instabilities into two types: parcel instability and wave instability. The simplest example of a parcel instability is the convective overturning that occurs when a fluid parcel is displaced vertically in a statically unstable fluid (see Section 2.7.3). Another example is inertial instability, which occurs when a parcel is displaced radially in an axisymmetric vortex with negative (positive) absolute vorticity in the Northern (Southern) Hemisphere. This instability was discussed in Section 7.5.1. A more general type of parcel instability, called *symmetric instability*, may also be significant in weather disturbances; this is discussed in Section 9.3.

Most of the instabilities of importance in meteorology, however, are associated with wave propagation; they cannot be easily related to the behavior of individual fluid parcels. The wave instabilities important for synoptic-scale meteorology generally are associated with zonally asymmetric perturbations to a zonally symmetric basic flow field. In general the basic flow is a jet stream that has both horizontal and vertical mean flow shears. *Barotropic instability* is a wave instability associated with the horizontal shear in a jetlike current. Barotropic instabilities grow by extracting kinetic energy from the mean flow field. *Baroclinic instability*, on the other hand, is associated with vertical shear of the mean flow. Baroclinic instabilities grow by converting potential energy associated with the mean horizontal temperature gradient that must exist to provide thermal wind balance for the vertical shear in the basic state flow. In neither of these instability types does the parcel method provide a satisfactory stability criterion. A more rigorous approach is required in which a linearized version of the governing equations is analyzed to determine the structure and amplification rate for the various wave modes supported by the system.

As indicated in Problem 7.2, the traditional approach to instability analysis is to assume that a small perturbation consisting of a single Fourier wave mode of the form $\exp[ik(x - ct)]$ is introduced into the flow and to determine the conditions for which the phase velocity c has an imaginary part. This technique, which is called the *normal modes* method, will be applied in the next section to analyze the stability of a baroclinic current.

An alternative method of instability analysis is the *initial value* approach. This method is motivated by the recognition that in general the perturbations from which storms develop cannot be described as single normal mode disturbances, but may have a complex structure. The initial growth of such disturbances may strongly depend on the potential vorticity distribution in the initial disturbance. On the time scale of a day or two such growth can be quite different from that of a normal mode of similar scale, although in the absence of nonlinear interactions the fastest-growing normal mode disturbance must eventually dominate.

A strong dependence of cyclogenesis on initial conditions occurs when a large-amplitude upper-level potential vorticity anomaly is advected into a region where there is a preexisting meridional temperature gradient at the surface. In that case, as shown schematically in Fig. 8.1, the circulation induced by the upper-level anomaly (which extends downward, as discussed in Section 6.3) leads to temperature advection at the surface; this induces a potential vorticity anomaly near the surface, which in turn reinforces the upper-level anomaly. Under some conditions the surface and upper-level potential vorticity anomalies can become locked in phase, so that the induced circulations produce a very rapid amplification of the anomaly pattern. Detailed discussion of the initial value approach to cyclogenesis is beyond the scope of this text. Here we concentrate instead on the simplest normal mode instability models.

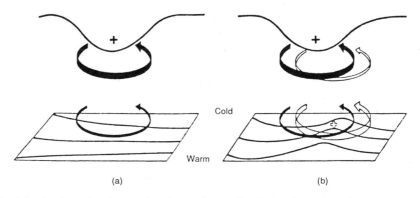

(a)   (b)

Fig. 8.1 A schematic picture of cyclogenesis associated with the arrival of an upper-level positive vorticity perturbation over a lower-level baroclinic region. (a) The lower-level cyclonic vorticity induced by the upper-level vorticity anomaly. The circulation induced by the vorticity anomaly is shown by the solid arrows, and potential temperature contours are shown at the lower boundary. The advection of potential temperature by the induced lower-level circulation leads to a warm anomaly slightly east of the upper-level vorticity anomaly. This in turn will induce a cyclonic circulation as shown by the open arrows in (b). The induced upper-level circulation will reinforce the original upper-level anomaly and can lead to amplification of the disturbance. (After Hoskins *et al.*, 1985.)

8.2 Baroclinic Instability: A Two-Layer Model

Even for a highly idealized mean flow profile the mathematical treatment of baroclinic instability in a continuously stratified atmosphere is rather complicated. Thus, before considering such a model we first focus on the simplest model that can incorporate baroclinic processes. The atmosphere is represented by two discrete layers bounded by surfaces numbered 0, 2, and 4 (generally taken to be the 0-, 500-, and 1000-mb surfaces, respectively) as shown in Fig. 8.2. The quasi-geostrophic vorticity equation for the midlatitude β plane is applied at the levels designated by 1 and 3 in Fig. 8.2, and the thermodynamic energy equation is applied at level 2.

Before writing out the specific equations of the two-layer model, it is convenient to define a *geostrophic streamfunction*, $\psi \equiv \Phi/f_0$. Then the geostrophic wind (6.7) and the geostrophic vorticity (6.15) can be expressed respectively as

$$\mathbf{V}_\psi = \mathbf{k} \times \nabla \psi, \qquad \zeta_g - \nabla^2 \psi \tag{8.1}$$

The quasi-geostrophic vorticity equation (6.19) and the hydrostatic thermodynamic energy equation (6.13) can then be written in terms of ψ and ω as

$$\frac{\partial}{\partial t} \nabla^2 \psi + \mathbf{V}_\psi \cdot \nabla(\nabla^2 \psi) + \beta \frac{\partial \psi}{\partial x} = f_0 \frac{\partial \omega}{\partial p} \tag{8.2}$$

$$\frac{\partial}{\partial t} \left(\frac{\partial \psi}{\partial p} \right) = -\mathbf{V}_\psi \cdot \nabla \left(\frac{\partial \psi}{\partial p} \right) - \frac{\sigma}{f_0} \omega \tag{8.3}$$

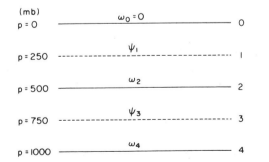

Fig. 8.2 Arrangement of variables in the vertical for the two-level baroclinic model.

We now apply the vorticity equation (8.2) at the two levels designated as 1 and 3, which are at the middle of the two layers. To do this we must estimate the divergence term $\partial \omega / \partial p$ at these levels using finite difference approximations to the vertical derivatives:

$$\left(\frac{\partial \omega}{\partial p}\right)_1 \approx \frac{\omega_2 - \omega_0}{\delta p}, \qquad \left(\frac{\partial \omega}{\partial p}\right)_3 \approx \frac{\omega_4 - \omega_2}{\delta p} \tag{8.4}$$

where δp is the pressure interval between levels 0-2 and 2-4 and subscript notation is used to designate the vertical level for each dependent variable. The resulting vorticity equations are

$$\frac{\partial}{\partial t} \nabla^2 \psi_1 + \mathbf{V}_1 \cdot \nabla (\nabla^2 \psi_1) + \beta \frac{\partial \psi_1}{\partial x} = \frac{f_0}{\delta p} \omega_2 \tag{8.5}$$

$$\frac{\partial}{\partial t} \nabla^2 \psi_3 + \mathbf{V}_3 \cdot \nabla (\nabla^2 \psi_3) + \beta \frac{\partial \psi_3}{\partial x} = -\frac{f_0}{\delta p} \omega_2 \tag{8.6}$$

where we have used the fact that $\omega_0 = 0$ and assumed that $\omega_4 = 0$, which is approximately true for a level lower boundary surface.

We next write the thermodynamic energy equation (8.3) at level 2. Here we must evaluate $\partial \psi / \partial p$ using the difference formula

$$(\partial \psi / \partial p)_2 \approx (\psi_3 - \psi_1) / \delta p$$

The result is

$$\frac{\partial}{\partial t} (\psi_1 - \psi_3) = -\mathbf{V}_2 \cdot \nabla (\psi_1 - \psi_3) + \frac{\sigma \delta p}{f_0} \omega_2 \tag{8.7}$$

The first term on the right-hand side in (8.7) is the advection of the 250–750-mb thickness by the wind at 500-mb. However, ψ_2, the 500-mb streamfunction, is not a predicted field in this model. Therefore, ψ_2 must be obtained by linearly interpolating between the 250- and 750-mb levels:

$$\psi_2 = (\psi_1 + \psi_3)/2 \tag{8.8}$$

If this interpolation formula is used, (8.5)–(8.7) become a closed set of prediction equations in the variables ψ_1, ψ_3, and ω_2.

8.2.1 LINEAR PERTURBATION ANALYSIS

To keep the analysis as simple as possible we assume that the stream-functions ψ_1 and ψ_3 consists of basic state parts that depend linearly on y alone, plus perturbations that depend only on x and t. Thus, we let

$$\psi_1 = -U_1 y + \psi_1'(x, t)$$

$$\psi_3 = -U_3 y + \psi_3'(x, t) \qquad (8.9)$$

$$\omega_2 = \omega_2'(x, t)$$

The zonal velocities at levels 1 and 3 are then constants with the values U_1 and U_3, respectively. Hence, the perturbation field has meridional and vertical velocity components only.

Substituting from (8.9) into (8.5)–(8.7) and linearizing yields the perturbation equations

$$\left(\frac{\partial}{\partial t} + U_1 \frac{\partial}{\partial x}\right) \frac{\partial^2 \psi_1'}{\partial x^2} + \beta \frac{\partial \psi_1'}{\partial x} = \frac{f_0}{\delta p} \omega_2' \qquad (8.10)$$

$$\left(\frac{\partial}{\partial t} + U_3 \frac{\partial}{\partial x}\right) \frac{\partial^2 \psi_3'}{\partial x^2} + \beta \frac{\partial \psi_3'}{\partial x} = -\frac{f_0}{\delta p} \omega_2' \qquad (8.11)$$

$$\left(\frac{\partial}{\partial t} + U_m \frac{\partial}{\partial x}\right)(\psi_1' - \psi_3') - U_T \frac{\partial}{\partial x}(\psi_1' + \psi_3') = \frac{\sigma \delta p}{f_0} \omega_2' \qquad (8.12)$$

where we have linearly interpolated to express $\mathbf{V}_2$ in terms of ψ_1 and ψ_3 and have defined

$$U_m \equiv (U_1 + U_3)/2, \qquad U_T \equiv (U_1 - U_3)/2$$

Thus, U_m and U_T are, respectively, the vertically averaged mean zonal wind and the mean thermal wind for the interval $\delta p/2$.

The dynamical properties of this system are more clearly expressed if (8.10)–(8.12) are combined to eliminate ω_2'. We first note that (8.10) and (8.11) can be rewritten as

$$\left[\frac{\partial}{\partial t} + (U_m + U_T) \frac{\partial}{\partial x}\right] \frac{\partial^2 \psi_1'}{\partial x^2} + \beta \frac{\partial \psi_1'}{\partial x} = \frac{f_0}{\delta p} \omega_2' \qquad (8.13)$$

$$\left[\frac{\partial}{\partial t} + (U_m - U_T) \frac{\partial}{\partial x}\right] \frac{\partial^2 \psi_3'}{\partial x^2} + \beta \frac{\partial \psi_3'}{\partial x} = -\frac{f_0}{\delta p} \omega_2' \qquad (8.14)$$

We now define the barotropic and baroclinic perturbations as

$$\psi_m \equiv (\psi_1' + \psi_3')/2, \qquad \psi_T \equiv (\psi_1' - \psi_3')/2 \tag{8.15}$$

Adding (8.13) and (8.14) and using the definitions in (8.15) yields

$$\left[\frac{\partial}{\partial t} + U_m \frac{\partial}{\partial x} \right] \frac{\partial^2 \psi_m}{\partial x^2} + \beta \frac{\partial \psi_m}{\partial x} + U_T \frac{\partial}{\partial x} \left(\frac{\partial^2 \psi_T}{\partial x^2} \right) = 0 \tag{8.16}$$

while subtracting (8.14) from (8.13) and combining with (8.12) to eliminate ω_2' yields

$$\left[\frac{\partial}{\partial t} + U_m \frac{\partial}{\partial x} \right] \left(\frac{\partial^2 \psi_T}{\partial x^2} - 2\lambda^2 \psi_T \right) + \beta \frac{\partial \psi_T}{\partial x} + U_T \frac{\partial}{\partial x} \left(\frac{\partial^2 \psi_m}{\partial x^2} + 2\lambda^2 \psi_m \right) = 0 \tag{8.17}$$

where $\lambda^2 \equiv f_0^2/[\sigma(\delta p)^2]$. Equations (8.16) and (8.17) govern the evolution of the barotropic (vertically averaged) and baroclinic (thermal) perturbation vorticities, respectively.

As in Chapter 7 we assume that wavelike solutions exist of the form

$$\psi_m = A e^{ik(x - ct)}, \qquad \psi_T = B e^{ik(x - ct)} \tag{8.18}$$

Substituting these assumed solutions into (8.16) and (8.17) and dividing through by the common exponential factor, we obtain a pair of simultaneous linear algebraic equations for the coefficients A, B:

$$ik[(c - U_m)k^2 + \beta]A - ik^3 U_T B = 0 \tag{8.19}$$

$$ik[(c - U_m)(k^2 + 2\lambda^2) + \beta]B - ikU_T(k^2 - 2\lambda^2)A = 0 \tag{8.20}$$

Since this set is homogeneous, nontrivial solutions will exist only if the determinant of the coefficients of A and B is zero. Thus the phase speed c must satisfy the condition

$$\begin{vmatrix} (c - U_m)k^2 + \beta & -k^2 U_T \\ -U_T(k^2 - 2\lambda^2) & (c - U_m)(k^2 + 2\lambda^2) + \beta \end{vmatrix} = 0$$

which gives a quadratic dispersion equation in c:

$$(c - U_m)^2 k^2(k^2 + 2\lambda^2) + 2(c - U_m)\beta(k^2 + \lambda^2)$$

$$+ [\beta^2 + U_T^2 k^2(2\lambda^2 - k^2)] = 0 \tag{8.21}$$

which is analogous to the linear wave dispersion equations developed in Chapter 7. The dispersion relationship (8.21) yields for the phase speed

$$c = U_m - \frac{\beta(k^2 + \lambda^2)}{k^2(k^2 + 2\lambda^2)} \pm \delta^{1/2} \qquad (8.22)$$

where

$$\delta \equiv \frac{\beta^2 \lambda^4}{k^4(k^2 + 2\lambda^2)^2} - \frac{U_T^2(2\lambda^2 - k^2)}{(k^2 + 2\lambda^2)}$$

We have now shown that (8.18) is a solution for the system (8.16)–(8.17) only if the phase speed satisfies (8.22). Although (8.22) appears to be rather complicated, it is immediately apparent that if $\delta < 0$ the phase speed will have an imaginary part and the perturbations will amplify exponentially. Before discussing the general physical conditions required for exponential growth it is useful to consider two special cases.

As the first special case we let $U_T = 0$ so that the basic state thermal wind vanishes and the mean flow is barotropic. The phase speeds in this case are

$$c_1 = U_m - \beta k^{-2} \qquad (8.23)$$

and

$$c_2 = U_m - \beta(k^2 + 2\lambda^2)^{-1} \qquad (8.24)$$

These are real quantities that correspond to the free (normal mode) oscillations for the two-level model with a barotropic basic state current. The phase speed c_1 is simply the dispersion relationship for a barotropic Rossby wave with no y dependence (see Section 7.7). Substituting the expression (8.23) in place of c in (8.19)–(8.20) we see that in this case $B = 0$ so that the perturbation is barotropic in structure. The expression (8.24), on the other hand, may be interpreted as the phase speed for an internal baroclinic Rossby wave. Note that c_2 is a dispersion relationship analogous to the Rossby wave speed for a homogeneous ocean with a free surface, which was given in Problem 7.15. But, in the two-level model, the factor $2\lambda^2$ appears in the denominator in place of the f_0^2/gH for the oceanic case. In each of these cases there is vertical motion associated with the Rossby wave so that static stability modifies the wave speed. It is left as a problem for the reader to show that if c_2 is substituted into (8.19)–(8.20), the resulting fields of ψ_1 and ψ_3 are 180° out of phase so that the perturbation is baroclinic, although the basic state is barotropic. Furthermore, the ω_2' field is 1/4 cycle out of phase with the 250-mb geopotential field, with the maximum upward motion occurring west of the 250-mb trough.

This vertical motion pattern may be understood if we note that $c_2 - U_m < 0$, so that the disturbance pattern moves westward *relative to the mean wind*. Now, viewed in a coordinate system moving with the mean wind the vorticity changes are due only to the planetary vorticity advection and the convergence terms while the thickness changes must be caused solely by the adiabatic heating or cooling due to vertical motion. Hence, there must be rising motion west of the 250-mb trough in order to produce the thickness changes required by the westward motion of the system.

Comparing (8.23) and (8.24) we see that the phase speed of the baroclinic mode is generally much less than that of the barotropic mode, since for average midlatitude tropospheric conditions $\lambda^2 \approx 2 \times 10^{-12}\, m^{-2}$, which is comparable in magnitude to k for zonal wavelength of ~ 4500 km.[1]

As the second special case, we assume that $\beta = 0$. This case corresponds, for example, to a laboratory situation in which the fluid is bounded above and below by rotating horizontal planes so that the gravity and rotation vectors are everywhere parallel. In such a situation

$$c = U_m \pm U_T \left(\frac{k^2 - 2\lambda^2}{k^2 + 2\lambda^2} \right)^{1/2} \tag{8.25}$$

For waves with zonal wave numbers satisfying $k^2 < 2\lambda^2$, (8.25) has an imaginary part. Thus, all waves longer than the critical wavelenth $L_c = \sqrt{2}\pi / \lambda$ will amplify. From the definition of λ we can write

$$L_c = \delta p \pi (2\sigma)^{1/2} / f_0$$

For typical tropospheric conditions $(2\sigma)^{1/2} \approx 2 \times 10^{-3}\, N^{-1}\, m^3\, s^{-1}$. Thus, with $\delta p = 50$ kPa and $f_0 = 10^{-4}\, s^{-1}$ we find that $L_c = 3000$ km. It is also clear from this formula that the critical wavelength for baroclinic instability increases with the static stability. The role of static stability in stabilizing the shorter waves can be understood qualitatively as follows: For a sinusoidal perturbation, the relative vorticity, and hence the differential vorticity advection, increases with the square of the wave number. But, as shown in Chapter 6, a secondary vertical circulation is required to maintain hydrostatic temperature changes and geostrophic vorticity changes in the presence of differential vorticity advection. Thus, for a geopotential perturbation of fixed amplitude the relative strength of the accompanying vertical circulation

---

[1] The presence of the free internal Rossby wave should actually be regarded as a weakness of the two-level model. Lindzen *et al.* (1968) have shown that this mode does not correspond to any free oscillation of the real atmosphere. Rather, it is a spurious mode resulting from the use of the upper boundary condition $\omega = 0$ at $p = 0$, which formally turns out to be equivalent to putting a lid at the top of the atmosphere.

must increase as the wavelength of the disturbance decreases. Since static stability tends to resist vertical displacements, the shortest wavelengths will thus be stabilized.

It is also of interest that with $\beta = 0$ the critical wavelength for instability does not depend on the magnitude of the basic state thermal wind U_T. The growth rate, however, does depend on U_T. According to (8.18) the time dependence of the disturbance solution has the form $\exp(-ikct)$. Thus, the exponential growth rate is $\alpha = kc_i$, where c_i designates the imaginary part of the phase speed. In the present case

$$\alpha = kU_T\left(\frac{2\lambda^2 - k^2}{2\lambda^2 + k^2}\right)^{1/2} \tag{8.26}$$

so that the growth rate increases linearly with the mean thermal wind.

Returning to the general case where all terms are retained in (8.22), the stability criterion is most easily understood by computing the *neutral curve*, which connects all values of U_T and k for which $\delta = 0$ so that the flow is *marginally stable*. From (8.22), the condition $\delta = 0$ implies that

$$\frac{\beta^2\lambda^4}{k^4(2\lambda^2 + k^2)} = U_T^2(2\lambda^2 - k^2) \tag{8.27}$$

This complicated relationship between U_T and k can best be displayed by solving (8.27) for $k^4/2\lambda^4$, yielding

$$k^4/(2\lambda^4) = 1 \pm [1 - \beta^2/(4\lambda^4 U_T^2)]^{1/2}$$

In Fig. 8.3 the nondimensional quantity $k^2/2\lambda^2$, which is a measure of the zonal wavelength, is plotted against the nondimensional parameter $2\lambda^2 U_T/\beta$, which is proportional to the thermal wind. As indicated in the figure, the neutral curve separates the unstable region of the U_T, k plane from the stable region. It is clear that the inclusion of the β effect serves to stabilize the flow, for now unstable roots exist only for $|U_T| > \beta/(2\lambda^2)$. In addition the minimum value of U_T required for unstable growth depends strongly on k. Thus, the β effect strongly stabilizes the long-wave end of the wave spectrum ($k \to 0$). Again, the flow is always stable for waves shorter than the critical wavelength $L_c = \sqrt{2}\pi/\lambda$.

This long-wave stabilization associated with the β effect is caused by the rapid westward propagation of long waves (i.e., Rossby wave propagation), which occurs only when the β effect is included in the model. It can be shown that baroclinically unstable waves always propagate at a speed which lies between the maximum and minimum mean zonal wind speeds. Thus,

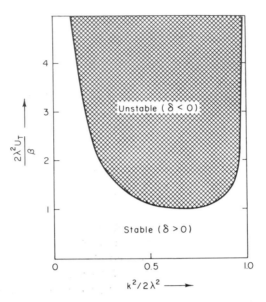

Fig. 8.3   Neutral stability curve for the two-level baroclinic model.

for our two-level model in the usual midlatitude case where $U_1 > U_3 > 0$, the real part of the phase speed satisfies the inequality $U_3 < c_r < U_1$ for unstable waves. In a continuous atmosphere this would imply that there must be a level where $U = c_r$. Such a level is called a *critical level* by theoreticians and a *steering level* by synopticians. For long waves and weak basic state wind shear, solution of (8.22) will have $c_r < U_3$, there is no steering level, and unstable growth cannot occur.

Differentiating (8.27) with respect to k and setting $dU_T/dk = 0$, we find that the minmum value of U_T for which unstable waves may exist occurs when $k^2 = \sqrt{2}\lambda^2$. This wave number corresponds to the wave of maximum instability. Wave numbers of observed disturbances should be close to the wave number of maximum instability, for if U_T were gradually raised from zero the flow would first become unstable for perturbations of wave number $k = 2^{1/4}\lambda$. Those perturbations would then amplify and in the process remove energy from the mean thermal wind, thereby decreasing U_T and stabilizing the flow. Under normal conditions of static stability the wavelength of maximum instability is about 4000 km, which is close to the average wavelength for midlatitude synoptic systems. Furthermore, the thermal wind required for marginal stability at this wavelength is only about $U_T \approx 4$ m s^{-1}, which implies a shear of 8 m s^{-1} between 250 and 750-mb. Shears greater than this are certainly common in middle latitudes for the zonally averaged flow. Therefore, the observed behavior of midlatitude synoptic systems is consistent with the hypothesis that such systems can originate from infinitesimal perturbations of a baroclinically unstable basic current. Of course, in the real atmosphere many other factors may influence the development of synoptic systems—for example, instabilities due to lateral shear in the jet stream, nonlinear interactions of finite-amplitude perturbations, and the release of latent heat in precipitating systems. However, observational studies, laboratory simulations, and numerical models all suggest that baroclinic instability is a primary mechanism for synoptic-scale wave development in middle latitudes.

8.2.2 VERTICAL MOTION IN BAROCLINIC WAVES

Since the two-level model is a special case of the quasi-geostrophic system, the physical mechansms responsible for forcing vertical motion should be those that were discussed in Section 6.4. Thus, the forcing of vertical motion can be expressed in terms of the sum of the forcing by thermal advection (evaluated at level 2) plus the differential vorticity advection (evaluated as the difference between the vorticity advection at level 1 and that at level 3). Alternatively, the forcing of vertical motion can be expressed in terms of the divergence of the $\mathbf{Q}$ vector.

The $\mathbf{Q}$-vector form of the omega equation for the two-level model can be derived simply from (6.35). We first estimate the second term on the left-hand side by finite differencing in p. Using (8.4) and again letting $\omega_0 = \omega_4 = 0$, we obtain

$$\frac{\partial^2 \omega}{\partial p^2} \approx \frac{(\partial \omega / \partial p)_3 - (\partial \omega / \partial p)_1}{\delta p} \approx -\frac{2\omega_2}{(\delta p)^2}$$

and observe that temperature in the two-level model is represented as

$$\frac{RT}{p} = -\frac{\partial \Phi}{\partial p} \approx \frac{f_0}{\delta p}(\psi_1 - \psi_3)$$

Thus, (6.35) becomes

$$\sigma(\nabla^2 - 2\lambda^2)\omega_2 = -2\nabla \cdot \mathbf{Q} \tag{8.28}$$

where

$$\mathbf{Q} = \frac{f_0}{\delta p}\left[-\frac{\partial \mathbf{V}_2}{\partial x} \cdot \nabla(\psi_1 - \psi_3), -\frac{\partial \mathbf{V}_2}{\partial y} \cdot \nabla(\psi_1 - \psi_3)\right]$$

In order to examine the forcing of vertical motion in baroclinically unstable waves we linearize (8.28) by specifying the same basic state and perturbation variables as in (8.9). For this situation, in which the mean zonal wind and the perturbation streamfunctions are independent of y, the $\mathbf{Q}$ vector has only an x-component:

$$Q_1 = \frac{f_0}{\delta p}\left[\frac{\partial^2 \psi_2'}{\partial x^2}(U_1 - U_3)\right] = \frac{2f_0}{\delta p} U_T \zeta_2'$$

The pattern of the $\mathbf{Q}$ vector in this case is similiar to that of Fig. 6.13 with eastward (westward) pointing $\mathbf{Q}$ centered at the 500-mb trough (ridge). This is consistent with the fact that $\mathbf{Q}$ represents the change of temperature gradient forced by geostrophic motion alone. In this simple model the temperature gradient is entirely due to the vertical shear of the mean zonal wind $[U_T \propto -\partial \bar{T}/\partial y]$ and the shear of the perturbation meridional velocity tends to advect warm air poleward east of the 500-mb trough and cold air equatorward west of the 500-mb trough so that there is a tendency to produce a component of temperature gradient directed eastward at the trough.

The forcing of vertical motion by the $\mathbf{Q}$ vector in the linearized model is from (8.28)

$$\left(\frac{\partial^2}{\partial x^2} - 2\lambda^2\right)\omega_2' = -\frac{4f_0}{\sigma\delta p} U_T \frac{\partial \zeta_2'}{\partial x} \tag{8.29}$$

Observing that

$$\left(\frac{\partial^2}{\partial x^2} - 2\lambda^2\right)\omega_2' \propto -\omega_2'$$

we may interpret (8.29) physically by noting that

$$w_2' \propto -\omega_2' \propto -U_T \frac{\partial \zeta_2'}{\partial x} \propto -v_2' \frac{\partial \bar{T}}{\partial y}$$

Thus, sinking (rising) motion is forced by negative (positive) advection of disturbance vorticity by the basic state thermal wind or alternatively by cold (warm) advection of the basic state thermal field by the perturbation meridional wind.

We now have the information required to diagram the structure of a baroclinically unstable disturbance in the two-level model. In the lower part of Fig. 8.4 we show schematically the phase relatonship between the

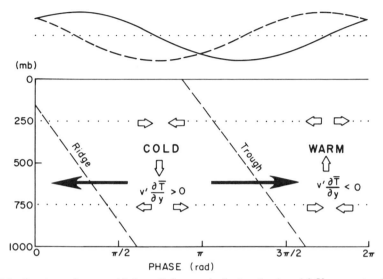

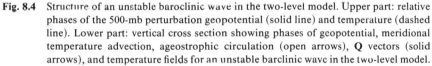

Fig. 8.4 Structure of an unstable baroclinic wave in the two-level model. Upper part: relative phases of the 500-mb perturbation geopotential (solid line) and temperature (dashed line). Lower part: vertical cross section showing phases of geopotential, meridional temperature advection, ageostrophic circulation (open arrows), **Q** vectors (solid arrows), and temperature fields for an unstable baroclinic wave in the two-level model.

geopotential field and the divergent secondary motion field for the usual midlatitude situation where $U_T > 0$. Linear interpolation has been used between levels so that the trough and ridge axes are straight lines tilted back toward the west with height. In this example the ψ_1 field lags the ψ_3 field by about 65° in phase so that the trough at 250-mb lies 65° in phase west of the 750-mb trough. At 500-mb the perturbation thickness field lags the geopotential field by one-quarter wavelength as shown in the top part of Fig. 8.4 and the thickness and vertical motion fields are in phase. Note that the temperature advection by the perturbation meridional wind is in phase with the 500-mb thickness field so that the advection of the basic state temperature by the perturbation wind acts to intensify the perturbation thickness field. This tendency is also illustrated by the zonally oriented **Q** vectors shown at the 500-mb level in the figure.

The pattern of vertical motion forced by the divergence of the **Q** vector as shown in Fig. 8.4 is associated with a divergence–convergence pattern that contributes a positive (negative) vorticity tendency near the 250-mb trough (ridge) and a negative (positive) vorticity tendency near the 750 mb ridge (trough). Since in all cases these vorticity tendencies tend to increase the extreme values of vorticity at the troughs and ridges, this secondary circulation system will act to increase the strength of the disturbance.

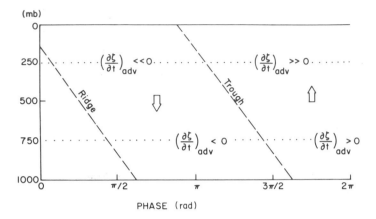

Fig. 8.5 Vertical cross section showing phase of vorticity change due to vorticity advection for an unstable baroclinic wave in the two-level model.

The total vorticity change at each level is, or course, determined by the sum of vorticity advection and vortex stretching due to the divergent circulation. The relative contributions of these processes are indicated schematically in Figs. 8.5 and 8.6, respectively. As can be seen in Fig. 8.5, vorticity advection leads the vorticity field by one-quarter wavelength. Since in this case the basic state wind increases with height, the vorticity advection at 250-mb is larger than that at 750 mb. If no other processes influenced the vorticity field, the effect of this differential vorticity advection would be to move the upper level trough and ridge pattern eastward more rapidly than

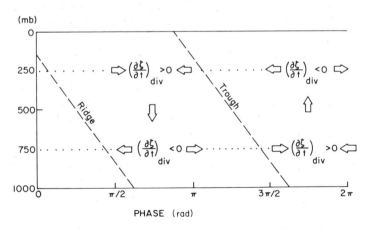

Fig. 8.6 Vertical cross section showing phase of vorticity change due to divergence–convergence for an unstable baroclinic wave in the two-level model.

the lower-level pattern. Thus, the westward tilt of the trough–ridge pattern would quickly be destroyed. The maintenance of this tilt in the presence of differential vorticity advection is due to the concentration of vorticity by vortex stretching associated with the divergent secondary circulation. Referring to Fig. 8.6, we see that concentration of vorticity by the divergence effect lags the vorticity field by about 65° at 250 mb and leads the vorticity field by about 65° at 750 mb. As a result, the net vorticity tendencies ahead of the vorticity maxima and minima are less than the advective tendencies at the upper level and greater than the advective tendencies at the lower level:

$$\left|\frac{\partial \zeta_1'}{\partial t}\right| < \left|\frac{\partial \zeta_1'}{\partial t}\right|_{adv}, \qquad \left|\frac{\partial \zeta_3'}{\partial t}\right| > \left|\frac{\partial \zeta_3'}{\partial t}\right|_{adv}$$

Furthermore, vorticity concentration by the divergence effect will tend to amplify the vorticity perturbations in the troughs and ridges at both the 250-mb and 750-mb levels, as is required for a growing disturbance.

8.3 The Energetics of Baroclinic Waves

In the previous section we showed that under suitable conditions a vertically sheared geostrophically balanced basic state flow is unstable to small wavelike perturbations with horizontal wavelengths in the range of observed synoptic-scale systems. Such baroclinically unstable perturbations will amplify exponentially by drawing energy from the mean flow. In this section we consider the energetics of linearized baroclinic disturbances and show that it is the potential energy of the mean flow that is the energy source for baroclinically unstable perturbations.

8.3.1 AVAILABLE POTENTIAL ENERGY

Before discussing the energetics of baroclinic waves, it is necessary to consider the energy of the atmosphere from a more general point of view. For all practical purposes, the total energy of the atmosphere is the sum of internal energy, gravitational potential energy, and kinetic energy. However, it is not necessary to consider separately the variations of internal and gravitational potential energy because in a hydrostatic atmosphere these two forms of energy are proportional and may be combined into a single term called the *total potential energy*. The proportionality of internal and gravitational potential energy can be demonstrated by considering these forms of energy for a column of air of unit horizontal cross section extending from the surface to the top of the atmosphere.

If we let dE_I be the internal energy in a vertical section of the column of height dz, then from the definition of internal energy [see (2.4)]

$$dE_I = \rho c_v T\, dz$$

so that the internal energy for the entire column is

$$E_I = c_v \int_0^\infty \rho T\, dz \qquad (8.30)$$

On the other hand, the gravitational potential energy for a slab of thickness dz at a height z is just

$$dE_P = \rho g z\, dz$$

so that the gravitational potential energy in the entire column is

$$E_P = \int_0^\infty \rho g z\, dz = -\int_{p_0}^0 z\, dp \qquad (8.31)$$

where we have substituted from the hydrostatic equation to obtain the last integral in (8.31). Integrating (8.31) by parts and using the ideal gas law, we obtain

$$E_P = \int_0^\infty p\, dz = R \int_0^\infty \rho T\, dz \qquad (8.32)$$

Comparing (8.30) and (8.32) we see that $c_v E_P = R E_I$. Thus, the total potential energy may be expressed as

$$E_P + E_I = (c_p/c_v)E_I = (c_p/R)E_P \qquad (8.33)$$

Therefore, in a hydrostatic atmosphere the total potential energy can be obtained by computing either E_I or E_P alone.

The total potential energy is not a very useful measure of energy in the atmosphere because only a very small fraction of the total potential energy is available for conversion to kinetic energy in storms. To demonstrate qualitatively why most of the total potential energy is unavailable we consider a simple model atmosphere that initially consists of two equal masses of dry air separated by a vertical partition as shown in Fig. 8.7. The two air masses are at uniform potential temperatures θ_1 and θ_2, respectively, with $\theta_1 < \theta_2$. The ground level pressure on each side of the partition is taken

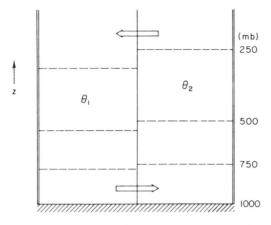

Fig. 8.7   Two air masses of differing potential temperature separated by a vertical partition. Dashed lines indicate isobaric surfaces. Arrows show direction of motion when partition is removed.

to be 1000-mb. We wish to compute the maximum kinetic energy that can be realized by an adiabatic rearrangement of mass within the same volume when the partition is removed.

Now for an adiabatic process, total energy is conserved:

$$E_K + E_P + E_I - \text{constant}$$

where E_K denotes the kinetic energy. If the air masses are initially at rest $E_K = 0$. Thus, if we let primed quantities denote the final state

$$E'_K + E'_P + E'_I = E_P + E_I$$

so that with the aid of (8.33) the kinetic energy realized by removal of the partition may be expressed as

$$E'_K = (c_p/c_v)(E_I - E'_I)$$

Since θ is conserved for an adiabatic process, the two air masses cannot mix. It is clear that E'_I will be a minimum (designated by E''_I) when the masses are rearranged so that the air at potential temperature θ_1 lies entirely beneath the air at potential temperature θ_2 with the 500-mb surface as the horizontal boundary between the two masses. In that case the total potential energy $(c_p/c_v)E''_I$ is not available for conversion to kinetic energy because no adiabatic process can further reduce E''_I.

The *available potential energy* (APE) can now be defined as the difference between the total potential energy of a closed system and the minimum total potential energy that could result from an adiabatic redistribution of mass. Thus, for the idealized model given above, the APE, which is designated by the symbol P, is

$$P = (c_p/c_v)(E_I - E_I'') \qquad (8.34)$$

which is equivalent to the maximum kinetic energy that can be realized by an adiabatic process.

Lorenz (1960) showed that available potential energy is given approximately by the volume integral over the entire atmosphere of the variance of potential temperature on isobaric surfaces. Thus, letting $\bar{\theta}$ designate the average potential temperature for a given pressure surface and θ' the local deviation from the average, the average available potential energy per unit volume satisfies the proportionality

$$\bar{P} \propto V^{-1} \int (\overline{\theta'^2}/\bar{\theta}^2) \, dV$$

where V designates the total volume. For the quasi-geostrophic model, this proportionality is an exact measure of the available potential energy, as we will show in the following subsection.

Observations indicate that for the atmosphere as a whole

$$\bar{P}/[(c_p/c_v)\bar{E}_I] \sim 5 \times 10^{-3}, \qquad \bar{K}/\bar{P} \sim 10^{-1}$$

Thus only about 0.5% of the total potential energy of the atmosphere is available, and of the available portion only about 10% is actually converted to kinetic energy. From this point of view the atmosphere is a rather inefficient heat engine.

8.3.2   ENERGY EQUATIONS FOR TWO-LAYER MODEL

In the two-layer model of Section 8.2, the perturbation temperature field is proportional to $\psi_1' - \psi_3'$, the 250–750-mb thickness. Thus, in view of the discussion in the previous section, we anticipate that the available potential energy in this case will be proportional to $(\psi_1' - \psi_3')^2$. To show that this in fact must be the case, we derive the energy equations for the system in the following manner: We first multiply (8.10) by $-\psi_1'$, (8.11) by $-\psi_3'$, and (8.12) by $(\psi_1' - \psi_3')$. We then integrate the resulting equations over one wavelength of the perturbation in the zonal direction. The resulting zonally averaged[2] terms will be denoted by overbars as done previously in

---

[2] A zonal average generally designates the average around an entire circle of latitude. However, for a disturbance consisting of a single sinusoidal wave of wave number $k = m/(a \cos \phi)$, where m is an integer, the average over a wavelength is identical to a zonal average.

Chapter 7:

$$(\overline{}) = L^{-1} \int_0^L (\ ) \, dx$$

where L is the wavelength of the perturbation. Thus, for the first term in (8.10) we have after multiplying by $-\psi_1'$, averaging, and differentiating by parts:

$$\overline{-\psi_1' \frac{\partial}{\partial t} \left(\frac{\partial^2 \psi_1'}{\partial x^2} \right)} = -\overline{\frac{\partial}{\partial x} \left[\psi_1' \frac{\partial}{\partial x} \left(\frac{\partial \psi_1'}{\partial t} \right) \right]} + \overline{\frac{\partial \psi_1'}{\partial x} \frac{\partial}{\partial t} \left(\frac{\partial \psi_1'}{\partial x} \right)}$$

The first term on the right-hand side vanishes because it is the integral of a perfect differential in x over a complete cycle. The second term on the right can be rewritten in the form

$$\frac{1}{2} \frac{\partial}{\partial t} \overline{\left(\frac{\partial \psi_1'}{\partial x} \right)^2}$$

which is just the rate of change of the perturbation kinetic energy per unit mass averaged over a wavelength. Similarly, $-\psi_1'$ times the advection term on the left in (8.10) can be written after integration in x as

$$-U_1 \overline{\psi_1' \frac{\partial^2}{\partial x^2} \left(\frac{\partial \psi_1'}{\partial x} \right)} = -U_1 \overline{\frac{\partial}{\partial x} \left[\psi_1' \frac{\partial}{\partial x} \left(\frac{\partial \psi_1'}{\partial x} \right) \right]}$$

$$+ U_1 \overline{\frac{\partial \psi_1'}{\partial x} \frac{\partial^2 \psi_1'}{\partial x^2}} = \frac{U_1}{2} \overline{\frac{\partial}{\partial x} \left(\frac{\partial \psi_1'}{\partial x} \right)^2} = 0$$

Thus, the advection of kinetic energy vanishes when integrated over a wavelength. Evaluating the various terms in (8.11) and (8.12) in the same manner after multiplying through by $-\psi_1'$ and $(\psi_1' - \psi_3')$, respectively, we obtain the following set of perturbation energy equations:

$$\frac{1}{2} \frac{\partial}{\partial t} \overline{\left(\frac{\partial \psi_1'}{\partial x} \right)^2} = -\frac{f_0}{\delta p} \overline{\omega_2' \psi_1'} \tag{8.35}$$

$$\frac{1}{2} \frac{\partial}{\partial t} \overline{\left(\frac{\partial \psi_3'}{\partial x} \right)^2} = +\frac{f_0}{\delta p} \overline{\omega_2' \psi_3'} \tag{8.36}$$

$$\frac{1}{2} \frac{\partial}{\partial t} \overline{(\psi_1' - \psi_3')^2} = U_T \overline{(\psi_1' - \psi_3') \frac{\partial}{\partial x} (\psi_1' + \psi_3')} + \frac{\sigma \delta p}{f_0} \overline{\omega_2'(\psi_1' - \psi_3')} \tag{8.37}$$

where as before $U_T \equiv (U_1 - U_3)/2$.

Defining the perturbation kinetic energy to be the sum of the kinetic energies of the 250- and 750-mb levels:

$$K' \equiv (1/2)[\overline{(\partial \psi'_1/\partial x)^2 + (\partial \psi'_3/\partial x)^2}]$$

we find by adding (8.35) and (8.36) that

$$dK'/dt = -(f_0/\delta p)\overline{\omega'_2(\psi'_1 - \psi'_3)} = -(2f_0/\delta p)\overline{\omega'_2 \psi_T} \tag{8.38}$$

Thus, the rate of change of perturbation kinetic energy is proportional to the correlation between the perturbation thickness and vertical motion. If we now define the perturbation available potential energy as

$$P' \equiv \lambda^2 \overline{(\psi'_1 - \psi'_3)^2}/2$$

we obtain from (8.37)

$$dP'/dt = \lambda^2 U_T \overline{(\psi'_1 - \psi'_3)\partial(\psi'_1 + \psi'_3)/\partial x} + (f_0/\delta p)\overline{\omega'_2(\psi'_1 - \psi'_3)}$$

$$= 4\lambda^2 U_T \overline{\psi_T \partial \psi_m/\partial x} + (2f_0/\delta p)\overline{\omega'_2 \psi_T} \tag{8.39}$$

The last term in (8.39) is just equal and opposite to the kinetic energy source term in (8.38). This term clearly must represent a conversion between potential and kinetic energy. If on the average the vertical motion is positive $(\omega'_2 < 0)$ where the thickness is greater than average $(\psi'_1 - \psi'_3 > 0)$ and vertical motion is negative where thickness is less than average, then

$$\overline{\omega'_2(\psi'_1 - \psi'_3)} = 2\overline{\omega'_2 \psi_T} < 0$$

and the perturbation potential energy is being converted to kinetic energy. Physically, this correlation represents an overturning in which cold air aloft is replaced by warm air from below, a situation that clearly tends to lower the center of mass and hence the potential energy of the perturbation. However, the available potential energy and kinetic energy of a disturbance can still grow simultaneously, provided that the potential energy generation-owing to the first term in (8.39) exceeds the rate of potential energy conversion to kinetic energy.

The potential energy generation term in (8.39) depends on the correlation between the perturbation thickness ψ_T and the meridional velocity at 500-mb, $\partial \psi_m/\partial x$. In order to understand the role of this term, it is helpful to consider a particular sinusoidal wave disturbance. Suppose that the barotropic and baroclinic parts of the disturbance can be written respectively as

$$\psi_m = A_m \cos k(x - ct) \quad \text{and} \quad \psi_T = A_T \cos k(x + x_0 - ct) \tag{8.40}$$

where x_0 designates the phase difference. Since ψ_m is proportional to the 500-mb geopotential and ψ_T is proportional to the 500-mb temperature (or 250–750 mb thickness), the phase angle kx_0 gives the phase difference between the geopotential and temperature fields at 500 mb. Furthermore, A_m and A_T are measures of the amplitudes of the 500-mb disturbance geopotential and thickness fields, respectively. Using the expressions in (8.40) we obtain

$$
\overline{\psi_T \frac{\partial \psi_m}{\partial x}} = -\frac{k}{L} \int_0^L A_T A_m \cos k(x + x_0 - ct) \sin k(x - ct)\, dx
$$

$$
= \frac{kA_T A_m \sin kx_0}{L} \int_0^L [\sin k(x - ct)]^2\, dx \tag{8.41}
$$

$$
= (A_T A_m k \sin kx_0)/2
$$

Substituting from (8.41) into (8.39) we see that for the usual midlatitude case of a westerly thermal wind ($U_T > 0$) the correlation in (8.41) must be positive if the perturbation potential energy is to increase. Thus, kx_0 must satisfy the inequality $0 < kx_0 < \pi$. Furthermore, the correlation will be a positive maximum for $kx_0 = \pi/2$, that is, when the temperature wave lags the geopotential wave by 90° in phase at 500-mb.

This case is shown schematically in Fig. 8.4. Clearly, when the temperature wave lags the geopotential by one-quarter cycle, the northward advection of warm air by the geostrophic wind east of the 500-mb trough and the southward advection of cold air west of the 500-mb trough are both maximized. As a result, cold advection is strong below the 250-mb trough and warm advection is strong below the 250-mb ridge. In that case, as discussed previously in Section 6.3.1, the upper-level disturbance will intensify. It should also be noted here that if the temperature wave lags the geopotential wave, the trough and ridge axes will tilt westward with height, which, as mentioned in Section 6.1, is observed to be the case for amplifying mid-latitude synoptic systems.

Referring again to Fig. 8.4 and recalling the vertical motion pattern implied by the omega equation (8.29), we see that the signs of the two terms on the right in (8.39) cannot be the same. In the westward-tilting perturbation of Fig. 8.4, the vertical motion must be downward in the cold air behind the trough at 500-mb. Hence, the correlation between temperature and vertical velocity must be positive in this situation; that is,

$$
\overline{\omega_2' \psi_T} < 0
$$

Thus, for quasi-geostrophic perturbations, a westward tilt of the perturbation with height implies both that the horizontal temperature advection will increase the available potential energy of the perturbation and that the vertical circulation will convert perturbation available potential energy to perturbation kinetic energy. Conversely, an eastward tilt of the system with height would change the signs of both terms on the right in (8.39).

Although the signs of the potential energy generation term and the potential energy conversion term in (8.39) are always opposite for a developing baroclinic wave, it is only the potential energy generation rate that determines the growth of the total energy $P' + K'$ of the disturbance. This may be proved by adding (8.38) and (8.39) to obtain

$$d(P' + K')/dt = 4\lambda^2 U_T \overline{\psi_T \partial \psi_m / \partial x}$$

Thus, provided the correlation between the meridional velocity and temperature is positive and $U_T > 0$, the total energy of the perturbation will increase. Note that the vertical circulation merely converts disturbance energy between the available potential and kinetic forms without affecting the total energy of the perturbation. The rate of increase of the total energy of the perturbation depends on the magnitude of the basic state thermal wind U_T. This is, of course, proportional to the zonally averaged meridional temperature gradient. Since the generation of perturbation energy requires systematic poleward transport of warm air and equatorward transport of cold air, it is clear that baroclinically unstable disturbances tend to reduce the meridional temperature gradient and hence the available potential energy of the mean flow. This latter process cannot be mathematically described in terms of the linearized equations. However, from Fig. 8.8 we can see qualitatively that parcels that move poleward (equatorward) and upward (downward) with slopes less than the slope of the zonal mean potential temperature surface will become warmer (colder) than their surroundings. For such parcels the correlations between disturbance meridional velocity and temperature and between disturbance vertical velocity and temperature will both be positive as required for baroclinically unstable disturbances. For parcels that have trajectory slopes greater than the mean potential temperature slope, however, both of these correlations will be negative. Such parcels must then convert disturbance kinetic energy to disturbance available potential energy, which is in turn converted to zonal mean available potential energy. Therefore, in order that perturbations be able to extract potential energy from the mean flow the perturbation parcel trajectories in the meridional plane must have slopes less than the slopes of the potential temperature surfaces, and a permanent rearrangement of air must take place for there to be a net heat transfer. Since we have

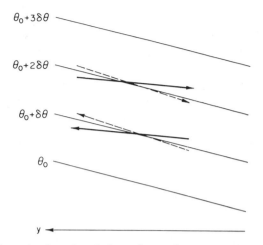

Fig. 8.8   Slopes of parcel trajectories relative to the zonal mean potential temperature surfaces for a baroclinically unstable disturbance (solid arrows) and for a baroclinically stable disturbance (dashed arrows).

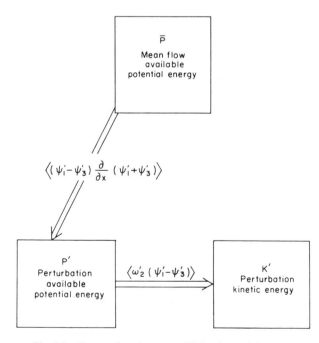

Fig. 8.9   Energy flow in an amplifying baroclinic wave.

previously seen that poleward-moving air must rise and equatorward-moving air must sink, it is clear that the rate of energy generation can be greater for an atmosphere in which the meridional slope of the potential temperature surfaces is large. We can also see more clearly why baroclinic instability has a short-wave cutoff. As previously mentioned, the intensity of the vertical circulation must increase as the wavelength decreases. Thus, the slopes of the parcel trajectories must increase with decreasing wavelength, and for some critical wavelength the trajectory slopes will become greater than the slopes of the potential temperature surfaces. Unlike convective instability, where the most rapid amplification occurs for the smallest possible scales, baroclinic instability is most effective at an intermediate range of scales.

The energy flow for quasi-geostrophic perturbations is summarized in Fig. 8.9 by means of a block diagram. In this type of energy diagram each block represents a reservoir of a particular type of energy and the arrows designate the direction of energy flow. The complete energy cycle cannot be derived in terms of linear perturbation theory but will be discussed qualitatively in Chapter 10.

8.4 Baroclinic Instability of a Continuously Stratified Atmosphere

In the two previous sections some basic aspects of baroclinic instability were elucidated in the context of a simple two-layer model. The dependence of growth rate on vertical shear and the existence of a short-wave cutoff were clearly demonstrated. The two-layer model, however, does have one severe constraint; it assumes that the altitude dependence of large-scale systems can be adequately represented with only two degrees of freedom in the vertical (i.e., the streamfunctions at the 250- and 750-mb levels). Although most synoptic-scale systems in midlatitudes are observed to have vertical scales comparable to the depth of the troposphere, observed vertical structures do differ. Disturbances that are concentrated near the ground or near the tropopause can hardly be accurately represented in the two-layer model.

An analysis of the structure of baroclinic modes for realistic mean zonal wind profiles is quite complex and indeed can only be done by numerical methods. However, without obtaining specific normal mode solutions, it is possible to obtain necessary conditions for baroclinic or barotropic instability from an integral theorem first developed by Rayleigh. This theorem, which is discussed in Section 8.4.2, also shows how baroclinic instability is intimately related to the mean meridional gradient of potential vorticity and the mean meridional temperature gradient at the surface.

If a number of simplifying assumptions are made it is possible to pose the stability problem for a continuously stratified atmosphere in a fashion that leads to a second-order differential equation for the vertical structure that can be solved by standard methods. This problem was originally studied by the British meteorologist Eady (1949) and, although mathematically similar to the two-layer model, provides additional insights. It will be developed in Section 8.4.3.

8.4.1 LOG-PRESSURE COORDINATES

Derivation of the Rayleigh theorem and the Eady stability model is facilitated if we transform from the standard isobaric coordinates to a vertical coordinate based on the logarithm of pressure. In the *log-pressure coordinates*, the vertical coordinate is defined as

$$z^* \equiv -H \ln (p/p_s) \tag{8.42}$$

where p_s is a standard reference pressure (usually taken to be 1000 mb) and H is a standard scale height, $H \equiv RT_s/g$, with T_s a global average temperature. For the special case of an isothermal atmosphere at temperature T_s, z^* is exactly equal to geometric height, and the density profile is given by the reference density

$$\rho_0(z^*) = \rho_s \exp (-z^*/H)$$

where ρ_s is the density at $z^* = 0$.

For an atmosphere with a realistic temperature profile, z^* is only approximately equivalent to the height, but in the troposphere the difference is usually quite small.

The vertical velocity in this coordinate system is

$$w^* \equiv Dz^*/dt$$

The horizontal momentum equation in the log-pressure system is the same as that in the isobaric system:

$$DV/Dt + f\mathbf{k} \times \mathbf{V} = -\nabla \Phi \tag{8.43}$$

However, the operator D/Dt is now defined as

$$D/Dt \equiv \partial/\partial t + \mathbf{V} \cdot \nabla + w^* \, \partial/\partial z^*$$

The hydrostatic equation $\partial\Phi/\partial p = -\alpha$ can be transformed to the log-pressure system by multiplying through by p and using the ideal gas law to get

$$\partial\Phi/\partial \ln p = -RT$$

which upon dividing through by $-H$ and using (8.42) gives

$$\partial\Phi/\partial z^* = +RT/H \tag{8.44}$$

The log-pressure form of the continuity equation can be obtained by transforming from the isobaric coordinate form (3.5). We first note that

$$w^* \equiv -(H/p)Dp/Dt = -H\omega/p$$

so that

$$\frac{\partial\omega}{\partial p} = -\frac{\partial}{\partial p}\left(\frac{pw^*}{H}\right) = \frac{\partial w^*}{\partial z^*} - \frac{w^*}{H} = \frac{1}{\rho_0}\frac{\partial(\rho_0 w^*)}{\partial z^*}$$

Thus, in log-pressure coordinates the continuity equation becomes simply

$$\frac{\partial u}{\partial x} + \frac{\partial v}{\partial y} + \frac{1}{\rho_0}\frac{\partial(\rho_0 w^*)}{\partial z^*} = 0 \tag{8.45}$$

It is left as a problem for the reader to show that the first law of thermodynamics (3.6) can be expressed in log-pressure form as

$$\left(\frac{\partial}{\partial t} + \mathbf{V}\cdot\nabla\right)\frac{\partial\Phi}{\partial z^*} + w^* N^2 = \frac{\kappa J}{H} \tag{8.46}$$

where

$$N^2 \equiv (R/H)(\partial T/\partial z^* + \kappa T/H)$$

is the buoyancy frequency squared (see Section 2.7.3) and $\kappa \equiv R/c_p$. Unlike the static stability parameter in the isobaric form of the thermodynamic equation (3.6), the parameter N^2 varies only weakly with height in the troposphere; it can be assumed to be constant without serious error. This is a major advantage of the log-pressure formulation.

The quasi-geostrophic potential vorticity equation (6.25) has the same form as in the isobaric system, but with q defined as

$$q \equiv \nabla^2\psi + f + \frac{1}{\rho_0}\frac{\partial}{\partial z^*}\left(\varepsilon\rho_0\frac{\partial\psi'}{\partial z^*}\right) \tag{8.47}$$

where $\varepsilon \equiv f_0^2/N^2$.

8.4.2 BAROCLINIC INSTABILITY: THE RAYLEIGH THEOREM

We now examine the stability problem for a continuously stratified atmosphere on the midlatitude β plane. The linearized form of the quasi-geostrophic potential vorticity equation (6.25) can be expressed in log-pressure coordinates as

$$\left(\frac{\partial}{\partial t} + \bar{u}\frac{\partial}{\partial x}\right)q' + \frac{\partial \bar{q}}{\partial y}\frac{\partial \psi'}{\partial x} = 0 \tag{8.48}$$

where

$$q' \equiv \nabla^2 \psi' + \frac{1}{\rho_0}\frac{\partial}{\partial z^*}\left(\varepsilon\rho_0\frac{\partial \psi'}{\partial z^*}\right) \tag{8.49}$$

and

$$\frac{\partial \bar{q}}{\partial y} = \beta - \frac{\partial^2 \bar{u}}{\partial y^2} - \frac{1}{\rho_0}\frac{\partial}{\partial z^*}\left(\varepsilon\rho_0\frac{\partial \bar{u}}{\partial z^*}\right) \tag{8.50}$$

As in the two-layer model, boundary conditions are required at lower and upper boundary pressure surfaces. Assuming that the vertical motion w^* vanishes at these boundaries, the linearized form of the thermodynamic energy equation (8.3) valid at horizontal boundary surfaces is simply

$$\left(\frac{\partial}{\partial t} + \bar{u}\frac{\partial}{\partial x}\right)\frac{\partial \psi'}{\partial z^*} - \frac{\partial \psi'}{\partial x}\frac{\partial \bar{u}}{\partial z^*} = 0 \tag{8.51}$$

The sidewall boundary conditions are

$$\partial \psi'/\partial x = 0, \quad \text{hence } \psi' = 0 \quad \text{at} \quad y = \pm L \tag{8.52}$$

We now assume that the perturbation consists of a single zonal Fourier component propagating in the x direction:

$$\psi'(x, y, z, t) = \text{Re}\{\Psi(y, z)\exp[ik(x - ct)]\} \tag{8.53}$$

where $\Psi(y, z) = \Psi_r + \Psi_i$ is a complex amplitude, k is the zonal wave number, and $c = c_r + c_i$ is a complex phase speed. Note that (8.53) can alternatively be expressed as

$$\psi'(x, y, z, t) = e^{kc_i t}[\Psi_r \cos k(x - c_r t) - \Psi_i \sin k(x - c_r t)]$$

Thus, the relative magnitudes of Ψ_r and Ψ_i determine the phase of the wave for any y, z^*.

Substituting from (8.53) into (8.48) and (8.51) yields

$$(\bar{u} - c)\left[\frac{\partial^2 \Psi}{\partial y^2} - k^2 \Psi + \frac{1}{\rho_0}\frac{\partial}{\partial z^*}\left(\varepsilon\rho_0\frac{\partial \Psi}{\partial z^*}\right)\right] + \frac{\partial \bar{q}}{\partial y}\Psi = 0 \qquad (8.54)$$

and

$$(\bar{u} - c)\frac{\partial \Psi}{\partial z^*} - \frac{\partial \bar{u}}{\partial z^*}\Psi = 0 \qquad \text{at } z^* = 0 \qquad (8.55)$$

If the upper boundary is taken to be a rigid lid at a finite height, as sometimes done in theoretical studies, the condition (8.55) is appropriate at that boundary as well. Alternatively, the upper boundary condition can be specified by requiring Ψ to remain finite as $z^* \to \infty$.

Equation (8.54) together with its boundary conditions constitutes a linear boundary value problem for $\Psi(y, z^*)$. However, it is generally not simple to obtain solutions for realistic mean zonal wind profiles. Nevertheless, we can obtain some useful information on stability properties simply by analyzing the energetics of the system.

Dividing (8.54) by $(\bar{u} - c)$ and separating the resulting equation into real and imaginary parts, we obtain

$$\frac{\partial^2 \Psi_r}{\partial y^2} + \frac{1}{\rho_0}\frac{\partial}{\partial z^*}\left(\varepsilon\rho_0\frac{\partial \Psi_r}{\partial z^*}\right) - [k^2 - \delta_r(\partial\bar{q}/\partial y)]\Psi_r - \delta_i\,\partial\bar{q}/\partial y\,\Psi_i = 0 \quad (8.56)$$

$$\frac{\partial^2 \Psi_i}{\partial y^2} + \frac{1}{\rho_0}\frac{\partial}{\partial z^*}\left(\varepsilon\rho_0\frac{\partial \Psi_i}{\partial z^*}\right) - [k^2 - \delta_r(\partial\bar{q}/\partial y)]\Psi_i + \delta_i\,\partial\bar{q}/\partial y\,\Psi_r = 0 \quad (8.57)$$

where

$$\delta_r = \frac{\bar{u} - c_r}{(\bar{u} - c_r)^2 + c_i^2} \quad \text{and} \quad \delta_i = \frac{c_i}{(\bar{u} - c_r)^2 + c_i^2}$$

Similarly, dividing (8.55) through by $(\bar{u} - c)$ and separating into real and imaginary parts gives for the boundary condition at $z^* = 0$:

$$\frac{\partial \Psi_r}{\partial z^*} + \frac{\partial \bar{u}}{\partial z^*}(\delta_i\Psi_i - \delta_r\Psi_r) = 0, \qquad \frac{\partial \Psi_i}{\partial z^*} - \frac{\partial \bar{u}}{\partial z^*}(\delta_r\Psi_i + \delta_i\Psi_r) = 0 \qquad (8.58)$$

Multiplying (8.56) by Ψ_i, (8.57) by Ψ_r and subtracting the latter from the former yields

$$\rho_0\left[\Psi_i\frac{\partial^2\Psi_r}{\partial y^2}-\Psi_r\frac{\partial^2\Psi_i}{\partial y^2}\right]+\left[\Psi_i\frac{\partial}{\partial z^*}\left(\varepsilon\rho_0\frac{\partial\Psi_r}{\partial z^*}\right)-\Psi_r\frac{\partial}{\partial z^*}\left(\varepsilon\rho_0\frac{\partial\Psi_i}{\partial z^*}\right)\right]$$

$$-\rho_0\delta_i\partial\bar{q}/\partial y(\Psi_i^2+\Psi_r^2)=0 \qquad (8.59)$$

Using the chain rule of differentiation, (8.59) can be expressed in the form

$$\rho_0\frac{\partial}{\partial y}\left[\Psi_i\frac{\partial\Psi_r}{\partial y}-\Psi_r\frac{\partial\Psi_i}{\partial y}\right]+\frac{\partial}{\partial z^*}\left[\varepsilon\rho_0\left(\Psi_i\frac{\partial\Psi_r}{\partial z^*}-\Psi_r\frac{\partial\Psi_i}{\partial z^*}\right)\right]$$

$$-\rho_0\delta_i\,\partial\bar{q}/\partial y\,(\Psi_i^2+\Psi_r^2)=0 \qquad (8.60)$$

The first term in brackets in (8.60) is a perfect differential in y; the second term is a perfect differential in z^*. Thus, if (8.60) is integrated over the y, z^* domain the result can be expressed as

$$\int_0^\infty\left[\Psi_i\frac{\partial\Psi_r}{\partial y}-\Psi_r\frac{\partial\Psi_i}{\partial y}\right]_{-L}^{+L}\rho_0\,dz^*+\int_{-L}^{+L}\left[\varepsilon\rho_0\left(\Psi_i\frac{\partial\Psi_r}{\partial z^*}-\Psi_r\frac{\partial\Psi_i}{\partial z^*}\right)\right]_0^\infty dy$$

$$=\int_{-L}^{+L}\int_0^\infty\rho_0\delta_i\frac{\partial\bar{q}}{\partial y}(\Psi_i^2+\Psi_r^2)\,dy\,dz^* \qquad (8.61)$$

But from (8.52) $\Psi_i=\Psi_r=0$ at $y=\pm L$ so that the first integral in (8.61) vanishes. Furthermore, if Ψ remains finite as $z^*\to\infty$ the contribution to the second integral of (8.61) at the upper boundary vanishes. If we then use (8.58) to eliminate the vertical derivatives in this term at the lower boundary, (8.61) can be expressed as

$$c_i\left[\int_{-L}^{+L}\int_0^\infty\frac{\partial\bar{q}}{\partial y}\frac{\rho_0|\Psi|^2}{|\bar{u}-c|^2}\,dy\,dz^*-\int_{-L}^{+L}\varepsilon\frac{\partial\bar{u}}{\partial z^*}\frac{\rho_0|\Psi|^2}{|\bar{u}-c|^2}\bigg|_{z^*=0}dy\right]=0$$

$$(8.62)$$

where $|\Psi|^2=\Psi_r^2+\Psi_i^2$ is the disturbance amplitude squared.

Equation (8.62) has important implications for the stability of quasi-geostrophic perturbations. For unstable modes c_i must be nonzero, and thus the quantity in square brackets in (8.62) must vanish. Since $|\Psi|^2/|\bar{u}-c|^2$ is nonnegative, instability is possible only when $\partial \bar{u}/\partial z^*$ at the lower boundary and $\partial \bar{q}/\partial y$ in the whole domain satisfy certain constraints:

(a) If $\partial \bar{u}/\partial z^* = 0$ at $z^* = 0$ (which by thermal wind balance implies that the meridional temperature gradient vanishes at the boundary), the second integral in (8.62) vanishes. Thus, the first integral must also vanish for instability to occur. This can occur only if $\partial \bar{q}/\partial y$ changes sign within the domain (i.e., $\partial \bar{q}/\partial y = 0$ somewhere). This is referred to as the *Rayleigh necessary condition* and is another demonstration of the fundamental role played by potential vorticity. Since $\partial \bar{q}/\partial y$ is normally positive, it is clear that, in the absence of temperature gradients at the lower boundary, a region of negative meridional potential vorticity gradients must exist in the interior for instability to be possible.

(b) If $\partial \bar{q}/\partial y \geq 0$ everywhere, it is necessary that $\partial \bar{u}/\partial z^* > 0$ somewhere at the lower boundary for $c_i > 0$.

(c) If $\partial \bar{u}/\partial z^* < 0$ everywhere at $z^* = 0$, it is necessary that $\partial \bar{q}/\partial y < 0$ somewhere for instability to occur. Thus, there is an asymmetry between westerly and easterly shear at the lower boundary, with the former more favorable for baroclinic instability.

The basic state potential vorticity gradient (8.50) can be written in the form

$$\frac{\partial \bar{q}}{\partial y} = \beta - \frac{\partial^2 \bar{u}}{\partial y^2} + \frac{\varepsilon}{H}\frac{\partial \bar{u}}{\partial z^*} - \varepsilon \frac{\partial^2 \bar{u}}{\partial z^{*2}} - \frac{\partial \varepsilon}{\partial z^*}\frac{\partial \bar{u}}{\partial z^*}$$

Thus, since β is positive everywhere, if ε is constant a negative basic state potential vorticity gradient can occur only for strong positive mean flow curvature (i.e., $\partial^2 \bar{u}/\partial y^2$ or $\partial^2 \bar{u}/\partial z^{*2} \gg 0$) or strong negative vertical shear ($\partial \bar{u}/\partial z^* \ll 0$). Strong positive meridional curvature can occur at the core of an easterly jet or on the flanks of a westerly jet. Instability associated with such horizontal curvature is referred to as *barotropic instability*. The normal baroclinic instability in midlatitudes is associated with mean flows in which $\partial \bar{q}/\partial y > 0$ and $\partial \bar{u}/\partial z^* > 0$ at the ground. Hence, a mean meridional temperature gradient at the ground is essential for the existence of such instability. Baroclinic instability can also be excited at the tropopause owing to the rapid decrease of ε with height if there is a sufficiently strong easterly mean wind shear to cause a local reversal in the mean potential vorticity gradient.

8.4.3 THE EADY STABILITY PROBLEM

In this section we analyze the structures (eigenfunctions) and growth rates (eigenvalues) for unstable modes in the simplest possible model that satisfies the necessary conditions for instability in a continuous atmosphere given in the previous subsection. For simplicity we make the following assumptions:

(i) Basic state density constant (Boussinesq approximation).
(ii) f-plane geometry ($\beta = 0$).
(iii) $\partial \bar{u}/\partial z^* = \Lambda = $ constant.
(iv) Rigid lids at $z^* = 0$ and H.

These conditions are only a crude model of the atmosphere, but they provide a first approximation for study of the dependence of vertical structure on horizontal scale and stability. Despite the zero mean potential vorticity in the domain, the Eady model satisfies the necessary conditions for instability discussed in the previous subsection because vertical shear of the basic state mean flow at the upper boundary provides an additional term in (8.62) that is equal and opposite to the lower boundary integral.

Using the above approximations, the quasi-geostrophic potential vorticity equation and thermodynamic energy equation are

$$\left(\frac{\partial}{\partial t} + \bar{u}\frac{\partial}{\partial x}\right)\left(\nabla^2 \psi' + \varepsilon \frac{\partial^2 \psi'}{\partial z^{*2}}\right) = 0 \qquad (8.63)$$

and

$$\left(\frac{\partial}{\partial t} + \bar{u}\frac{\partial}{\partial x}\right)\frac{\partial \psi'}{\partial z^*} - \frac{\partial \psi'}{\partial x}\frac{\partial \bar{u}}{\partial z^*} = 0 \qquad (8.64)$$

respectively, where again $\varepsilon \equiv f_0^2/N^2$.

Letting

$$\psi'(x, y, z^*, t) = \Psi(z^*)\cos ly \exp[ik(x - ct)]$$
$$\bar{u}(z^*) = \Lambda z^* \qquad (8.65)$$

where as in the previous subsection $\Psi(z^*)$ is a complex amplitude and c a complex phase speed, and substituting from (8.65) into (8.63), we find that the vertical structure is given by the solution of the standard second-order differential equation

$$d^2\Psi/dz^{*2} - \alpha^2\Psi = 0 \qquad (8.66)$$

where $\alpha^2 = (k^2 + l^2)/\varepsilon$. A similar substitution into (8.64) yields the boundary condition

$$(\Lambda z^* - c)\,d\Psi/dz^* - \Psi\Lambda = 0 \qquad \text{at } z^* = 0, H \qquad (8.67)$$

valid for rigid horizontal boundaries at the surface ($z^* = 0$) and the tropopause ($z^* = H$).

The general solution of (8.66) can be written in the form

$$\Psi(z^*) = A \sinh \alpha z^* + B \cosh \alpha z^* \qquad (8.68)$$

Substituting from (8.68) into the boundary conditions (8.67) for $z^* = 0$ and H yields a set of two linear homogeneous equations in the amplitude coefficients A and B:

$$-c\alpha A - B\Lambda = 0$$

$$\alpha(\Lambda H - c)(A \cosh \alpha H + B \sinh \alpha H) - \Lambda(A \sinh \alpha H + B \cosh \alpha H) = 0$$

As in the two-layer model, a nontrivial solution exists only if the determinant of the coefficients of A and B vanishes. Again, this leads to a quadratic equation in the phase speed c. The solution (see Problem 8.12) has the form

$$c = \frac{\Lambda H}{2} \pm \frac{\Lambda H}{2}\left[1 - \frac{4 \cosh \alpha H}{\alpha H \sinh \alpha H} + \frac{4}{\alpha^2 H^2}\right]^{1/2} \qquad (8.69)$$

Thus

$$c_i \neq 0 \quad \text{if} \quad 1 - \frac{4 \cosh \alpha H}{\alpha H \sinh \alpha H} + \frac{4}{\alpha^2 H^2} < 0$$

and the flow is then baroclinically unstable. When the quantity in square brackets in (8.69) is equal to zero, the flow is said to be *neutrally stable*. This condition occurs for $\alpha = \alpha_c$ where

$$\alpha_c^2 H^2/4 - \alpha_c H(\tanh \alpha_c H)^{-1} + 1 = 0 \qquad (8.70)$$

Using the identity

$$\tanh \alpha_c H = \frac{2 \tanh(\alpha_c H/2)}{1 + \tanh^2(\alpha_c H/2)}$$

we can factor (8.70) to yield

$$\left[\frac{\alpha_c H}{2} - \tanh\left(\frac{\alpha_c H}{2}\right)\right]\left[\frac{\alpha_c H}{2} - \coth\left(\frac{\alpha_c H}{2}\right)\right] = 0 \qquad (8.71)$$

Thus, the critical value of α is given by $\alpha_c H/2 = \coth(\alpha_c H/2)$, which implies $\alpha_c H \cong 2.4$. Hence instability requires $\alpha < \alpha_c$, or

$$(k^2 + l^2) < (\alpha_c^2 f_0^2 / N^2) \approx 5.76 / L_R^2$$

where $L_R \equiv NH/f_0 \approx 1000$ km is the Rossby *radius of deformation* for a continuously stratified fluid [compare λ defined just below (8.17)]. For waves with equal zonal and meridional wave numbers ($k = l$) the wavelength of maximum growth rate turns out to be

$$L_m = 2\sqrt{2}\,\pi L_R / (H\alpha_m) \cong 5500 \text{ km}$$

where α_m is the value of α for which kc_i is a maximum.

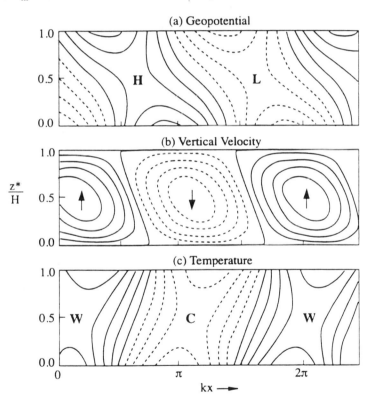

Fig. 8.10 Properties of the most unstable Eady wave. (a) Contours of perturbation geopotential height; H and L designate ridge and trough axes, respectively. (b) Contours of vertical velocity; up and down arrows designate axes of maximum upward and downward motion, respectively. (c) Contours of perturbation temperature; W and C designate axes of warmest and coldest temperatures, respectively. In all panels 1 and 1/4 wavelengths are shown for clarity.

Substituting this value of α into the solution for the vertical structure of the streamfunction (8.68) and using the lower boundary condition to express the coefficient B in terms of A, we can determine the vertical structure of the most unstable mode. As shown in Fig. 8.10, trough and ridge axes slope westward with height, in agreement with the requirements for extraction of available potential energy from the mean flow. The axes of the warmest and coldest air, however, tilt eastward with height, a result that could not be determined from the two-layer model where temperature was given at a single level. Furthermore, panels (a) and (b) of Fig. 8.10 show that east of the upper-level trough axis, where the perturbation meridional velocity is positive, the vertical velocity is also positive. Thus, parcel motion is poleward and upward in the region where $\theta' > 0$. Conversely, west of the upper level trough axis parcel motion is equatorward and downward where $\theta' < 0$. Both cases are thus consistent with the energy-converting parcel trajectory slopes shown in Fig. 8.8.

Problems

8.1. Show using Eq. (8.26) that the maximum growth rate for baroclinic instability when $\beta = 0$ occurs for

$$k^2 = 2\lambda^2(\sqrt{2} - 1)$$

How long does it take the most rapidly growing wave to amplify by a factor of e^1 if $\lambda = 2 \times 10^{-6} \, m^{-1}$ and $U_T = 20 \, m \, s^{-1}$?

8.2. Solve for ψ_3' and ω_2' in terms of ψ_1' for a baroclinic Rossby wave whose phase speed satisfies (8.24). Explain the phase relationship between ψ_1', ψ_3', and ω_2' in terms of the quasi-geostrophic theory. (Note that $U_T = 0$ in this case.)

8.3. For the case $U_1 = -U_3$ and $k^2 = \lambda^2$ solve for ψ_3' and ω_2' in terms of ψ_1' for marginally stable waves [i.e., $\delta = 0$ in (8.22)].

8.4. For the case $\beta = 0$, $k^2 = \lambda^2$, and $U_m = U_T$ solve for ψ_3' and ω_2' in terms of ψ_1'. Explain the phase relationships between ω_2', ψ_1', and ψ_3' in terms of the energetics of quasi-geostrophic waves for the amplifying wave.

8.5. Suppose that a baroclinic fluid is confined between two rigid horizontal lids in a rotating tank in which $\beta = 0$ but friction is presented in the form of linear drag proportional to the velocity (i.e., $\mathbf{Fr} = -\mu \mathbf{V}$). Show that the two-level model perturbation vorticity equations in Cartesian coordinates can be written as

$$\left(\frac{\partial}{\partial t} + U_1 \frac{\partial}{\partial x} + \mu\right)\frac{\partial^2 \psi_1'}{\partial x^2} - \frac{f}{\delta p}\omega_2' = 0$$

$$\left(\frac{\partial}{\partial t} + U_3 \frac{\partial}{\partial x} + \mu\right)\frac{\partial^2 \psi_3'}{\partial x^2} + \frac{f}{\delta p}\omega_2' = 0$$

where perturbations are assumed in the form given in (8.9). Assuming solutions of the form (8.15), show that the phase speed satisfies a relationship similar to (8.22) with β replaced everywhere by $i\mu k$ and that as a result the condition for baroclinic instability becomes

$$U_T > \mu(2\lambda^2 - k^2)^{-1/2}$$

8.6. For the case $\beta = 0$ determine the phase difference between the 250-mb and 750-mb geopotential fields for the most unstable baroclinic wave (see Problem 8.1). Show that the 500-mb geopotential and thickness fields are 90° out of phase.

8.7. For the conditions of Problem 8.6, given that the amplitude of ψ_m is $A = 10^7 \, \text{m}^2 \, \text{s}^{-1}$ solve the system (8.19)–(8.20) to obtain B. Let $\lambda^2 = 2 \times 10^{-12} \, \text{m}^{-2}$ and $U_T = 15 \, \text{m s}^{-1}$.

8.8. For the situation of Problem 8.7 compute ω_2' using expression (8.29).

8.9. Compute the total potential energy per unit cross-sectional area for an atmosphere with an adiabatic lapse rate given that the temperature and pressure at the ground are $p = 10^5 \, \text{Pa}$ and $T = 300 \, \text{K}$, respectively.

8.10. Consider two air masses at the uniform potential temperatures $\theta_1 = 320 \, \text{K}$ and $\theta_2 = 340 \, \text{K}$ which are separated by a vertical partition as shown in Fig. 8.5. Each air mass occupies a horizontal area of $10^4 \, \text{m}^2$ and extends from the surface ($p_0 = 10^5 \, \text{Pa}$) to the top of the atmosphere. What is the available potential energy for this system? What fraction of the total potential energy is available in this case?

8.11. For the unstable baroclinic wave which satisfies the conditions given in Problems 8.7 and 8.8 compute the energy conversion terms in (8.39) and (8.40) and hence obtain the instantaneous rate of change of the perturbation kinetic and available potential energies.

8.12. Starting with (8.63) and (8.64), derive the phase speed c for the Eady wave given in (8.69).

8.13. Unstable baroclinic waves play an important role in the global heat budget by transferring heat poleward. Show that for the Eady wave solution the poleward heat flux averaged over a wavelength,

$$\overline{v'T'} = \frac{1}{L} \int_0^L v'T' \, dx$$

is independent of height and is positive for a growing wave. How does the magnitude of the heat flux at a given instant change if the mean wind shear is doubled?

Suggested References

Charney (1947) is the classic paper on baroclinic instability. The mathematical level is advanced, but Charney's paper contains an excellent qualitative summary of the main results which is very readable.

Hoskins, McIntyre, and Robertson (1985) discuss cyclogenesis and baroclinic instability from a potential vorticity perspective.

Pedlosky, *Geophysical Fluid Dynamics* (2nd ed.), contains a thorough treatment of baroclinic instability at a mathematically advanced level.

Chapter

9 Mesoscale Circulations

The previous chapters have focused primarily on the dynamics of synoptic and planetary-scale circulations. Such large-scale motions are strongly influenced by the rotation of the earth so that outside the equatorial zone the Coriolis force dominates over inertia (i.e., the Rossby number is small). To a first approximation, as we saw in Chapter 6, they can be modeled by quasi-geostrophic theory.

The study of quasi-geostrophic motions has been a central theme of dynamic meteorology for many years. Not all important circulations fit into the quasi-geostrophic classification, however. Some have Rossby numbers of order unity, and some are hardly at all influenced by the rotation of the earth. Such circulations include a wide variety of phenomena. They all, however, are characterized by horizontal scales that are smaller than the synoptic scale (i.e., the *macroscale* of motion) but larger than the scale of an individual fair weather cumulus cloud (i.e., the *microscale*). Hence, they can be conveniently classified as *mesoscale* circulations. Most severe weather is associated with mesoscale motion systems. Thus, understanding of the mesoscale is of both scientific and practical importance.

9.1 Energy Sources for Mesoscale Circulations

Mesoscale dynamics is generally defined to include the study of motion systems that have horizontal scales in the range of about 10–1000 km. It includes circulations ranging from thunderstorms and internal gravity waves at the small end of the scale to fronts and hurricanes at the large end. Given the diverse nature of mesoscale systems, it is not surprising that there is no single conceptual framework, equivalent to the quasi-geostrophic theory, which can provide a unified model for the dynamics of the mesoscale. Indeed, the dominant dynamical processes vary enormously depending on the type of mesoscale circulation system involved.

Possible sources of mesoscale disturbances include instabilities that occur intrinsically on the mesoscale, forcing by mesoscale thermal or topographic sources, nonlinear transfer of energy from either macroscale or microscale motions, and interaction of cloud physical and dynamical processes.

Although instabilities associated with the mean velocity or thermal structure of the atmosphere are a rich source of atmospheric disturbances, most instabilities have their maximum growth rates either on the large scale (baroclinic and most barotropic instability) or on the small scale (convection and Kelvin–Helmholtz instability.) Only symmetric instability (to be discussed in Section 9.3) appears to be an intrinsically mesoscale instability.

Mountain waves created by flow over individual peaks are generally regarded as small-scale phenomena. But flow over large mountain ranges can produce orographic disturbances in the 10–100-km mesoscale range. Flow over mountain ranges, such as the Front Range of the Colorado Rockies, can under some conditions of mean flow and static stability lead to strong downslope wind storms. Such mesoscale features have their scales determined by the horizontal scales of the mountains that force the waves.

Energy transfer from small scales to the mesoscale is a primary energy source for mesoscale convective systems. These may start as individual convective cells, which grow and combine to form thunderstorms, convective complexes such as squall lines and mesocyclones, and even hurricanes. Conversely, energy transfer from the large scale associated with temperature and vorticity advection in synoptic-scale circulations is responsible for the development of frontal circulations.

9.2 Fronts and Frontogenesis

In the discussion of baroclinic instability in Chapter 8 the mean thermal wind U_T was taken to be a constant independent of the y coordinate. That assumption was necessary to obtain a mathematically simple model that retained the basic instability mechanism. It was pointed out in Section 6.1, however, that baroclinicity is not uniformly distributed in the atmosphere. Rather, horizontal temperature gradients tend to be concentrated in baroclinic zones associated with tropospheric jet streams. Not surprisingly, the development of baroclinic waves is also concentrated in such regions.

We showed in Section 8.3 that the energetics of baroclinic waves require that they remove available potential energy from the mean flow. Thus, on average, baroclinic wave development tends to weaken the meridional temperature gradient (that is, reduce the mean thermal wind). The mean pole-to-equator temperature gradient is of course continually restored by differential solar heating, which maintains the time-averaged temperature gradient pattern. In addition, there are transient dynamical processes, which produce zones with greatly enhanced temperature gradients within individual baroclinic eddies. Such zones, which are particularly intense at the surface, are referred to as *fronts*. Processes that generate fronts are called *frontogenetic*. Frontogenesis usually occurs in association with developing baroclinic waves, which in turn are concentrated in the storm tracks associated with the time-mean jet streams. Thus, even though on average baroclinic disturbances transport heat down the mean temperature gradient and tend to weaken the temperature difference between the polar and tropical regions, locally the flow associated with baroclinic disturbances may actually enhance the temperature gradient.

9.2.1 THE KINEMATICS OF FRONTOGENESIS

A complete discussion of the dynamics of frontogenesis is beyond the scope of this text. A qualitative description of frontogenesis can be obtained, however, by considering the evolution of the temperature gradient when temperature is treated as a passive tracer in a specified horizontal flow field. Such an approach is referred to as *kinematic*; it considers the effects of advection on a field variable without reference to the underlying physical forces or to any influence of the advected tracer on the flow field.

The influence of a purely geostrophic flow on the temperature gradient was given in terms of the $\mathbf{Q}$ vector in (6.37). If for simplicity we focus on the meridional temperature gradient, then from (6.37)

$$\frac{D_g}{Dt}\left(\frac{\partial T}{\partial y}\right) = -\left[\frac{\partial u_g}{\partial y}\frac{\partial T}{\partial x} - \frac{\partial u_g}{\partial x}\frac{\partial T}{\partial y}\right] \tag{9.1}$$

where we have used the fact that the geostrophic wind is nondivergent so that $\partial v_g/\partial y = -\partial u_g/\partial x$. The two terms within the brackets on the right in (9.1) can be interpreted as the forcing of the meridional temperature gradient by horizontal shear deformation and stretching deformation, respectively.

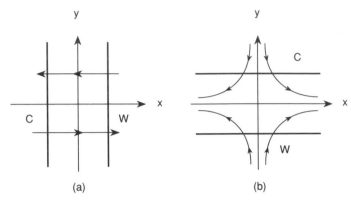

(a)   (b)

Fig. 9.1   Frontogenetic flow configurations: (a) horizontal shearing deformation; (b) horizontal stretching deformation.

Horizontal shear has two effects on a fluid parcel; it tends to rotate the parcel (owing to shear vorticity) and to deform the parcel through stretching parallel to the shear vector (i.e., along the x axis in Fig. 9.1a) and shrinking along the horizontal direction perpendicular to the shear vector. Thus, the x-directed temperature gradient in Fig. 9.1a is both rotated into the positive y direction and intensified by the shear. Horizontal shear is an important frontogenetic mechanism in both cold and warm fronts. For example, in the schematic surface pressure chart of Fig. 9.2 the geostrophic wind has a

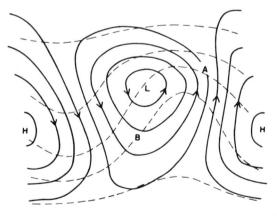

Fig. 9.2   Schematic surface isobars (solid lines) and isotherms (dashed lines) for a baroclinic wave disturbance. Arrows show direction of geostrophic wind. Horizontal stretching deformation intensifies the temperature gradient at A; horizontal shear deformation intensifies the gradient at B. (After Hoskins and Bretherton, 1972. Reproduced with permission of the American Meteorological Society.)

northerly component west of point B and a southerly component east of point B. The resulting cyclonic shear will tend to rotate the isotherms and to concentrate them along the line of maximum shear passing through B. (Note the strong cold advection northwest of B and the weak thermal advection southeast of B.)

Horizontal *stretching deformation* tends to advect the temperature field so that the isotherms become concentrated along the axis of *dilation* (the x axis in Fig. 9.1b), provided that the initial temperature field has a finite gradient along the axis of *contraction* (the y axis in Fig. 9.1b). That this effect is represented by the second term on the right in (9.1) can be verified by noting from Fig. 9.1b that $\partial T/\partial y < 0$ and $\partial u_g/\partial x > 0$.

The velocity field shown in Fig. 9.1b is a pure deformation field. That is, a parcel advected by it will have its shape changed without undergoing rotation or change in horizontal area. This deformation field has a stream-function given by $\psi = -Kxy$, where K is a constant. Such a field may be characterized by the rate at which advection changes the shape of a horizontal area element. This may be illustrated by considering the rectangular element with sides δx and δy. The shape may be represented by the ratio $\delta x/\delta y$, and the fractional rate of change of shape can thus be expressed as

$$\frac{1}{\delta x/\delta y}\frac{D(\delta x/\delta y)}{Dt} = \frac{1}{\delta x}\frac{D\delta x}{Dt} - \frac{1}{\delta y}\frac{D\delta y}{Dt} \approx \frac{\partial u}{\partial x} - \frac{\partial v}{\partial y}$$

It is easily verified that the fractional rate of change of $\delta x/\delta y$ for the velocity field in Fig. 9.1b equals $+2K$. Thus, a square parcel with sides parallel to the x and y axes would be deformed into a rectangle as the sides parallel to the x axis stretch and those parallel to the y axis contract in time at a constant rate.

A pure stretching deformation field is both irrotational and nondivergent. Thus, a parcel advected by a pure deformation field will merely have its shape changed in time, without any rotation or change in horizontal area. Horizontal deformation at low levels is an important mechanism for the development of both cold and warm fronts. In the example of Fig. 9.2 the flow near point A has $\partial u/\partial x > 0$ and $\partial v/\partial y < 0$, so there is a stretching deformation field present with its axis of contraction nearly orthogonal to the isotherms. This deformation field leads to strong warm advection south of point A and weak warm advection north of point A.

Although, as shown in Fig. 9.2, the low-level flow in the vicinity of a developing warm front may resemble a pure deformation field, the total flow in the upper troposphere in baroclinic disturbances seldom resembles that of a pure deformation field owing to the presence of strong mean westerlies. Rather, a combination of mean flow plus horizontal stretching deformation produces a *confluent* flow as shown in Fig. 9.3. Such confluence acts to concentrate the cross-stream temperature gradient as parcels move downstream. Confluent regions are always present in the tropospheric jet stream owing to the influence of quasi-stationary planetary-scale waves on the position and intensity of the jet. In fact, even a monthly mean 500-mb chart (see Fig. 6.3) reveals two regions of large-scale confluence immediately to the east of the continents of Asia and North America. Observationally, these two regions are known to be regions of intense baroclinic wave development and frontogenesis.

The mechanisms of horizontal shear and horizontal stretching deformation operate to concentrate the pole–equator temperature gradient on the synoptic scale (~1000 km). These processes alone cannot cause the rapid frontogenesis often observed in extratropical systems, in which the temperature gradient can become concentrated in a zone of characteristic width ~50 km on a time scale of 1–2 d. This rapid reduction in scale is caused primarily by the frontogenetic character of the secondary circulation driven by the quasi-geostrophic synoptic-scale flow (moist processes may also be important in rapid frontogenesis).

The nature of the secondary flow may be deduced from the pattern of **Q** vectors illustrated in Fig. 9.3a. As discussed in Section 6.4.2, the divergence of **Q** forces a secondary ageostrophic circulation. For the situation of Fig. 9.3 this circulation is in the cross-frontal plane as illustrated in Fig. 9.3b. Advection of the temperature field by this ageostrophic circulation tends

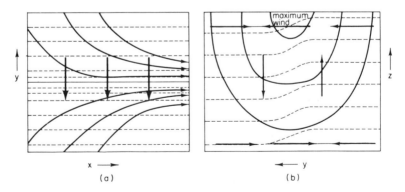

Fig. 9.3 (a) Horizontal streamlines, **Q** vectors (thick arrows), and isotherms in a frontogenetic confluence. (b) Vertical section across the confluence, showing isotachs (solid), isotherms (dashed), and vertical and transverse motions (arrows). (Adapted from Sawyer, 1956.)

to increase the horizontal temperature gradient at the surface on the warm side of the jet axis. Temperature advection by the upper-level secondary circulation, on the other hand, tends to concentrate the temperature gradient on the cold side of the jet axis. As a result, the frontal zone tends to slope toward the cold-air side with height. The differential vertical motion associated with the ageostrophic circulation tends to weaken the front in the midtroposphere owing to adiabatic temperature changes (adiabatic warming on the cold side of the front and adiabatic cooling on the warm side). For this reason fronts are most intense in the lower troposphere and near the tropopause.

The secondary circulation associated with frontogenesis is required to maintain the thermal wind balance between the along-front flow and the cross-front temperature gradient in the presence of advective processes that tend to destroy this balance. The concentration of the isotherms by the horizontal advection increases the cross-stream pressure gradient at upper levels and thus requires an increase in the vertical shear of the along-jet flow in order to maintain thermal wind balance. The required upper-level acceleration of the jet is produced by the Coriolis force caused by the cross-jet ageostrophic wind, which develops in response to the increased cross-stream gradient in the jet stream core. As the jet accelerates, cyclonic vorticity must be generated on the cold side of the jet axis and anticyclonic vorticity on the warm side. These vorticity changes require that the horizontal flow at the jet stream level be convergent on the cold side of the jet axis and divergent on the warm side of the jet axis. The vertical circulation and low-level secondary ageostrophic motion required by mass continuity are indicated in Fig. 9.3b.

9.2.2 SEMIGEOSTROPHIC THEORY

To analyze the dynamics of the frontogenetic motion fields discussed in the previous subsection it is convenient to use the Boussinesq approximation introduced in Section 7.4, in which density is replaced by a constant reference value ρ_0 except where it appears in the buoyancy force. This approximation simplifies the equations of motion without affecting the main features of the results. It is also useful to replace the total pressure and density fields with deviations from their standard atmosphere values. Thus, we let $\Phi(x, y, z, t) = (p - p_0)/\rho_0$ designate the pressure deviation normalized by density and $\Theta = \theta - \theta_0$ designate the potential temperature deviation, where p_0 and θ_0 are the height-dependent standard atmosphere values of pressure and potential temperature, respectively.

With the above definitions the horizontal momentum equations, thermodynamic energy equation, hydrostatic approximation, and continuity equation become

$$\frac{Du}{Dt} - fv + \frac{\partial \Phi}{\partial x} = 0 \tag{9.2}$$

$$\frac{Dv}{Dt} + fu + \frac{\partial \Phi}{\partial y} = 0 \tag{9.3}$$

$$\frac{D\Theta}{Dt} + w \frac{d\theta_0}{dz} = 0 \tag{9.4}$$

$$\frac{g\Theta}{\theta_0} = \frac{\partial \Phi}{\partial z} \tag{9.5}$$

$$\frac{\partial u}{\partial x} + \frac{\partial v}{\partial y} + \frac{\partial w}{\partial z} = 0 \tag{9.6}$$

where

$$\frac{D}{Dt} = \frac{\partial}{\partial t} + u \frac{\partial}{\partial x} + v \frac{\partial}{\partial y} + w \frac{\partial}{\partial z}$$

From the discussion of the previous subsection it should be clear that the horizontal scale of variations parallel to a front is much larger than the cross-frontal scale. This scale separation suggests that to a first approximation we can model fronts as two-dimensional structures. For convenience we choose a coordinate system in which the front is stationary and take the cross-frontal direction to be parallel to the y axis. Then $L_x \gg L_y$, where L_x and L_y designate the along-front and cross-front length scales. Similarly, $U \gg V$, where U and V respectively designate the along-front and cross-front velocity scales. Figure 9.4 shows these scales relative to the front.

Letting $U \sim 10 \text{ m s}^{-1}$, $V \sim 1 \text{ m s}^{-1}$, $L_x \sim 1000 \text{ km}$, and $L_y \sim 100 \text{ km}$, we find that it is possible to utilize the differing scales of the along-front and cross-front motion to simplify the dynamics. Assuming that $D/Dt \sim V/L_y$ (the cross-front advection time scale) and defining a Rossby number, $\text{Ro} \equiv V/fL_y \ll 1$, the magnitude of the ratios of the inertial and Coriolis terms in

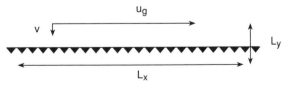

Fig. 9.4   Velocity and length scales relative to a front parallel to the x axis.

the x and y components of the momentum equation can be expressed as

$$\frac{|Du/Dt|}{|fv|} \sim \frac{UV/L_y}{fV} \sim \text{Ro}\left(\frac{U}{V}\right) \sim 1$$

$$\frac{|Dv/Dt|}{|fu|} \sim \frac{V^2/L_y}{fU} \sim \text{Ro}\left(\frac{V}{U}\right) \sim 10^{-2}$$

The along-front velocity is in geostrophic balance with the cross-front pressure gradient with error of order 1%, but geostrophy does not hold even approximately for the cross-front velocity. Therefore, if the geostrophic wind components are defined by

$$fu_g = -\partial\Phi/\partial y, \qquad fv_g = \partial\Phi/\partial x$$

and we separate the horizontal velocity field into geostrophic and ageostrophic parts, to a good approximation $u = u_g$, but $v = v_g + v_a$, where v_g and v_a are the same order of magnitude.

The x component of the horizontal momentum equation (9.2), the thermodynamic energy equation (9.4), and the continuity equation (9.6) for frontal scaling can thus be expressed as

$$\frac{Du_g}{Dt} - fv_a = 0 \tag{9.7}$$

$$\frac{D\Theta}{Dt} + w\frac{d\theta_0}{dz} = 0 \tag{9.8}$$

$$\frac{\partial v_a}{\partial y} + \frac{\partial w}{\partial z} = 0 \tag{9.9}$$

Since the along-front velocity is in geostrophic balance, u_g and Θ are related by the thermal wind relationship:

$$f\frac{\partial u_g}{\partial z} = -\frac{g}{\theta_0}\frac{\partial\Theta}{\partial y} \tag{9.10}$$

Note that (9.7) and (9.8) differ from their quasi-geostrophic analogs; although zonal momentum is still approximated geostrophically and advection parallel to the front is geostrophic, the advection of momentum and temperature across the front is due not only to the geostrophic wind but also to the ageostrophic (v_a, w) circulation:

$$\frac{D}{Dt} = \frac{D_g}{Dt} + \left(v_a\frac{\partial}{\partial y} + w\frac{\partial}{\partial z}\right)$$

where D_g/Dt was defined in (6.8). Replacement of momentum by its geostrophic value in (9.7) is referred to as the *geostrophic momentum approximation*, and the resulting set of prediction equations are called the *semigeostrophic* equations.[1]

9.2.3 CROSS-FRONTAL CIRCULATION

Equations (9.7)–(9.10) form a closed set, which can be used to determine the cross-frontal ageostrophic circulation in terms of the distribution of zonal wind or temperature. Suppose that the large-scale geostrophic flow is acting to intensify the north–south temperature gradient through deformation as shown in Fig. 9.3. As the temperature gradient increases, the vertical shear of the zonal wind must also increase to maintain geostrophic balance. This requires an increase of u_g in the upper troposphere, which must be produced by the Coriolis force associated with a cross-frontal ageostrophic circulation [see (9.7)]. The structure of this secondary circulation can be computed by deriving an equation analogous to the omega equation discussed in Section 6.4.2 expressed in the Boussinesq approximation.

We first differentiate (9.8) with respect to y and use the chain rule to express the result as

$$\frac{D}{Dt}\left(\frac{g}{\theta_0}\frac{\partial\Theta}{\partial y}\right) = Q_2 - \frac{\partial v_a}{\partial y}\frac{g}{\theta_0}\frac{\partial\Theta}{\partial y} - \frac{\partial w}{\partial y}\left(N^2 + \frac{g}{\theta_0}\frac{\partial\Theta}{\partial z}\right) \qquad (9.11)$$

where $N^2 \equiv (g/\theta_0)\, d\theta_0/dz$ and

$$Q_2 = -\frac{\partial u_g}{\partial y}\frac{g}{\theta_0}\frac{\partial\Theta}{\partial x} - \frac{\partial v_g}{\partial y}\frac{g}{\theta_0}\frac{\partial\Theta}{\partial y} \qquad (9.12)$$

is just the y component of the **Q** vector discussed in Section 6.4.2 expressed in the Boussinesq approximation.

Next we differentiate (9.7) with respect to z, again use the chain rule to rearrange terms, and use the thermal wind equation (9.10) to replace $\partial u_g/\partial z$ by $\partial\Theta/\partial y$ on the right-hand side. The result can then be written as

$$\frac{D}{Dt}\left(f\frac{\partial u_g}{\partial z}\right) = Q_2 + \frac{\partial v_a}{\partial z}f\left(f - \frac{\partial u_g}{\partial y}\right) + \frac{\partial w}{\partial z}\frac{g}{\theta_0}\frac{\partial\Theta}{\partial y} \qquad (9.13)$$

---

[1] Some authors reserve this name for a version of the equations written in a transformed set of coordinates called geostrophic coordinates (e.g., Hoskins, 1975).

Again, as we saw in Section 6.4.2, the geostrophic forcing (given by Q_2) tends to destroy thermal wind balance by changing the temperature gradient and vertical shear parts of the thermal wind equation in equal but opposite senses. This tendency of geostrophic advection to destroy geostrophic balance is counteracted by the cross-frontal secondary circulation.

In this case the secondary circulation is a two-dimensional overturning in the y, z plane. It can thus be represented in terms of a streamfunction ψ defined so that

$$v_a = -\partial\psi/\partial z, \qquad w = \partial\psi/\partial y \qquad (9.14)$$

which identically satisfies the continuity equation (9.9). Adding (9.11) and (9.13), using the thermal wind balance (9.10) to eliminate the time derivative, and using (9.14) to eliminate v_a and w, we obtain the *Sawyer–Eliassen equation*

$$N_s^2 \frac{\partial^2\psi}{\partial y^2} + F^2 \frac{\partial^2\psi}{\partial z^2} + 2S^2 \frac{\partial^2\psi}{\partial y\,\partial z} = 2Q_2 \qquad (9.15)$$

where

$$N_s^2 \equiv N^2 + \frac{g}{\theta_0}\frac{\partial\Theta}{\partial z}, \qquad F^2 \equiv f\left(f - \frac{\partial u_g}{\partial y}\right), \qquad S^2 \equiv -\frac{g}{\theta_0}\frac{\partial\Theta}{\partial y} \qquad (9.16)$$

Equation (9.15) can be compared with the quasi-geostrophic version obtained by neglecting advection by the ageostrophic circulation in (9.7) and (9.8). This has the form

$$N^2 \frac{\partial^2\psi}{\partial y^2} + f^2 \frac{\partial^2\psi}{\partial z^2} = 2Q_2 \qquad (9.17)$$

Thus, in the quasi-geostrophic case the coefficients in the differential operator on the left depend only on the standard atmosphere static stability, N, and the planetary vorticity, f, while in the semigeostrophic case they depend on the deviation of potential temperature from its standard profile through the N_s and S terms and the absolute vorticity through the F term.

An equation of the form (9.17), in which the coefficients of the derivatives on the left are positive, is referred to as an *elliptic boundary value problem*. It has a solution ψ that is uniquely determined by Q_2 plus the boundary conditions. For the situation pictured in Fig. 9.5 the forcing term Q_2 is negative in the frontal region since both $\partial v_g/\partial y$ and $\partial\Theta/\partial y$ are negative. The streamfunction describes an elliptical circulation with warm air rising

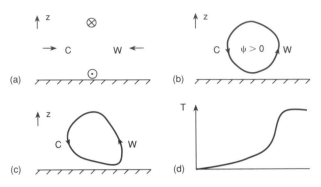

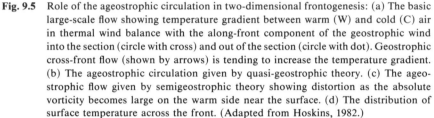

Fig. 9.5   Role of the ageostrophic circulation in two-dimensional frontogenesis: (a) The basic
large-scale flow showing temperature gradient between warm (W) and cold (C) air
in thermal wind balance with the along-front component of the geostrophic wind
into the section (circle with cross) and out of the section (circle with dot). Geostrophic
cross-front flow (shown by arrows) is tending to increase the temperature gradient.
(b) The ageostrophic circulation given by quasi-geostrophic theory. (c) The ageostrophic flow given by semigeostrophic theory showing distortion as the absolute
vorticity becomes large on the warm side near the surface. (d) The distribution of
surface temperature across the front. (Adapted from Hoskins, 1982.)

and cold air sinking as shown in Fig. 9.5b. The semigeostrophic case (9.15)
is also an elliptic boundary value problem provided that $N^2 F^2 - S^4$ is
positive. This is nearly always the case for an unsaturated atmosphere. The
spatial variation of the coefficients in (9.16) and the presence of the cross-
derivative term produce a distortion of the secondary circulation, as shown
in Fig. 9.5c. There is an intensification of the cross-frontal flow near the
surface in the region of large absolute vorticity on the warm air side of the
front and a tilting of the circulation with height.

The influence of the ageostrophic circulation on the time scale for frontogenesis can be illustrated by comparing the processes included in quasi-
geostrophic and semigeostrophic frontogenesis. The rate of surface frontal
development owing to a steady geostrophic stretching deformation parallel
to the x axis is given by (9.1) as

$$\frac{D_g}{Dt}\left(\frac{\partial T}{\partial y}\right) = K \frac{\partial T}{\partial y}$$

where $K = \partial u_g / \partial x$ is the amplification rate for $|\partial T/\partial y|$. Thus, the e^1
amplification time scale for quasi-geostrophic frontogenesis is $K^{-1} \sim 10^5$ s
(assuming that u_g changes by 10 m s^{-1} in 1000 km). This implies that if
temperature is advected only by the geostrophic flow, more than 2 days
would be required to increase the temperature gradient by an order of
magnitude.

For semigeostrophic advection, on the other hand, there is a positive feedback that greatly reduces the time scale of frontogenesis. As the temperature contrast increases, Q_2 increases, and the secondary circulation must also increase, so that the amplification *rate* of $|\partial T/\partial y|$ increases with $|\partial T/\partial y|$ rather than remaining constant as in the quasi-geostrophic case. Owing to this feedback, in the absence of frictional effects, the semigeostrophic model can produce an infinite temperature gradient at the surface in less than 12 hours.

9.3 Symmetric Instability

Observations indicate that mesoscale bands of cloud and precipitation commonly occur in association with synoptic-scale systems. These are often aligned with fronts but are separate entities. In common with fronts, such features are generally associated with strong baroclinicity and have length scales parallel to the mean wind shear that are much larger than the scales across the wind shear. A plausible source for such features is a two-dimensional form of baroclinic instability known as *symmetric instability*.

For typical atmospheric conditions buoyancy tends to stabilize air parcels against vertical displacements and rotation tends to stabilize parcels with respect to horizontal displacements. Instability with respect to vertical displacements is referred to as hydrostatic (or simply static) instability (see Section 2.7.3). For an unsaturated atmosphere static instability requires that the lapse rate exceed the dry adiabatic value ($N^2 < 0$). Instability with respect to horizontal displacements, on the other hand, is referred to as inertial instability (see Section 7.5.1). Inertial instability requires that the product of the Coriolis parameter and the absolute vorticity be negative $[f(f + \zeta) < 0]$.

If parcels are displaced along slantwise paths rather than vertical or horizontal paths, it is possible under certain conditions for the displacements to be unstable even when the conditions for ordinary static and inertial stability are separately satisfied. Such instability can occur only in the presence of vertical shear of the mean horizontal wind and may be regarded as a special form of baroclinic instability in which the perturbations are independent of the coordinate parallel to the mean flow. Alternatively, as shown below, symmetric instability may be regarded as isentropic inertial instability.

For convenience in deriving the conditions for symmetric instability we use the Boussinesq equations of the previous section and again assume that the mean wind is directed along the x axis and is in thermal wind balance with the meridional temperature gradient:

$$f \, \partial u_g / \partial z = -(g/\theta_0) \, \partial \Theta / \partial y \qquad (9.18)$$

Following Sections 2.7.3 and 7.5.1, we measure the stability with respect to vertical displacements by the distribution of total potential temperature $\theta = \theta_0 + \Theta$ and that with respect to horizontal displacements by the distribution of the so-called absolute zonal momentum, which was defined in (7.53) as $M \equiv fy - u_g$, where the sign convention is taken so that $\partial M / \partial y = f - \partial u_g / \partial y$ yields the zonal mean absolute vorticity.

For a barotropic flow, potential temperature surfaces are horizontally oriented and absolute momentum surfaces are vertically oriented in the meridional plane. When the mean flow is westerly and increases with height, however, the potential temperature and absolute momentum surfaces both slope upward toward the pole (Fig. 9.6). The comparative strengths of the vertical and horizontal restoring forces in the midlatitude troposphere are given by the ratio $N^2/(f \, \partial M \, \partial y) \sim 10^4$. Thus, parcel motion in the plane orthogonal to the mean flow will remain much closer to θ surfaces than to

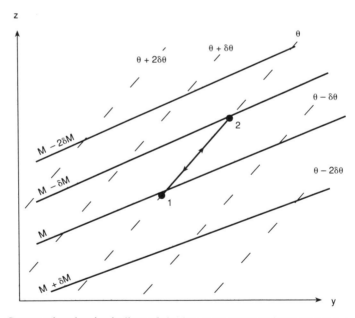

Fig. 9.6 Cross section showing isolines of absolute momentum and potential temperature for a symmetrically unstable basic state. Motion along the isentropic path between points labeled 1 and 2 is unstable since M decreases with latitude along the path. See text for details.

M surfaces. It is thus natural to utilize isentropic coordinates to analyze parcel displacements. The arguments of Section 7.5.1 still apply provided that derivatives with respect to y are taken at constant θ. The stability of such motions thus depends on the relative slope of the θ and M surfaces. Ordinarily the M surfaces will slope more than the θ surfaces and parcel displacements are stable, but when the θ surfaces slope more than the M surfaces so that

$$f(\partial M/\partial y)_\theta < 0 \qquad (9.19)$$

the flow is unstable with respect to displacements along the θ surfaces. This situation may occur in regions of very strong horizontal temperature gradient and weak vertical stability. The condition (9.19) is similar to the criterion (7.54) for inertial instability, except that here the derivative of M is taken along a sloping θ surface.

If (9.19) is multiplied by $-g(\partial\theta/\partial p)$ the criterion for symmetric instability can be expressed in terms of the distribution of Ertel potential vorticity (4.12) in the simple form

$$f\bar{P} < 0 \qquad (9.20)$$

where $\bar{P}$ is the potential vorticity of the basic state geostrophic flow. Thus, if the initial state potential vorticity in the Northern Hemisphere is everywhere positive, then symmetric instability cannot develop through adiabatic motions, since potential vorticity is conserved following the motion and will always remain positive.

As an alternative method of demonstrating that (9.19) is the condition for symmetric instability we consider the change in mean kinetic energy required for exchange of the tubes of fluid labeled 1 and 2 in Fig. 9.6. (These tubes are assumed to extend infinitely along the x axis so that the problem is two-dimensional.) Since the tubes lie on the same potential temperature surface, they have the same available potential energy. Thus a spontaneous exchange of parcels is possible if $\delta(\mathrm{KE})$, the kinetic energy of the zonal flow after exchange minus that in the initial state, is negative. Otherwise some external source of energy is required to furnish the kinetic energy of the meridional and vertical motions required for the exchange.

Initially the motion of the tubes is parallel to the x axis and in geostrophic balance so that the absolute momentum for the two tubes is

$$M_1 = fy_1 - u_1 = fy - u_g(y, z)$$
$$M_2 = fy_2 - u_2 = f(y + \delta y) - u_g(y + \delta y, z + \delta z) \qquad (9.21)$$

Absolute momentum is conserved by the tubes so that after exchange the perturbation zonal velocities are given by

$$M_1' = f(y + \delta y) - u_1' = M_1, \qquad M_2' = fy - u_2' = M_2 \qquad (9.22)$$

Eliminating M_1 and M_2 between (9.21) and (9.22) and solving for the disturbance zonal wind we obtain

$$u_1' = f(\delta y) + u_1, \qquad u_2' = -f(\delta y) + u_2$$

The difference in zonal kinetic energy between the final state and the initial state is given by

$$\delta(\mathrm{KE}) = \tfrac{1}{2}(u_1'^2 + u_2'^2) - \tfrac{1}{2}(u_1^2 + u_2^2)$$

$$= f\,\delta y(u_1 - u_2 + f\,\delta y) = f\,\delta y(M_2 - M_1) \qquad (9.23)$$

Thus, $\delta(\mathrm{KE})$ is negative, and unforced meridional motion may occur, provided that $f(M_2 - M_1) < 0$. This is equivalent to the condition (9.19) since the tubes lie along the same θ surface.

To estimate the likelihood that conditions for symmetric instability may be satisfied it is useful to express the stability criterion in terms of a mean flow *Richardson number*. To do this we first note that the slope of an M surface can be estimated from noting that on an M surface

$$\delta M = \frac{\partial M}{\partial y}\,\delta y + \frac{\partial M}{\partial z}\,\delta z = 0$$

so that the ratio of δz to δy at constant M is

$$\left(\frac{\delta z}{\delta y}\right)_M = \frac{-\partial M/\partial y}{\partial M/\partial z} = \frac{f - \partial u_g/\partial y}{\partial u_g/\partial z} \qquad (9.24)$$

Similarly, the slope of a potential temperature surface is

$$\left(\frac{\delta z}{\delta y}\right)_\theta = \frac{-\partial \theta/\partial y}{\partial \theta/\partial z} = \frac{f\,\partial u_g/\partial z}{(g/\theta_0)\,\partial \theta/\partial z} \qquad (9.25)$$

where we have used the thermal wind relationship to express the meridional temperature gradient in terms of the vertical shear of the zonal wind. The ratio of (9.24) to (9.25) is simply

$$\frac{(\delta z/\delta y)_M}{(\delta z/\delta y)_\theta} = \frac{f(f - \partial u_g/\partial y)((g/\theta_0)\,\partial\theta/\partial z)}{[f^2(\partial u_g/\partial z)^2]} = \frac{F^2 N_s^2}{S^4} \qquad (9.26)$$

where the notation in the last term is defined in (9.16).

Recalling that symmetric instability requires that the slopes of the θ surfaces exceed those of the M surfaces, the necessary condition for instability of geostrophic flow parallel to the x axis becomes

$$\frac{(\delta z/\delta y)_M}{(\delta z/\delta y)_\theta} = \frac{f(f - \partial u_g/\partial y)\text{Ri}}{f^2} = \frac{F^2 N_s^2}{S^4} < 1 \tag{9.27}$$

where the mean flow *Richardson number* Ri is defined as

$$\text{Ri} \equiv \frac{(g/\theta_0)\,\partial\theta/\partial z}{(\partial u_g/\partial z)^2}$$

Thus, if the relative vorticity of the mean flow vanishes $(\partial u_g/\partial y = 0)$, $\text{Ri} < 1$ is required for instability.

The condition (9.27) can be related to (9.20) by observing that (9.27) requires that $F^2 N_s^2 - S^4 < 0$ for symmetric instability. It can be shown (see Problem 9.1) that

$$F^2 N_s^2 - S^4 = (\rho f g/\theta_0)\bar{P} \tag{9.28}$$

Because the large-scale potential vorticity is normally positive (negative) in the Northern (Southern) Hemisphere, (9.28) is ordinarily positive; the condition for symmetric instability is rarely satisfied. If the atmosphere is saturated, however, the relevant static stability condition involves the lapse rate of equivalent potential temperature, and neutral conditions with respect to symmetric instability may easily occur (see Section 9.5).

Finally, it is worth noting the relationships among the condition for symmetric instability, the distribution of mean state potential vorticity, and the mathematical structure of the Sawyer–Eliassen equation (9.15). When $F^2 N_s^2 - S^4 > 0$ (i.e., when the product of f and the mean flow potential vorticity is positive) the flow is stable with respect to symmetric baroclinic perturbations and, as pointed out in Section 9.2.3, the Sawyer–Eliassen equation is an elliptic boundary value problem. In that case nonzero flow in the meridional plane must be forced by Q_2 [see (9.12)]. When the inequality (9.27) holds the Sawyer–Eliassen equation is no longer elliptic and unforced overturning may occur in the plane orthogonal to the mean flow.

9.4 Mountain Waves

In Section 7.5.2 we showed that stably stratified air forced to flow over sinusoidally varying surface topography creates oscillations, which can be either vertically propagating or vertically decaying, depending on whether the intrinsic wave frequency relative to the mean flow is less than or greater than the buoyancy frequency. Most topographic features on the surface of the earth do not, however, consist of regularly repeating lines of ridges. In general the distance between large topographic barriers is large compared to the horizontal scales of the barriers. Also, the static stability and the basic state flow are not usually constants, as was assumed in Section 7.5.2, but may vary strongly with height. Furthermore, nonlinear modifications of mountain waves are sometimes associated with strong surface winds along the lee slopes of ridges. Thus, mountain waves are significant features of mesoscale meteorology.

9.4.1 FLOW OVER ISOLATED RIDGES

Just as flow over a periodic series of sinusoidal ridges can be represented by a single sinusoidal function, i.e., by a single Fourier harmonic, flow over an isolated ridge can be approximated by the sum of a number of Fourier components (see Section 7.2.1). Thus, any zonally varying topography can be represented by a Fourier series of the form

$$h_M(x) = \sum_{s=1}^{\infty} \text{Re}[h_s \exp(ik_sx)] \tag{9.29}$$

where h_s is the amplitude of the sth Fourier component of the topography. We can then express the solution to the wave equation (7.46) as the sum of Fourier components:

$$w(x, z) = \sum_{s=1}^{\infty} \text{Re}\{W_s \exp[i(k_sx + m_sz)]\} \tag{9.30}$$

where $W_s = ik_s\bar{u}h_s$ and $m_s^2 = N^2/\bar{u}^2 - k_s^2$.

Individual Fourier modes will yield vertically propagating or vertically decaying contributions to the total solution (9.30) depending on whether m_s is real or imaginary. This in turn depends on whether k_s^2 is less than or greater than $N^2/\bar{u}^2$. Thus, each Fourier mode behaves just as the total solution (7.48) for periodic sinusoidal topography. For a narrow ridge Fourier components with wave numbers greater than $N/\bar{u}$ dominate in (9.29), and the resulting disturbance decays with height. For a broad ridge components with wave numbers less than $N/\bar{u}$ dominate and the disturbance propagates vertically. In the wide mountain limit where $m_s^2 \approx N^2/\bar{u}^2$ the flow is periodic in the vertical with vertical wavelength of $2\pi m_s^{-1}$, and phase lines tilt upstream with height as shown in Fig. 9.7.

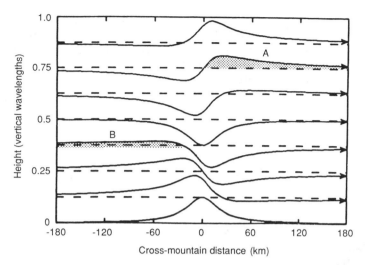

Fig. 9.7 Streamlines of flow over a broad isolated ridge, showing upstream phase tilt with height. The pattern is periodic in height and one vertical wavelength is shown. Orographic clouds may form in the shaded areas where streamlines are displaced upward from equilibrium either upstream or downstream of the ridge if sufficient moisture is present. (Courtesy of Dr. Dale Durran.)

Vertically propagating gravity waves generated by flow over broad topography can produce clouds both upstream and downstream of the topography depending on variations of the moisture distribution with altitude. In the example shown in Fig. 9.7 the positions labeled A and B indicate regions where streamlines are displaced upward downstream and upstream of the ridge, respectively. If sufficient moisture is present orographic clouds may then form in region A or B as suggested by the shading in Fig. 9.7.

9.4.2 LEE WAVES

If $\bar{u}$ and N are allowed to vary in height, then (7.46) must be replaced by

$$\left(\frac{\partial^2 w'}{\partial x^2} + \frac{\partial^2 w'}{\partial z^2}\right) + l^2 w' = 0 \tag{9.31}$$

where the *Scorer parameter, l,* is defined as

$$l^2 = N^2/\bar{u}^2 - \bar{u}^{-1}\, d^2\bar{u}/dz^2$$

and the condition for vertical propagation becomes $k_s^2 < l^2$.

If the mean cross-mountain wind speed increases strongly with height, or there is a low-level stable layer so that N decreases strongly with height,

there may be a layer near the surface in which vertically propagating waves
are permitted, which is topped by a layer in which the disturbance decays
in the vertical. In that case vertically propagating waves in the lower layer
are reflected when they reach the upper layer. Under some circumstances
the waves may be repeatedly reflected from the upper layer and the surface
downstream of the mountain, leading to a series of "trapped" lee waves as
shown in Fig. 9.8.

Vertical variations in the Scorer parameter can also modify the amplitude
of waves that are sufficiently long to propagate vertically through the entire
troposphere. Amplitude enhancement leading to wave breaking and turbulent mixing can occur if there is a *critical level* where the mean flow goes
to zero ($l \to \infty$).

9.4.3 DOWNSLOPE WINDSTORMS

Strong downslope surface winds are occasionally observed along the lee
slopes of mountain ranges. Although partial reflection of vertically propagating linear gravity waves may produce enhanced surface winds under some

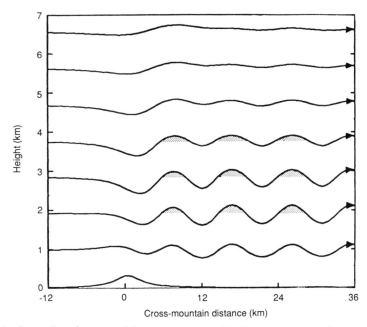

Fig. 9.8 Streamlines for trapped lee waves generated by flow over topography with vertical
variation of the Scorer parameter. Shading shows locations where lee wave clouds
may form. (Courtesy of Dr. Dale Durran.)

conditions, it appears that nonlinear processes are essential to account for observed windstorms associated with stable flow over topography.

To demonstrate the role of nonlinearity we assume that the troposphere has a stable lower layer of undisturbed depth h topped by a weakly stable upper layer and assume that the lower layer behaves as a barotropic fluid with a free surface $h(x, t)$. We assume that disturbances have zonal wavelengths much greater than the layer depth. The motion of the lower layer may then be described by the shallow-water equations of Section 7.3.2, but with the lower boundary condition replaced by

$$w(x, h_M) = Dh_M/Dt = u \, \partial h_M/\partial x$$

where h_M again denotes the height of the topography.

We first examine the linear behavior of this model by considering steady flow over small-amplitude topography. The linearized shallow-water equations (7.20) and (7.21) then become

$$\bar{u}\frac{\partial u'}{\partial x} + \frac{g\delta\rho}{\rho_1}\frac{\partial h'}{\partial x} = 0 \tag{9.32}$$

$$\bar{u}\frac{\partial(h'-h_M)}{\partial x} + H\frac{\partial u'}{\partial x} = 0 \tag{9.33}$$

Here $\delta\rho/\rho_1$ is the fractional change in density across the interface between the layers, $h' = h - H$, where H is the mean height of the interface, and $h' - h_M$ is the deviation from H of the thickness of the lower layer. The solutions for the set (9.32) and (9.33) can be expressed as

$$h' = -\frac{h_M(\bar{u}^2/c^2)}{(1-\bar{u}^2/c^2)}, \qquad u' = \frac{h_M}{H}\left(\frac{\bar{u}}{1-\bar{u}^2/c^2}\right) \tag{9.34}$$

where $c^2 \equiv (gH\,\delta\rho/\rho_1)$ is the shallow-water wave speed. The characteristics of the disturbance fields h' and u' depend on the magnitude of the mean flow *Froude number*, defined as $\mathrm{Fr}^2 \equiv \bar{u}^2/c^2$. When $\mathrm{Fr} < 1$, the flow is referred to as *subcritical*. In subcritical flow, the shallow-water gravity wave speed is greater than the mean flow speed, and the disturbance height and wind fields are out of phase. The interface height disturbance is negative and the velocity disturbance is positive over the topographic barrier as shown in Fig. 9.9a. When $\mathrm{Fr} > 1$, the flow is referred to as *supercritical*. In supercritical flow the mean flow exceeds the shallow-water gravity wave speed. Gravity waves cannot play a role in establishing the steady-state adjustment between

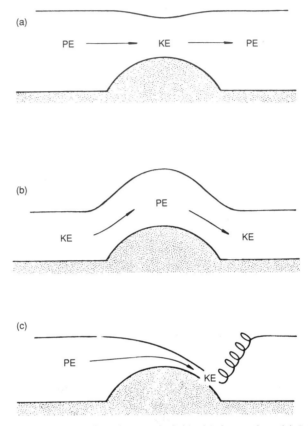

Fig. 9.9   Flow over an obstacle for a barotropic fluid with free surface. (a) Subcritical flow (Fr < 1 everywhere). (b) Supercritical flow (Fr > 1 everywhere). (c) Supercritical flow on lee slope with adjustment to subcritical flow at hydraulic jump near base of obstacle. (After Durran, 1990.)

the height and velocity disturbances because such waves are swept down-stream from the ridge by the mean flow. In this case the fluid thickens and slows as it ascends over the barrier (Fig. 9.9b). It is also clear from (9.34) that for Fr ~ 1 the perturbations are no longer small and the linear solution breaks down.

The nonlinear equations corresponding to the set (9.32) and (9.33) in the case $\delta\rho = \rho_1$ can be expressed as

$$u\frac{\partial u}{\partial x} + g\frac{\partial h}{\partial x} = 0 \qquad (9.35)$$

$$\frac{\partial}{\partial x}[u(h - h_M)] = 0 \tag{9.36}$$

Equation (9.35) may be integrated immediately to show that the sum of kinetic and potential energies, $u^2/2 + gh$, is constant following the motion. Thus, energy conservation requires that if u increases h must decrease, and vice versa. In addition, (9.36) shows that the mass flux $u(h - h_M)$ must also be conserved. The direction of the exchange between kinetic and potential energy in flow over a ridge is determined by the necessity that both (9.35) and (9.36) be satisfied.

If we multiply (9.35) by u and eliminate $\partial h/\partial x$ with the aid of (9.36) we find that

$$(1 - \mathrm{Fr}^2)\frac{\partial u}{\partial x} = \frac{ug}{c^2}\frac{\partial h_M}{\partial x} \tag{9.37}$$

where the shallow-water wave speed is now defined using the local thickness of the fluid:

$$c^2 \equiv g(h - h_M)$$

From (9.37) it is clear that the flow will accelerate on the upslope side of the ridge $(\partial u/\partial x > 0$ where $\partial h_M/\partial x > 0)$ if the Froude number is less than unity but will decelerate if the Froude number is greater than unity.

As a subcritical flow ascends the upslope side of a topographic barrier, Fr will tend to increase both from the increase in u and the decrease in c. If Fr = 1 at the crest, then from (9.37) the flow will become supercritical and continue to accelerate as it descends the lee side until it adjusts back to the ambient subcritical conditions in a turbulent hydraulic jump as illustrated in Fig. 9.9c. In this case very high velocities can occur along the lee slope since potential energy is converted into kinetic energy during the entire period that a fluid column traverses the barrier. Although conditions in the continuously stratified atmosphere are clearly more complex than in the shallow-water hydraulic model, numerical simulations have demonstrated that the hydraulic model provides a reasonable conceptual model for the primary processes occurring in downslope windstorms.

9.5 Cumulus Convection

Mesoscale storms associated with cumulus convection represent a large fraction of all meteorologically important mesoscale circulations. Before considering such systems it is necessary to examine a few of the essential thermodynamic and dynamical aspects of individual cumulus clouds. The subject of cumulus convection is extremely complex to treat theoretically. Much of this difficulty stems from the fact that cumulus clouds have a complex internal structure. They are generally composed of a number of short-lived individual rising towers, which are produced by elements of ascending buoyant air called *thermals*. Rising thermals *entrain* environmental air and thus modify the cloud air through mixing. The thermals are nonhydrostatic, nonsteady, and highly turbulent. The buoyancy of an individual thermal (i.e., the difference between its density and the density of the environment) depends on a number of factors including the environmental lapse rate, the rate of dilution by entrainment, and drag by the weight of liquid water in cloud droplets. Detailed discussion of the dynamics of thermal convection is beyond the scope of this text. In this section we utilize a simple one-dimensional cloud model and focus primarily on the thermodynamic aspects of moist convection. Convective storm dynamics will be considered in Section 9.6.

9.5.1 EQUIVALENT POTENTIAL TEMPERATURE

We have previously applied the parcel method to discuss the vertical stability of a dry atmosphere. We found that the stability of a dry parcel with respect to a vertical displacement depends on the lapse rate of potential temperature in the environment such that a parcel displacement is stable provided that $\partial\theta/\partial z > 0$ (i.e., the actual lapse rate is less than the adiabatic lapse rate). The same condition also applies to parcels in a moist atmosphere when the relative humidity is less than 100%. If, however, a parcel of moist air is forced to rise, it will eventually become saturated at a level called the lifting condensation level (LCL). A further forced rise will then cause condensation and latent heat release, and the parcel will then cool at the saturated adiabatic lapse rate. If the environmental lapse rate is greater than the saturated adiabatic lapse rate and the parcel is forced to continue to rise, it will reach a level at which it becomes buoyant relative to its surroundings. It can then freely accelerate upward. The level at which this occurs is called the level of free convection (LFC).

Discussion of parcel dynamics in a moist atmosphere is facilitated by defining a thermodynamic field called the *equivalent potential temperature*. Equivalent potential temperature, designated by θ_e, is the potential temperature that a parcel of air would have if all its moisture were condensed and the resulting latent heat used to warm the parcel. The temperature of an air parcel can be brought to its equivalent potential value by raising the parcel from its original level until all the water vapor in the parcel has condensed and fallen out, then compressing the parcel adiabatically to a pressure of 1000 mb. Since the condensed water is assumed to fall out, the temperature increase during the compression will be at the dry adiabatic rate and the parcel will arrive back at its original level with a temperature that is higher than its original temperature. Thus, the process is irreversible. Ascent of this type, in which all condensation products are assumed to fall out, is called *pseudoadiabatic* ascent. (It is not a truly adiabatic process because the liquid water that falls out carries a small amount of heat with it.)

A complete derivation of the mathematical expression relating θ_e to the other variables of state is rather involved and will be relegated to Appendix D. For most purposes, however, it is sufficient to use an approximate expression for θ_e that can be immediately derived from the entropy form of the first law of thermodynamics (2.46). If we let q_s denote the mass of water vapor per unit mass of dry air in a saturated parcel (q_s is called the saturation mixing ratio), then the rate of diabatic heating per unit mass is

$$J = -L_c \frac{Dq_s}{Dt}$$

where L_c is the latent heat of condensation. Thus, from the first law of thermodynamics

$$c_p \frac{D \ln \theta}{Dt} = -\frac{L_c}{T} \frac{Dq_s}{Dt} \tag{9.38}$$

For a saturated parcel undergoing pseudoadiabatic ascent the rate of change in q_s following the motion is much larger than the rate of change in T or L_c. Therefore,

$$d \ln \theta \approx -d(L_c q_s / c_p T) \tag{9.39}$$

Integrating (9.39) from the initial state (θ, q_s, T) to a state where $q_s \approx 0$ we obtain

$$\ln(\theta / \theta_e) \approx -L_c q_s / c_p T$$

where θ_e, the potential temperature in the final state, is approximately the equivalent potential temperature defined above. Thus, θ_e for a saturated parcel is given by

$$\theta_e \approx \theta \exp(L_c q_s / c_p T) \tag{9.40}$$

Expression (9.40) may also be used to compute θ_e for an unsaturated parcel provided that the temperature used in the formula is the temperature that the parcel would have if expanded adiabatically to saturation (i.e., T_{LCL}) and the saturation mixing ratio is replaced by the *actual* mixing ratio of the initial state. Thus, equivalent potential temperature is conserved for a parcel during both dry adiabatic and pseudoadiabatic displacements.

An alternative to θ_e, which is sometimes used in studies of convection, is the *moist static energy*, defined as $h \equiv s + L_c q$, where $s \equiv c_p T + gz$ is the *dry static energy*. It can be shown (Problem 9.3) that

$$c_p T \, d \ln \theta_e \approx dh \tag{9.41}$$

Hence, moist static energy is approximately conserved when θ_e is conserved.

9.5.2   THE PSEUDOADIABATIC LAPSE RATE

The first law of thermodynamics (9.38) can be used to derive a formula for the rate of change of temperature with respect to height for a saturated parcel undergoing pseudoadiabatic ascent. Using the definition of θ (2.44) we can rewrite (9.38) as

$$\frac{d \ln T}{dz} - \frac{R}{c_p} \frac{d \ln p}{dz} = -\frac{L_c}{c_p T} \frac{dq_s}{dz}$$

which upon nothing $q_s \equiv q_s(T, p)$ and applying the hydrostatic equation and equation of state can be expressed as

$$\frac{dT}{dz} + \frac{g}{c_p} = -\frac{L_c}{c_p} \left[\left(\frac{\partial q_s}{\partial T} \right)_p \frac{dT}{dz} - \left(\frac{\partial q_s}{\partial p} \right)_T \rho g \right]$$

Thus, following the ascending saturated parcel we have

$$\Gamma_s \equiv -\frac{dT}{dz} = \Gamma_d \frac{[1 - \rho L_c (\partial q_s/\partial p)_T]}{[1 + (L_c c_p)(\partial q_s/\partial T)_p]} \tag{9.42}$$

where $\Gamma_d \equiv g/c_p$ is the dry adiabatic lapse rate, and Γ_s is the *pseudoadiabatic lapse rate*, which is always less than Γ_d. Observed values of Γ_s range from ~ 4 K km^{-1} in warm humid air masses in the lower troposphere to ~ 6–7 K km^{-1} in the midtroposphere.

9.5.3 CONDITIONAL INSTABILITY

We showed in Section 2.7.3 that for dry adiabatic motions the atmosphere is statically stable provided that the lapse rate is less than the dry adiabatic lapse rate (i.e., the potential temperature increases with height). If the lapse rate Γ lies between the dry adiabatic and pseudoadiabatic values ($\Gamma_s < \Gamma < \Gamma_d$) the atmosphere is stably stratified with respect to dry adiabatic displacements but unstable with respect to pseudoadiabatic displacements. Such a situation is referred to as *conditional instability* (i.e., the instability is conditional to saturation of the air parcel).

The conditional stability criterion can also be expressed in terms of the gradient of a field variable θ_e^*, defined as the equivalent potential temperature of a hypothetically saturated atmosphere that has the thermal structure of the actual atmosphere.[2] Thus,

$$d \ln \theta_e^* = d \ln \theta + d(L_c q_s / c_p T) \qquad (9.43)$$

where T is the actual temperature, not the temperature after adiabatic expansion to saturation as in (9.40). To derive this condition we consider the motion of a saturated parcel in an environment in which the potential temperature is θ_0 at the level z_0. At the level $z_0 - \delta z$ the undisturbed environmental air thus has potential temperature

$$\theta_0 - (\partial \theta / \partial z) \delta z$$

Suppose a saturated parcel that has the environmental potential temperature at $z_0 - \delta z$ is raised to the level z_0. When it arrives at z_0 the parcel will have potential temperature

$$\theta_1 = \left(\theta_0 - \frac{\partial \theta}{\partial z} \delta z \right) + \delta \theta$$

where $\delta \theta$ is the change in parcel potential temperature owing to condensation during ascent through vertical distance δz. Assuming a pseudoadiabatic ascent, we see from (9.39) that

$$\frac{\delta \theta}{\theta} \approx -\delta \left(\frac{L_c q_s}{c_p T} \right) \approx -\frac{\partial}{\partial z} \left(\frac{L_c q_s}{c_p T} \right) \delta z$$

so that the buoyancy of the parcel when it arrives at z_0 is proportional to

$$\frac{\theta_1 - \theta_0}{\theta_0} \approx -\left[\frac{1}{\theta} \frac{\partial \theta}{\partial z} + \frac{\partial}{\partial z} \left(\frac{L_c q_s}{c_p T} \right) \right] \delta z \approx -\frac{\partial \ln \theta_e^*}{\partial z} \delta z$$

[2] Note that θ_e^* is *not* the same as θ_e except in a saturated atmosphere.

The saturated parcel will be warmer than its environment at z_0 provided that $\theta_1 > \theta_0$. Thus the conditional stability criterion for a saturated parcel is

$$\frac{\partial \theta_e^*}{\partial z} \begin{cases} <0 & \text{conditionally unstable} \\ =0 & \text{saturated neutral} \\ >0 & \text{conditionally stable} \end{cases} \qquad (9.44)$$

In Fig. 9.10 the vertical profiles of θ, θ_e, and θ_e^* for a typical sounding in the vicinity of an extratropical thunderstorm are shown. It is obvious from the figure that the sounding is conditionally unstable in the lower troposphere. However, this observed profile does not imply that convective

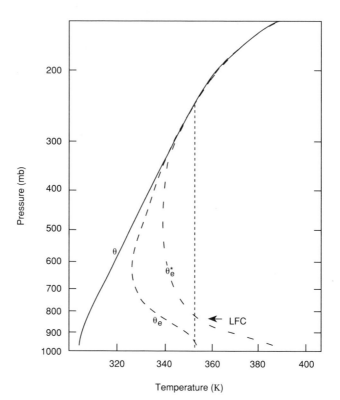

Fig. 9.10   Schematic sounding for a conditionally unstable environment characteristic of midwestern North America thunderstorm conditions, showing the vertical profiles of potential temperature θ, equivalent potential temperature θ_e, and the equivalent temperature θ_e^* of a hypothetically saturated atmosphere with the same temperature profile. Dotted line shows θ_e for a nonentraining parcel raised from the surface. Arrow denotes the LFC for the parcel.

overturning will occur spontaneously. The release of conditional instability requires not only that $\partial \theta_e^* / \partial z < 0$ but also parcel saturation at the environmental temperature of the level where the convection begins (i.e., the parcel must reach the LFC). The mean relative humidity in the troposphere is well below 100%, even in the boundary layer. Thus, low-level convergence with resulting forced layer ascent or vigorous vertical turbulent mixing in the boundary layer is required to produce saturation.

The amount of ascent necessary to raise a parcel to its LFC can be estimated simply from Fig. 9.10. A parcel rising pseudoadiabatically from a level $z_0 - \delta z$ will conserve the value of θ_e characteristic of the environment at $z_0 - \delta z$. But the buoyancy of a parcel depends only on the difference in density between the parcel and its environment. Thus, in order to compute the buoyancy of the parcel at z_0, it is not correct simply to compare θ_e of the environment at z_0 to $\theta_e(z_0 - \delta z)$ because if the environment is unsaturated the difference in θ_e between the parcel and the environment may be due primarily to the difference in mixing ratios, not to any temperature (density) difference. To estimate the buoyancy of the parcel $\theta_e(z_0 - \delta z)$ should instead be compared to $\theta_e^*(z_0)$, which is the equivalent potential temperature that the environment at z_0 would have if it were isothermally brought to saturation. The parcel will thus become buoyant where $\theta_e(z_0 - \delta z) > \theta_e^*(z_0)$, for then the parcel temperature will exceed the temperature of the environment. From Fig. 9.10 we see that θ_e for a parcel raised from about 960 mb will intersect the θ_e^* curve near 850 mb, whereas a parcel raised from any level much above 850 mb will not intersect θ_e^* no matter how far it is forced to ascend. It is for this reason that low-level convergence is usually required to initiate convective overturning over the oceans. Only air near the surface has a sufficiently high value of θ_e to become buoyant when it is forcibly raised. Convection over continental regions, on the other hand, can be initiated without significant boundary layer convergence since strong surface heating can produce positive parcel buoyancy all the way to the surface. Sustained deep convection, however, requires mean low-level moisture convergence.

9.5.4 CONVECTIVE AVAILABLE POTENTIAL ENERGY (CAPE)

Development of convective storms depends on the presence of environmental conditions favorable for the occurrence of deep convection. Several indices have been developed to measure the susceptibility of a given temperature and moisture profile to the occurrence of deep convection. A particularly useful measure is the *convective available potential energy* (CAPE). CAPE provides a measure of the maximum possible kinetic energy that a statically unstable parcel can acquire (neglecting effects of water vapor and condensed water on the buoyancy), assuming that the parcel ascends without mixing with the environment and instantaneously adjusts to the local environmental pressure.

The momentum equation for such a parcel is (2.51), which can be rewritten following the vertical motion of the parcel as

$$\frac{Dw}{Dt} = \frac{Dz}{Dt}\frac{Dw}{Dz} = w\frac{Dw}{Dz} = b' \tag{9.45}$$

where $b'(z)$ is the *buoyancy* given by

$$b' = g\frac{\rho_{env} - \rho_{parcel}}{\rho_{parcel}} = g\frac{T_{parcel} - T_{env}}{T_{env}} \tag{9.46}$$

and T_{env} designates the temperature of the environment. If (9.45) is integrated vertically from the level of free convection, z_{LFC}, to the level of neutral buoyancy, z_{LNB}, following the motion of the parcel the result is

$$\frac{w_{max}^2}{2} = \int_{z_{LFC}}^{z_{LNB}} g\left(\frac{T_{parcel} - T_{env}}{T_{env}}\right) dz \equiv B \tag{9.47}$$

Here B is the maximum kinetic energy per unit mass that a buoyant parcel could obtain by ascending from a state of rest at the level of free convection to the level of neutral buoyancy near the tropopause (see Fig. 9.10). This is an overestimate of the actual kinetic energy realized by a nonentraining parcel since the negative buoyancy contribution of liquid water reduces the effective buoyancy, especially in the tropics.

In a typical tropical oceanic sounding, parcel temperature excesses of 1–2 K may occur over a depth of 10–12 km. A typical value of CAPE is then $B \approx 500$ m^2 s^{-2}. In severe storm conditions in the midwest of North America, on the other hand, parcel temperature excesses can be 7–10 K (see Fig. 9.10) and $B \approx 2300$–3000 m^2 s^{-2}. Observed updrafts in the latter case (up to 50 m s^{-1}) are much stronger than in the former case (5–10 m s^{-1}). The small value of CAPE in the mean tropical environment is the major reason that updraft velocities in tropical cumulonimbus are observed to be much smaller than those in midlatitude thunderstorms.

9.5.5   ENTRAINMENT

In the previous subsection it was assumed that convective cells rise without mixing with environmental air so that they maintain constant θ_e during their rise. In reality, however, rising saturated air parcels tend to be diluted by entraining, or mixing in, some of the relatively dry environmental air. If the air in the environment is unsaturated, some of the liquid water in the rising parcel must be evaporated to maintain saturation in the convective cell as air from the environment is entrained. The evaporative cooling caused by entrainment will reduce the buoyancy of the convective parcel (i.e., lower its θ_e). Thus, the equivalent potential temperature in an entraining convection cell will decrease with height rather than remain constant. Similar considerations hold for any other conservable variable[3] in which environmental values differ from cloud values; entrainment will modify the in-cloud vertical profiles.

Denoting the amount of an arbitrary conservable variable per unit mass of air by A, the vertical dependence of A in an entraining convective cell can be estimated by assuming that to a first approximation the cell can be modeled as a steady-state jet as shown in Fig. 9.11. Thus, in a time increment δt a mass m of saturated cloud air with an amount of the arbitrary variable given by mA_{cld} mixes with a mass δm of entrained environmental air which has an amount of the arbitrary variable given by $\delta m A_{env}$. The change in the value of A within the cloud, δA_{cld}, is then given by the mass balance relationship

$$(m + \delta m)(A_{cld} + \delta A_{cld}) = mA_{cld} + \delta m A_{env} + (DA_{cld}/Dt)_S m \, \delta t \quad (9.48)$$

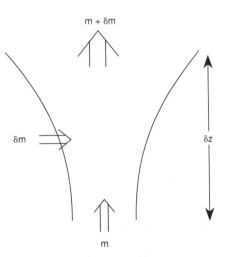

Fig. 9.11 An entraining jet model of cumulus convection. (See text for explanation.)

---

[3] A conservable variable is one that is conserved following the motion in the absence of sources and sinks (e.g., a chemical trace constituent).

where $(DA_{cld}/Dt)_S$ designates the rate of change of A_{cld} owing to sources and sinks unrelated to entrainment. Dividing through by δt in (9.48), neglecting the second-order term, and rearranging yields

$$\frac{\delta A_{cld}}{\delta t} = \left(\frac{DA_{cld}}{Dt}\right)_S - \frac{1}{m}\frac{\delta m}{\delta t}(A_{cld} - A_{env}) \qquad (9.49)$$

Noting that in the time increment δt an ascending parcel rises a distance $\delta z = w\,\delta t$, where w is the speed of ascent of the parcel, we can eliminate δt in (9.49) to obtain an equation for the vertical dependence of A_{cld} in a continuously entraining convective cell:

$$w\frac{dA_{cld}}{dz} = \left(\frac{DA_{cld}}{Dt}\right)_S - w\lambda(A_{cld} - A_{env}) \qquad (9.50)$$

where we have defined the *entrainment rate*, $\lambda \equiv d\ln m/dz$.

Letting $A_{cld} = \ln\theta_e$ and noting that θ_e is conserved in the absence of entrainment, (9.50) yields

$$\left(\frac{d\ln\theta_e}{dz}\right)_{cld} = -\lambda[(\ln\theta_e)_{cld} - (\ln\theta_e)_{env}]$$

$$\approx -\lambda\left[\frac{L_c}{c_pT}(q_s - q_{env}) + \ln\left(\frac{T_{cld}}{T_{env}}\right)\right] \qquad (9.51)$$

Thus an entraining convective cell is less buoyant than a nonentraining cell. Letting $A_{cld} = w$ in (9.50) and neglecting the pressure contribution to buoyancy, we find that the height dependence of kinetic energy per unit mass is given by

$$\frac{d}{dz}\left(\frac{w^2}{2}\right) = g\left(\frac{T_{cld} - T_{env}}{T_{env}}\right) - \lambda w^2 \qquad (9.52)$$

An entraining cell will undergo less acceleration than a nonentraining cell not only because the buoyancy is reduced but also because of the drag exerted by mass entrainment.

Equations (9.51) and (9.52) together with suitable relations for the cloud moisture variables can be used to determine the vertical profile of cloud variables. Such one-dimensional cloud models have been very popular in the past. Unfortunately, observed cloud properties, such as maximum cloud depth and cloud water concentration, cannot simultaneously be satisfactorily predicted by this type of model. In reality, pressure perturbations generated by the convective cells are important in the momentum budget, and entrainment does not occur instantaneously but in a sporadic manner which allows some cells to rise through most of the troposphere with very little dilution. Although more sophisticated one-dimensional models can partly overcome these deficiencies, some of the most important aspects of thunderstorm dynamics (e.g., the influence of vertical shear of the environmental wind) can be included only in multidimensional models.

9.6 Convective Storms

Convective storms can take a large variety of forms. These range in scale from isolated thunderstorms involving a single convective cloud (or *cell*) to mesoscale convective complexes consisting of ensembles of multicelled thunderstorms. Here we distinguish three primary types: the single-cell, the multicell, and the supercell storm. As we saw in the previous section, convective available potential energy measures whether thermodynamic conditions are favorable for the development of cumulus convection. CAPE, therefore, provides a guide to the strength of convection. It does not, however, provide any notion of the most likely type of mesoscale organization. It turns out, as suggested above, that storm type also depends on the vertical shear of the lower tropospheric environment.

When vertical shear is weak (<10 m s^{-1} below 4 km) single-cell storms occur. These tend to be short-lived (~30 min) and move with the meanwind in the lowest 8 km. When there is moderate vertical shear (~10–20 m s^{-1} below 4 km) multicell storms arise in which individual cells have lifetimes of ~30 min, but the storm lifetime may be many hours. In multicell storms the downdrafts induced by evaporation of precipitation form a dome of cold outflowing air at the surface. New cells tend to develop along the *gust front* where the cold outflow lifts conditionally unstable surface air. When vertical shear is large (>20 m s^{-1} below 4 km) the strong tilting of convective cells tends to delay storm development even in a thermodynamically favorable environment, so that an hour or more may be required for the initial cell to develop completely. This development may be followed by a split into two storms, which move to the left and right of the mean wind. Usually the left-moving storm dies rapidly while the right-moving storm slowly evolves into a rotating circulation with a single updraft core and trailing downdrafts, as will be discussed in the next section. Such supercell storms often produce heavy rain, hail, and damaging tornadoes. Ensembles of multicell or supercell storms are often organized along lines referred to as *squall lines*, which may move in a different direction than the individual thunderstorms.

9.6.1   Development of Rotation in Supercell Thunderstorms

The supercell thunderstorm is of particular dynamical interest because of its tendency to develop a rotating mesocyclone from an initially non-rotating environment. The dominance of cyclonic rotation in such systems might suggest that the Coriolis force plays a role in supercell dynamics. However, it can readily be shown that the rotation of the earth is not relevant to development of rotation in supercell storms.

Although a quantitative treatment of supercell dynamics requires that the density stratification of the atmosphere be taken into account, for the purpose of understanding the processes that lead to development of rotation in such systems and to the dominance of the right-moving cell, it is sufficient to use the Boussinesq approximation. The Euler momentum equation and continuity equation may then be expressed as follows:

$$\frac{D\mathbf{U}}{Dt} = \frac{\partial \mathbf{U}}{\partial t} + (\mathbf{U} \cdot \nabla)\mathbf{U} = -\frac{1}{\rho_0}\nabla p + b\mathbf{k}$$

$$\nabla \cdot \mathbf{U} = 0$$

Here $\mathbf{U} \equiv \mathbf{V} + \mathbf{k}w$, is the three-dimensional velocity, ∇ is the three-dimensional del operator, ρ_0 is the constant basic state density, p is the deviation of pressure from its horizontal mean, and $b \equiv -g\rho'/\rho_0$ is the total buoyancy.

It is convenient to rewrite the momentum equation using the vector identity

$$(\mathbf{U} \cdot \nabla)\mathbf{U} = \nabla\left(\frac{\mathbf{U} \cdot \mathbf{U}}{2}\right) - \mathbf{U} \times (\nabla \times \mathbf{U})$$

to obtain

$$\frac{\partial \mathbf{U}}{\partial t} = -\nabla\left(\frac{p}{\rho_0} + \frac{\mathbf{U} \cdot \mathbf{U}}{2}\right) + \mathbf{U} \times \boldsymbol{\omega} + b\mathbf{k} \tag{9.53}$$

Taking $\nabla \times (9.53)$ and recalling that the curl of the gradient vanishes, we obtain the three-dimensional vorticity equation

$$\frac{\partial \boldsymbol{\omega}}{\partial t} = \nabla \times (\mathbf{U} \times \boldsymbol{\omega}) + \nabla \times (b\mathbf{k}) \tag{9.54}$$

Letting $\zeta = \mathbf{k} \cdot \boldsymbol{\omega}$ be the vertical component of vorticity and taking $\mathbf{k} \cdot$ (9.54), we obtain an equation for the tendency of ζ in a nonrotating reference frame:

$$\frac{\partial \zeta}{\partial t} = \mathbf{k} \cdot \nabla \times (\mathbf{U} \times \boldsymbol{\omega}) \tag{9.55}$$

Notice that buoyancy affects only the horizontal vorticity components.

We now consider a flow consisting of a single convective updraft embedded in a basic state westerly flow that depends on z alone. Linearizing about this basic state by letting

$$\boldsymbol{\omega} = \mathbf{j} \, d\bar{u}/dz + \boldsymbol{\omega}'(x, y, z, t), \qquad \mathbf{U} = \mathbf{i}\bar{u} + \mathbf{U}'(x, y, z, t)$$

and noting that the linearized form of the right-hand side in (9.55) becomes

$$\mathbf{k} \cdot \nabla \times (\mathbf{U} \times \boldsymbol{\omega}) = -\mathbf{k} \cdot \nabla \times (\mathbf{i}w' \, d\bar{u}/dz + \mathbf{j}\bar{u}\zeta')$$

we find that the linearized vorticity tendency is

$$\frac{\partial \zeta'}{\partial t} = -\bar{u}\frac{\partial \zeta'}{\partial x} + \frac{\partial w'}{\partial y}\frac{d\bar{u}}{dz} \tag{9.56}$$

The first term on the right in (9.56) is just the advection by the basic state flow. The second term represents tilting of horizontal shear vorticity into the vertical by differential vertical motion.

Since $d\bar{u}/dz$ is positive, the vorticity tendency owing to this tilting will be positive to the south of the updraft core and negative to the north of the updraft core. As a result, a counterrotating vortex pair is established with cyclonic rotation to the south and anticyclonic rotation to the north of the initial updraft, as shown in Fig. 9.12a. Eventually, the development of negative buoyancy owing to precipitation loading generates an upper-level downdraft at the position of the initial updraft and the storm splits as shown in Fig. 9.12b. New updraft cores are established centered on the counterrotating vortex pair.

In order to understand the generation of updrafts in the vortices on the flanks of the storm we examine the perturbation pressure field. A diagnostic equation for the disturbance pressure is obtained by taking $\nabla \cdot$ (9.53) to yield

$$\nabla^2 \left(\frac{p}{\rho_0}\right) = -\nabla^2 \left(\frac{\mathbf{U} \cdot \mathbf{U}}{2}\right) + \nabla \cdot (\mathbf{U} \times \boldsymbol{\omega}) + \frac{\partial b}{\partial z} \tag{9.57}$$

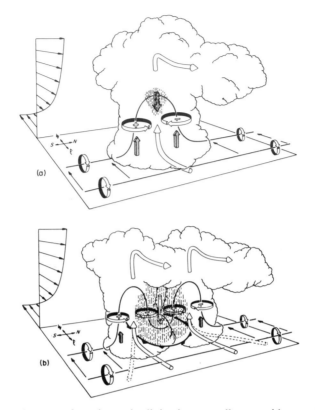

Fig. 9.12   Development of rotation and splitting in supercell storm with westerly mean wind shear (shown by storm relative wind arrows in the upper left corner of each panel). Cylindrical arrows show the direction of cloud relative air flow. Heavy solid lines show vortex lines with sense of rotation shown by circular arrows. Plus and minus signs indicate cyclonic and anticyclonic rotation caused by vortex tube tilting. Shaded arrows represent updraft and downdraft growth. Vertical dashed lines denote regions of precipitation. (a) Initial stage: the environmental shear vorticity is tilted and stretched into the vertical as it is swept into the updraft. (b) Splitting stage: downdraft forms between the new updraft cells. Barbed line at surface indicates downdraft outflow at surface. (After Klemp, 1987.)

The first two terms on the right in (9.57) represent dynamical forcing, while the last term represents buoyancy forcing. Observations and numerical models suggest that the buoyancy forcing in (9.57) produces pressure perturbations that tend partly to compensate the buoyancy force in the vertical momentum equation. Dynamically forced pressure perturbations, on the other hand, may generate substantial vertical accelerations.

In order to compute the dynamical contribution to the disturbance pressure gradient force in either the right or left side vortex we use cylindrical coordinates (r, λ, z) centered on the axis of rotation of either vortex and assume that to a first approximation the azimuthal velocity v_λ (positive for cyclonic flow) is independent of λ. In this system the storm relative horizontal motion and vertical component of vorticity are given approximately by

$$\mathbf{U}' \approx \mathbf{j}_\lambda v_\lambda, \qquad \mathbf{k} \cdot \boldsymbol{\omega}' = \zeta' \approx r^{-1} \partial(r v_\lambda)/\partial r$$

where $\mathbf{j}_\lambda$ is the unit vector in the azimuthal direction (positive counterclockwise), and r is the distance from the axis of the vortex. Letting $\mathbf{i}_\lambda$ be the unit vector in the radial direction we have

$$\mathbf{U} \times \boldsymbol{\omega} \approx \mathbf{i}_\lambda \frac{v_\lambda}{r} \frac{\partial}{\partial r} (r v_\lambda)$$

Assuming that the vertical scale is much larger than the radial scale, the Laplacian in cylindrical coordinates can be approximated by

$$\nabla^2 \approx \frac{1}{r} \frac{\partial}{\partial r} \left(r \frac{\partial}{\partial r} \right)$$

Thus, from (9.57) the dynamical component of the pressure perturbation in the vortices (designated p_{dyn}) can be expressed as

$$\frac{1}{r} \frac{\partial}{\partial r} \left(\frac{r}{\rho_0} \frac{\partial p_{dyn}}{\partial r} \right) \approx -\frac{1}{r} \frac{\partial}{\partial r} \left[r \frac{\partial(v_\lambda^2/2)}{\partial r} \right] + \frac{1}{r} \frac{\partial}{\partial r} \left[v_\lambda \frac{\partial(r v_\lambda)}{\partial r} \right] = \frac{1}{r} \frac{\partial v_\lambda^2}{\partial r} \quad (9.58)$$

Integrating (9.58) with respect to r we obtain the equation of cyclostrophic balance (see Section 3.2.4):

$$\rho_0^{-1} \partial p_{dyn}/\partial r \approx v_\lambda^2/r \qquad (9.59)$$

Hence, there is a pressure minimum at the vortex center irrespective of whether the rotation is cyclonic or anticyclonic. The strong midtropospheric rotation induced by vortex tube twisting and stretching creates a "centrifugal pump" effect, which causes a negative dynamical pressure perturbation centered in the vortices in the midtroposphere. This in turn produces an upward-directed dynamical contribution to the vertical component of the pressure gradient force and thus provides an upward acceleration, which produces updrafts in the cores of the counterrotating vortices as depicted in Fig. 9.12. These updrafts are separated by a downdraft that leads to a splitting of the storm and the development of two new centers of convection which move to the right and left of the original storm (Fig. 9.12b).

As discussed in this section, the tilting and stretching of horizontal vorticity associated with the vertical shear of the basic state wind can account for the development of mesoscale rotating supercells. This process does not, however, appear to be able to produce the large vorticities observed in the tornadoes that often accompany supercell thunderstorms. Numerical simulations suggest that these tend, rather, to involve tilting and stretching of especially strong horizontal vorticity produced by horizontal gradients in buoyancy that occur near the surface along the *gust front* where negatively buoyant outdrafts produced by convective downdrafts meet moist warm boundary layer air.

9.6.2 THE RIGHT-MOVING STORM

When the environmental wind shear is unidirectional, as in the case discussed above (see also Fig. 9.13a), the anticyclonic (left-moving) and cyclonic (right-moving) updraft cores are equally favored. In most severe storms in the central United States, however, the mean flow turns anticyclonically with height and this directional shear in the environment favors the right-moving storm center while inhibiting the left-moving center. Thus, right-moving storms are observed far more often than left-moving storms.

The dominance of the right-moving storm can be understood qualitatively by again considering the dynamical pressure perturbations. We define the basic state wind shear vector $\bar{\mathbf{S}} \equiv \partial \bar{\mathbf{V}}/\partial z$, which is assumed to turn clockwise with height. Noting that the basic state vorticity in this case is

$$\bar{\boldsymbol{\omega}} = \mathbf{k} \times \bar{\mathbf{S}} = -\mathbf{i}\, \partial \bar{v}/\partial z + \mathbf{j}\, \partial \bar{u}/\partial z$$

we see that there is a linear contribution to the dynamic pressure in (9.57) of the form

$$\nabla \cdot (\mathbf{U}' \times \bar{\boldsymbol{\omega}}) \approx -\nabla \cdot (w'\bar{\mathbf{S}})$$

From (9.57) the sign of the pressure perturbation owing to this effect may be determined by noting that

$$\nabla^2 p_{\text{dyn}} \sim -p_{\text{dyn}} \sim -\frac{\partial}{\partial x}(w'S_x) - \frac{\partial}{\partial y}(w'S_y) \tag{9.60}$$

which shows that there is a positive dynamical pressure perturbation upshear of the cell and a negative perturbation downshear (analogous to the positive pressure perturbation upwind and negative perturbation downwind of an obstacle). The resulting pattern of dynamical pressure perturbations is

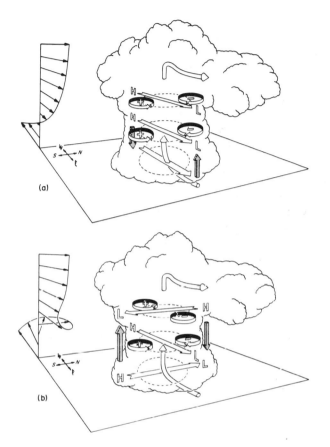

Fig. 9.13 Pressure and vertical vorticity perturbations produced by interaction of the updraft with environmental wind shear in a supercell storm. (a) Wind shear does not change direction with height. (b) Wind shear turns clockwise with height. Broad open arrows designate the shear vectors. *H* and *L* designate high and low dynamical pressure perturbations, respectively. Shaded arrows show resulting disturbance vertical pressure gradients. (After Klemp, 1987.)

shown in Fig. 9.13. In the case of unidirectional shear (Fig. 9.13a) the induced pressure pattern favors updraft growth on the leading edge of the storm. But, when the shear vector rotates clockwise with height as in Fig. 9.13b, (9.60) shows that a dynamical pressure disturbance pattern is induced in which there are an upward-directed vertical pressure gradient force on the flank of the cyclonically rotating cell and a downward-directed pressure gradient force on the flank of the anticyclonic cell. Thus, in the presence of clockwise rotation of the environmental shear stronger updrafts are favored in the right-moving cyclonic vortex to the south of the initial updraft.

9.7 Hurricanes

Hurricanes, which are also referred to as tropical cyclones and typhoons in some parts of the world, are intense vortical storms that develop over the tropical oceans in regions of very warm surface water. Although the winds and convective clouds observed in hurricanes are not truly axisymmetric about the vortex center, the fundamental aspects of hurricane dynamics can be modeled by idealizing the hurricane as an axisymmetric vortex. Typically hurricanes have radial scales of several hundred kilometers, similar to those of some midlatitude synoptic systems. But the horizontal scale of the region of intense convection and strong winds in a hurricane is typically only about 100 km in radius. Thus, it is reasonable to classify hurricanes as mesoscale systems.

Unlike the rotating convective storms treated in the previous section, the hurricane vortex cannot be understood without including the rotation of the earth in the vorticity balance. The rapid rotation observed in hurricanes is produced by concentration of the vertical component of absolute vorticity by vortex stretching, not by tilting horizontal vorticity into the vertical. Maximum tangential wind speeds in these storms range typically from 50 to 100 m s^{-1}. For such high velocities and relatively small scales, the centrifugal force term cannot be neglected compared to the Coriolis force. Thus, to a first approximation, the azimuthal velocity in a steady-state hurricane is in gradient wind balance with the radial pressure gradient force. Hydrostatic balance also holds on the hurricane scale, which implies that the vertical shear of the azimuthal velocity is a function of the radial temperature gradient.

9.7.1 DYNAMICS OF MATURE HURRICANES

An expression for the thermal wind in an axisymmetric hurricane can be easily derived starting from the gradient wind balance in cylindrical coordinates (r, λ, z). This can be written as

$$\frac{v_\lambda^2}{r} + f v_\lambda = \frac{\partial \Phi}{\partial r} \tag{9.61}$$

where r is the radial distance from the axis of the storm (positive outward) and v_λ is the tangential velocity (positive for anticlockwise flow). Alternatively, it is sometimes useful to use the absolute angular momentum $M_\lambda \equiv v_\lambda r + f r^2 / 2$ in place of v_λ since above the boundary layer M_λ is nearly conserved following the motion. (It thus plays the same role in cylindrically symmetric flow as the absolute momentum defined in Section 9.3 plays in linearly symmetric flow.) In terms of M_λ the gradient wind balance can be expressed as (see Problem 9.7)

$$\frac{M_\lambda^2}{r^3} - \frac{f^2 r}{4} = \frac{\partial \Phi}{\partial r} \tag{9.62}$$

We can eliminate Φ in (9.62) with the aid of the hydrostatic equation in log-pressure coordinates introduced in Section 8.4.1:

$$\frac{\partial \Phi}{\partial z^*} = \frac{RT}{H}$$

to obtain a relationship between the radial temperature gradient and the vertical shear of the absolute momentum:

$$\frac{1}{r^3} \frac{\partial M_\lambda^2}{\partial z^*} = \frac{R}{H} \frac{\partial T}{\partial r} \tag{9.63}$$

The cyclonic flow in a hurricane is observed to be a maximum near the top of the boundary layer. Above the boundary layer $\partial M_\lambda / \partial z^* < 0$, which by (9.63) implies that $\partial T / \partial r < 0$. Thus, a temperature maximum must occur at the center of the storm. This is consistent with the observation that hurricanes are *warm core* systems.

If we let the vertical scale of the disturbance be equal to the scale height H, the tangential velocity scale be $U \sim 50 \text{ m s}^{-1}$, the horizontal scale be $L \sim 100$ km, and assume that $f \sim 5 \times 10^{-5} \text{ s}^{-1}$ (corresponding to a latitude of about 20°), we find from (9.63) that the radial temperature fluctuation must have a magnitude

$$\delta T \sim (UL/R)(f + 2U/L) \sim 10°C$$

This strong radial variation in temperature is one of the most important thermodynamic characteristics of hurricanes.

The kinetic energy of hurricanes is maintained in the presence of boundary layer dissipation by conversion of latent heat energy acquired from the underlying ocean. This potential energy conversion is carried out by a transverse secondary circulation associated with the hurricane, as shown schematically in Fig. 9.14. This circulation consists of boundary layer inflow into a region of enhanced convection surrounding the storm center that is referred to as the *eyewall*, ascent within convective cloud towers that tend to be concentrated in the narrow outward-sloping eyewall, radial outflow in a thin layer near the tropopause, and gentle subsidence at large radius.

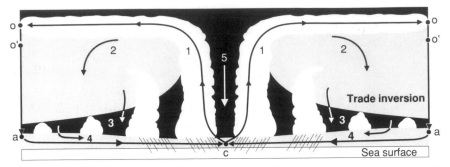

Fig. 9.14   Schematic cross section of the secondary meridional circulation in a mature hurricane. Air spirals in toward the eye (region 5) in the boundary layer (region 4),
ascends along constant-M surfaces in the eyewall cloud (region 1), and slowly
subsides and dries in regions 2 and 3. (After Emanuel, 1988.)

Observations show that evaporation of water from the sea surface into the
inward-flowing air in the boundary layer causes a large increase in θ_e as
the air approaches the eyewall region. Within the eyewall the θ_e and M_λ
surfaces coincide so that parcel ascent in the eyewall (along the path labeled
1 in Fig. 9.14) is neutral with respect to conditional symmetric instability
and thus does not require external forcing. The eyewall surrounds a central
eye of radius 5–50 km that is often calm and nearly cloud free.

The energetics of the steady-state hurricane can be viewed as an example
of a Carnot cycle heat engine in which heat is absorbed (in the form of
water vapor) from the ocean at temperature T_s and expelled by radiative
cooling to space at temperature T_0 at the top of the storm. Since $T_s \approx 300$ K
and $T_0 \approx 200$ K, the efficiency of the heat engine is very high.

9.7.2   HURRICANE DEVELOPMENT

The origin of tropical cyclones is still a matter of uncertainty. It is not
clear under what conditions a weak tropical disturbance can be transformed
into a hurricane. Although there are many tropical disturbances each year,
only rarely does one develop into a hurricane. Thus, the development of a
hurricane must require rather special conditions. Many theoretical investiga-
tions of this problem have assumed the initial existence of a small-amplitude
cylindrically symmetric disturbance and examined the conditions under
which unstable amplification of the disturbance can occur. As we saw in
Chapter 8, this sort of linear stability theory is quite successful in accounting
for the development of extratropical baroclinic disturbances.

In the tropics, however, the only well-documented linear instability is
conditional instability. This instability has its maximum growth rate on the

scale of an individual cumulus cloud. Therefore, it cannot explain the synoptic-scale organization of the motion. Observations indicate, moreover, that the mean tropical atmosphere is not saturated, even in the planetary boundary layer. Thus, a parcel must undergo a considerable amount of forced ascent before it reaches its LFC and becomes positively buoyant. Such forced parcel ascent can only be caused by small-scale motions, such as turbulent plumes in the boundary layer. The efficacy of boundary layer turbulence in producing parcel ascent to the LFC clearly depends on the temperature and humidity of the boundary layer environment. In the tropics it is difficult to initialize deep convection unless the boundary layer is brought toward saturation and destabilized, which may occur if there is large-scale (or mesoscale) ascent in the boundary layer. Thus, convection tends to be concentrated in regions of large-scale low-level convergence. This concentration arises not because the large-scale convergence directly "forces" the convection but rather because it preconditions the environment to be favorable for parcel ascent to the LFC.

Cumulus convection and the large-scale environmental motion may thus be viewed as cooperatively interacting. The diabatic heating owing to latent heat released by cumulus clouds produces a large-scale (or mesoscale) cyclonic disturbance; this disturbance in turn, through boundary layer pumping, drives the low-level moisture convergence necessary to maintain an environment favorable for development of cumulus convection. This interaction process is indicated schematically in Fig. 9.15.

There have been attempts to formalize these ideas into a linear stability theory (often referred to as *conditional instability of the second kind*, or *CISK*) which attributes hurricane growth to the organized interaction between the cumulus scale and the large-scale moisture convergence. These have not been very successful, since there is little evidence that such interaction leads to a growth rate maximum on the observed scale of hurricanes. In recent years a dramatically different view of the stability of the tropical atmosphere has come into vogue. This view, which is referred to as the *air–sea interaction theory*, is based on the fact emphasized in the previous subsection that the potential energy for hurricanes arises from the thermodynamic disequilibrium between the atmosphere and the underlying ocean.

The efficacy of air–sea interaction in providing potential energy to balance frictional dissipation depends on the rate of transfer of latent heat from the ocean to the atmosphere. This is a function of surface wind speed; strong surface winds, which produces a rough sea surface, can greatly increase the evaporation rate. Thus, hurricane development depends on the presence of a finite-amplitude initiating disturbance, such as an equatorial wave, to provide the winds required to produce strong evaporation. Given a suitable

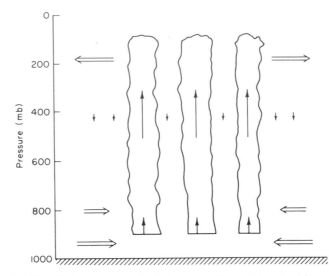

Fig. 9.15 Relationship between cumulonimbus convection and low-level large-scale convergence. Convergence moistens the environment and destabilizes it through layer ascent. This enables small-scale plumes to easily reach their levels of free convection and to produce cumulonimbus clouds. Diabatic heating owing to the resulting precipitation drives the large-scale circulation and thus maintains the large-scale convergence.

initial disturbance, a feedback may occur in which an increase in inward-spiraling surface winds increases the rate of moisture transfer from the ocean, which by bringing the boundary layer toward saturation increases the intensity of the convection, which further increases the secondary circulation.

The air–sea interaction theory is consistent with observations that hurricanes can develop only in the presence of very warm ocean surface temperatures. For surface temperatures less than 26°C, the converging flow in the boundary layer apparently cannot acquire a high enough equivalent potential temperature to sustain the intense transverse circulation needed to maintain the hurricane. Thus, it appears that hurricanes do not arise from linear instability of the tropical atmosphere, but develop from preexisting large-scale disturbances under rather special conditions.

Problems

9.1. Show by transforming from θ coordinates to height coordinates that the Ertel potential vorticity P is proportional to $F^2 N_s - S^4$. [See Eq. (9.28).]

9.2. Starting with the linearized Boussinesq equations for steady waves in a basic state zonal flow that is a function of height, derive (9.31) and verify the form given for the Scorer parameter.

9.3. An air parcel at 920 mb with temperature 20°C is saturated (mixing ratio 16 g kg^{-1}). Compute θ_e for the parcel.

9.4. Suppose that the mass of air in an entraining cumulus updraft increases exponentially with height so that $m = m_0 e^{z/H}$ where $H = 8$ km and m_0 is the mass at a reference level. If the updraft speed is 3 m s^{-1} at 2-km height, what is its value at a height of 8 km assuming that the updraft has zero net buoyancy?

9.5. Verify the approximate relationship between moist static energy and θ_e given by (9.41).

9.6. The azimuthal velocity component in some hurricanes is observed to have a radial dependence given by $v_\lambda = V_0(r_0/r)^2$ for distances from the center given by $r \geq r_0$. Letting $V_0 = 50$ m s^{-1} and $r_0 = 50$ km find the total geopotential difference between the far field ($r \to \infty$) and $r = r_0$, assuming gradient wind balance and $f_0 = 5 \times 10^{-5}$ s^{-1}. At what distance from the center does the Coriolis force equal the centrifugal force?

9.7. Starting with (9.61) derive the angular momentum form of the gradient wind balance for an axisymmetric vortex given by (9.62).

Suggested References

Hoskins (1982) discusses the semigeostrophic theory of frontogenesis and summarizes results from numerical models.

Durran (1990) reviews the dynamics of mountain waves and downslope windstorms.

Houze and Hobbs (1982) review observational aspects of convective storms.

Klemp (1987) describes the dynamics of tornadic thunderstorms.

Emanuel (1988) reviews the air–sea interaction theory for hurricanes.

Chapter

10 | The General Circulation

In its broadest sense the *general circulation* of the atmosphere is usually considered to include the totality of motions that characterizes the global-scale atmospheric flow. Specifically, the study of the general circulation is concerned with the dynamics of climate, that is, with the temporally averaged structures of the fields of wind, temperature, humidity, precipitation, and other meteorological variables. The general circulation may thus be considered to consist of the flow averaged in time over a period sufficiently long to remove the random variations associated with individual weather systems, but short enough to retain monthly and seasonal variations.

In the past, both observational and theoretical studies of the general circulation concentrated on the dynamics of the zonally averaged flow. The time-averaged circulation is, however, highly dependent on longitude owing to longitudinally asymmetric forcing by orography and land–sea heating contrasts. The longitudinally dependent components of the general circulation may be separated into *quasi-stationary circulations*, which vary little in time, *monsoonal circulations*, which are seasonally reversing, and various subseasonal and interannual components, which together account for the *low-frequency variability*. A complete understanding of the physical basis for the general circulation requires an explanation for not only the zonally averaged circulation but the longitudinally and time-varying components as well.

Nevertheless, to introduce the study of the general circulation it proves useful to isolate those processes that maintain the zonal mean flow (i.e., the flow averaged around latitude circles). This approach follows naturally from the linear wave studies of previous chapters in which flow fields were split into zonal mean and longitudinally dependent eddy components. In the present chapter, however, we concentrate not on the development and motion of the eddies but on the influence of the eddies on the structure of the zonal mean circulation. Focusing on the zonal mean allows us to isolate those features of the circulation that are not dependent on continentality and should thus be common to all thermally driven rotating fluid systems. In particular, we discuss the angular momentum and energy budgets of the zonally averaged flow. We also show that the mean meridional circulation (i.e., the circulation consisting of the zonal mean vertical and meridional velocity components) satisfies a diagnostic equation analogous to the omega equation of Section 6.4.1, but with the forcing determined by the distributions of diabatic heating and eddy heat and momentum fluxes.

Following our discussion of the zonal mean circulation we consider the longitudinally varying time-averaged circulation. In this chapter the primary emphasis is on extratropical aspects of the circulation; these can be discussed within the framework of quasi-geostrophic theory. The general circulation of the tropics will be considered in Chapter 11.

10.1 The Nature of the Problem

Theoretical speculation on the nature of the general circulation has quite a long history. Perhaps the most important early work on the subject was that of the eighteenth-century Englishman George Hadley. Hadley, in seeking a cause for the trade wind circulation, realized that this circulation must be a form of thermal convection driven by the difference in solar heating between the equatorial and polar regions. He visualized the general circulation as consisting of a zonally symmetric overturning in which the heated equatorial air rises and flows poleward, where it cools, sinks, and flows equatorward again. At the same time, the Coriolis force deflects the poleward-moving air at the upper levels to the east and the equatorward-moving air near the surface to the west. The latter is, of course, consistent with the observed northeasterly (southeasterly) surface flow in the Northern (Southern) Hemisphere trade wind zone. This type of circulation is now called a *Hadley circulation* or Hadley cell.

Although a circulation consisting of Hadley cells extending from equator to pole in each hemisphere is mathematically possible in the sense that such a circulation would not violate the laws of physics, the observed Hadley circulation is confined to the tropics. If the midlatitudes were dominated by Hadley cells, they would have zonal wind regimes characterized by upper-level westerlies and surface easterlies just as in the tropics. Extratropical upper tropospheric winds are indeed westerly. However, the surface winds in midlatitudes are also westerly; these are inconsistent with the momentum balance for a Hadley circulation.

Evidence from a number of studies indicates that for conditions existing in the earth's atmosphere a symmetric hemisphere-wide Hadley circulation would be baroclinically unstable. If such a circulation were to become established by some mechanism, it would quickly break down outside the tropics as baroclinic eddies developed and modified the zonal mean circulation through their heat and momentum fluxes.

The observed general circulation thus cannot be understood purely in terms of zonally symmetric processes. Rather, it can be thought of qualitatively as developing through three-dimensional interactions among radiative and dynamical processes. In the mean the net solar energy absorbed by the atmosphere and the earth must equal the infrared energy radiated back to space by the planet. The annually averaged solar heating is, however, strongly dependent on latitude, with a maximum at the equator and minima at the poles. The outgoing infrared radiation, on the other hand, is only weakly latitude dependent. Thus, there is a net radiation surplus in the equatorial region and deficit in the polar region. This differential heating warms the equatorial atmosphere relative to higher latitudes and creates a pole-to-equator temperature gradient. Hence it produces a growing store of zonal mean available potential energy. At some point however, the westerly thermal wind (which must develop if the motion is to be geostrophically balanced in the presence of the pole-to-equator temperature gradient) becomes baroclinically unstable. As shown in Chapter 8, the resulting baroclinic waves transport heat poleward. These waves will intensify until their heat transport (together with the heat transported by planetary waves and ocean currents) is sufficient to balance the radiation deficit in the polar regions so that the pole-to-equator temperature gradient ceases to grow. At the same time, these perturbations convert potential energy into kinetic energy, thereby maintaining the kinetic energy of the atmosphere against the effects of frictional dissipation.

From a thermodynamic point of view, the atmosphere may be regarded as a "heat engine" which absorbs net heat at relatively warm temperatures in the tropics (primarily in the form of latent heat owing to evaporation from the sea surface) and gives up heat at relatively cool temperatures in the extratropics. In this manner net radiation generates available potential energy, which is in turn partially converted to kinetic energy, which does work to maintain the circulation against frictional dissipation. Only a small fraction of the solar energy input actually gets converted to kinetic energy. Thus, from an engineer's viewpoint the atmosphere is a rather inefficient heat engine. However, if due account is taken of the many constraints operating on atmospheric motions, it appears that the atmosphere may in fact generate kinetic energy about as efficiently as dynamically possible.

The above qualitative discussion suggests that the gross features of the general circulation outside the tropics can be understood on the basis of quasi-geostrophic theory since, as we have previously seen, baroclinic instability is contained within the quasi-geostrophic framework. In view of this fact, and to keep the equations as simple as possible, our discussion of the zonally averaged and longitudinally varying components of the circulation outside the tropics will concentrate on those aspects that can be qualitatively represented by the quasi-geostrophic equations on the midlatitude β plane.

It should be recognized that a quasi-geostrophic model cannot provide a complete theory of the general circulation because in the quasi-geostrophic theory a number of assumptions are made concerning scales of motion that restrict in advance the possible types of solutions. Quantitative modeling of the general circulation requires complicated numerical models based on the primitive equations in spherical coordinates. The ultimate objective of these modeling efforts is to simulate the general circulation so faithfully that the climatological consequences of any change in the external parameters (such as the atmospheric concentration of carbon dioxide) can be accurately predicted. Present models can provide fairly accurate simulations of the current climate and plausible predictions of the response of the climate system to changes in external conditions. But uncertainties in the representations of a number of physical processes, particularly clouds and precipitation, limit the confidence that can be placed in quantitative climate change predictions based on such models.

10.2   The Zonally Averaged Circulation

The observed global distribution of the longitudinally averaged zonal wind for the two solstice seasons was shown in Fig. 6.1. Although there are important interhemispheric differences, the flow in both hemispheres is characterized by westerly jets with maximum zonal velocities at about 30°–35° latitude near the tropopause. These westerlies vary in longitude, especially in the Northern Hemisphere (see Fig. 6.2), but there is a significant zonally symmetric component, which we will refer to as the *mean zonal wind*.

Although departures of the time-averaged flow from zonal symmetry are important aspects of the general circulation, especially in the Northern Hemisphere, it is useful to first obtain an understanding of the dynamics of the zonally symmetric component before investigating the three-dimensional time mean circulation. In this section we examine the dynamics of zonally symmetric motions using quasi-geostrophic theory and the log-pressure coordinate system introduced in Chapter 8 and show that the meridional circulation associated with an axially symmetric vortex is dynamically analogous to the secondary divergent circulation in a baroclinic wave.

The component momentum equations, the hydrostatic approximation, the continuity equation, and the thermodynamic energy equation in the log-pressure system can be written as follows:

$$Du/Dt - fv + \partial\Phi/\partial x = X \tag{10.1}$$

$$Dv/Dt + fu + \partial\Phi/\partial y = Y \tag{10.2}$$

$$\partial\Phi/\partial z = H^{-1}RT \tag{10.3}$$

$$\partial u/\partial x + \partial v/\partial y + \rho_0^{-1}\partial(\rho_0 w)/\partial z = 0 \tag{10.4}$$

$$DT/Dt + (\kappa T/H)w = J/c_p \tag{10.5}$$

where

$$\frac{D}{Dt} \equiv \frac{\partial}{\partial t} + u\frac{\partial}{\partial x} + v\frac{\partial}{\partial y} + w\frac{\partial}{\partial z}$$

and X and Y designate the zonal and meridional components of drag owing to small-scale eddies.

For convenience in this and the following chapters we have dropped the asterisk notation used in Chapter 8 to distinguish the log-pressure coordinate from geometric height. Thus, z here designates the log-pressure variable defined in Section 8.4.1.

Analysis of the zonally averaged circulation involves study of the inter-action of longitudinally varying disturbances (referred to as eddies and denoted by primed variables) with the longitudinally averaged flow (referred to as the mean flow and denoted by overbars). Thus, any variable A is expanded in the form $A = \bar{A} + A'$. This sort of average is an *Eulerian mean* since it is evaluated at fixed latitude, height, and time. The Eulerian mean equations are obtained by taking zonal averages of (10.1)–(10.5). Such averaging is facilitated by using (10.4) to expand the material derivative for any variable A in flux form as follows:

$$\rho_0 \frac{DA}{Dt} = \rho_0 \left(\frac{\partial}{\partial t} + \mathbf{V} \cdot \nabla + w \frac{\partial}{\partial z} \right) A + A \left[\nabla \cdot (\rho_0 \mathbf{V}) + \frac{\partial}{\partial z} (\rho_0 w) \right] \quad (10.6)$$

Using the chain rule of differentiation and observing that $\rho_0 = \rho_0(z)$, this result can be rewritten as

$$\rho_0 \frac{DA}{Dt} = \frac{\partial}{\partial t} (\rho_0 A) + \frac{\partial}{\partial x} (\rho_0 A u) + \frac{\partial}{\partial y} (\rho_0 A v) + \frac{\partial}{\partial z} (\rho_0 A w)$$

Applying the zonal averaging operator then gives

$$\rho_0 \frac{\overline{DA}}{Dt} = \frac{\partial}{\partial t} (\rho_0 \bar{A}) + \frac{\partial}{\partial y} [\rho_0 (\bar{A}\bar{v} + \overline{A'v'})] + \frac{\partial}{\partial z} [\rho_0 (\bar{A}\bar{w} + \overline{A'w'})] \quad (10.7)$$

Here we have used the fact that $\partial(\bar{\ })/\partial x = 0$ since quantities with the overbar are independent of x. We have also used the fact that for any variables a and b

$$\overline{ab} = \overline{(\bar{a} + a')(\bar{b} + b')} = \overline{\bar{a}\bar{b}} + \overline{\bar{a}b'} + \overline{a'\bar{b}} + \overline{a'b'} = \bar{a}\bar{b} + \overline{a'b'}$$

which follows from that facts that $\bar{a}$ and $\bar{b}$ are independent of x and $\overline{a'} = \overline{b'} = 0$ so that, for example, $\overline{\bar{a}b'} = \bar{a}\overline{b'} = 0$.

The zonal-mean terms on the right in (10.7) can be rewritten in advective form if we first take the zonal mean of (10.4) to obtain the continuity equation for the mean meridional circulation:

$$\partial \bar{v}/\partial y + \rho_0^{-1} \partial (\rho_0 \bar{w})/\partial z = 0 \quad (10.8)$$

Applying the chain rule of differentiation to the mean terms on the right in (10.7) and substituting from (10.8) we can rewrite (10.7) as

$$\rho_0 \frac{\overline{DA}}{Dt} = \frac{\bar{D}}{Dt} (\rho_0 \bar{A}) + \frac{\partial}{\partial y} [\rho_0 (\overline{A'v'})] + \frac{\partial}{\partial z} [\rho_0 (\overline{A'w'})] \quad (10.9)$$

where

$$\frac{\bar{D}}{Dt} \equiv \frac{\partial}{\partial t} + \bar{v} \frac{\partial}{\partial y} + \bar{w} \frac{\partial}{\partial z} \quad (10.10)$$

is the rate of change following the mean meridional motion $(\bar{v}, \bar{w})$.

10.2.1   THE CONVENTIONAL EULERIAN MEAN

Applying the averaging scheme of (10.9) to (10.1) and (10.5) we obtain
the zonal mean zonal momentum and thermodynamic energy equations for
quasi-geostrophic motions on the midlatitude β plane:

$$\partial \bar{u}/\partial t - f_0 \bar{v} = -\partial(\overline{u'v'})/\partial y + \bar{X} \tag{10.11}$$

$$\partial \bar{T}/\partial t + N^2 HR^{-1}\bar{w} = -\partial(\overline{v'T'})/\partial y + \bar{J}/c_p \tag{10.12}$$

where N is the buoyancy frequency defined by

$$N^2 \equiv \frac{R}{H}\left(\frac{\kappa T_0}{H} + \frac{dT_0}{dz}\right)$$

In (10.11) and (10.12), consistent with quasi-geostrophic scaling, we have
neglected advection by the ageostrophic mean meridional circulation and
vertical eddy flux divergences. It is easily confirmed that for quasi-geostrophic scales these terms are small compared to the retained terms (see
Problem 10.4). We have included the zonally averaged turbulent drag in
(10.11) because stresses owing to unresolved eddies may be important not
only in the boundary layer, but also in the upper troposphere and lower
stratosphere (see Section 10.8.4).

Similar scaling shows that the zonally averaged meridional momentum
equation can be accurately approximated by geostrophic balance,

$$f_0\bar{u} = -\partial\bar{\Phi}/\partial y$$

This can be combined with the hydrostatic relationship (10.3) to give the
thermal wind relation

$$f_0 \, \partial \bar{u}/\partial z + RH^{-1}\,\partial \bar{T}/\partial y = 0 \tag{10.13}$$

This relationship between the zonal mean wind and potential temperature
distributions imposes a strong constraint on the ageostrophic mean
meridional circulation $(\bar{v}, \bar{w})$. In the absence of a mean meridional circulation the eddy momentum flux divergence in (10.11) and eddy heat flux
divergence in (10.12) would tend separately to change the mean zonal wind
and temperature fields and hence would destroy thermal wind balance. The
pressure gradient force that results from any small departure of the mean
zonal wind from geostrophic balance will, however, drive a mean meridional
circulation, which adjusts the mean zonal wind and temperature fields so

that (10.13) remains valid. In many situations this compensation allows the mean zonal wind to remain unchanged even in the presence of large eddy heat and momentum fluxes. The mean meridional circulation thus plays exactly the same role in the zonal mean circulation that the secondary divergent circulation plays in the synoptic-scale quasi-geostrophic system. In fact, for steady-state mean flow conditions the $(\bar{v}, \bar{w})$ circulation must just balance the eddy forcing plus diabatic heating so that the balances in (10.11) and (10.12) are as follows:

Coriolis force $f_0\bar{v} \approx$ divergence of eddy momentum fluxes

Adiabatic cooling $\approx$ diabatic heating plus convergence of
eddy heat fluxes

Analysis of observations shows that outside the tropics these balances appear to be approximately true above the boundary layer. Thus, changes in the zonal mean flow arise from small imbalances between the forcing terms and the mean meridional circulation.

The Eulerian mean meridional circulation can be determined in terms of the forcing from an equation similar to the omega equation of Section 6.4. Before deriving this equation it is useful to observe that the mean meridional mass circulation is nondivergent in the meridional plane. Thus, it can be represented in terms of a meridional mass transport streamfunction, which identically satisfies the continuity equation (10.8), by letting

$$\rho_0\bar{v} = -\frac{\partial \bar{\chi}}{\partial z}, \qquad \rho_0\bar{w} = \frac{\partial \bar{\chi}}{\partial y} \tag{10.14}$$

The relationship of the sign of the streamfunction $\bar{\chi}$ to the sense of the mean meridional circulation is shown schematically in Fig. 10.1.

The diagnostic equation for $\bar{\chi}$ is derived by first taking

$$f_0\frac{\partial}{\partial z}(10.11) + \frac{R}{H}\frac{\partial}{\partial y}(10.12)$$

and then using (10.13) to eliminate the time derivatives and (10.14) to express the mean meridional circulation in terms of $\bar{\chi}$. The resulting elliptic equation has the form

$$\frac{\partial^2 \bar{\chi}}{\partial y^2} + \frac{f_0^2}{N^2}\rho_0\frac{\partial}{\partial z}\left(\frac{1}{\rho_0}\frac{\partial \bar{\chi}}{\partial z}\right)$$

$$= \frac{\rho_0}{N^2}\left[\frac{\partial}{\partial y}\left(\frac{\kappa \bar{J}}{H} - \frac{R}{H}\frac{\partial(\overline{v'T'})}{\partial y}\right) - f_0\left(\frac{\partial^2(\overline{u'v'})}{\partial z\,\partial y} - \frac{\partial \bar{X}}{\partial z}\right)\right] \tag{10.15}$$

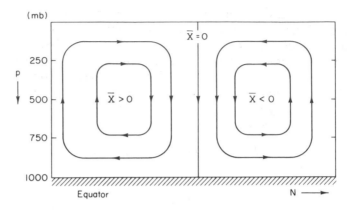

Fig. 10.1   Relationship of the Eulerian mean meridional streamfunction to the vertical and meridional motion.

Equation (10.15) can be used to diagnose qualitatively the mean meridional circulation. Since $\bar{\chi}$ must vanish on the boundaries it can be represented by a double Fourier series in y and z. Hence, the elliptic operator on the left-hand side of (10.15) is approximately proportional to $-\bar{\chi}$ and (10.15) states qualitatively that

$$\bar{\chi} \propto -\frac{\partial}{\partial y}\,(\text{diabatic heating}) + \frac{\partial^2}{\partial y^2}\,(\text{large-scale eddy heat flux})$$

$$+ \frac{\partial^2}{\partial y\,\partial z}\,(\text{large-scale eddy momentum flux}) + \frac{\partial}{\partial z}\,(\text{zonal drag force})$$

Now, diabatic heating in the Northern Hemisphere decreases for increasing y. Thus, the first term on the right is positive and tends to force a mean meridional cell with $\bar{\chi} > 0$. This is referred to as a *thermally direct* cell since warm air is rising and cool air sinking. It is this process that primarily accounts for the Hadley circulation of the tropics as illustrated in Fig. 10.2. For an idealized Hadley cell in the absence of eddy sources the differential diabatic heating would be balanced only by adiabatic cooling near the equator and adiabatic warming at higher latitudes.

In the extratropical Northern Hemisphere poleward eddy heat fluxes owing to both transient synoptic-scale eddies and stationary planetary waves tend to transfer heat poleward, producing a maximum poleward heat flux

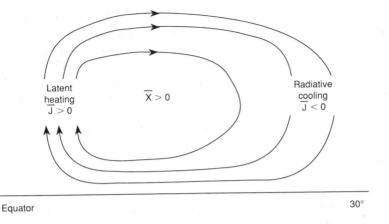

Fig. 10.2 Schematic Eulerian mean meridional circulation, showing the streamfunction for a thermally direct Hadley cell.

$\overline{v'T'}$ in the lower troposphere at about 50° latitude as shown in Fig. 10.3. Since $\bar{\chi}$ is proportional to the second derivative of $\overline{v'T'}$, which should be negative where $\overline{v'T'} > 0$, the heat flux forcing term should tend to produce a mean meridional cell with $\bar{\chi} < 0$ centered in the lower troposphere at midlatitudes. Thus, the eddy heat flux tends to drive an indirect meridional circulation.

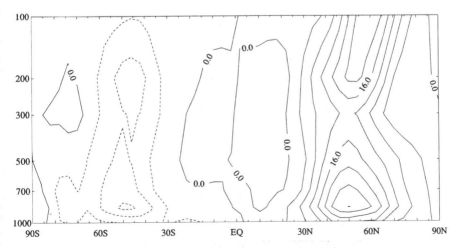

Fig. 10.3 Observed northward eddy heat flux distribution (°C m s^{-1}) for Northern Hemisphere winter. (Adapted from Schubert *et al.*, 1990.)

The existence of this indirect meridional circulation can be understood in terms of the need to maintain geostrophic and hydrostatic balance. North of the latitude where $\overline{v'T'}$ is a maximum there is a convergence of eddy heat flux, while equatorward of that latitude there is a divergence. Thus, the eddy heat transport tends to reduce the pole-to-equator mean temperature gradient. If the mean zonal flow is to remain geostrophic, the thermal wind must then also decrease. In the absence of eddy momentum transport, this decrease in the thermal wind can only be produced by the Coriolis torque owing to a mean meridional circulation with the sense of that in Fig. 10.4. At the same time, it is not surprising to find that the vertical mean motions required by continuity oppose the temperature tendency associated with the eddy heat flux by producing adiabatic warming in the region of eddy heat flux divergence and adiabatic cooling in the region of eddy heat flux convergence.

The next to last forcing term in (10.15) is proportional to the vertical gradient of the horizontal eddy momentum flux convergence. But it can be shown (Problem 10.5) that

$$-\frac{\partial^2 \overline{u'v'}}{\partial z\,\partial y} = +\frac{\partial \overline{v'\zeta'}}{\partial z}$$

Thus, this term is proportional to the vertical derivative of the meridional vorticity flux and plays the same role in forcing a mean meridional circulation as the differential vorticity advection term of the omega equation plays in forcing vertical motions in baroclinic disturbances. To interpret this eddy forcing physically, we suppose as shown in Fig. 10.5 that the momentum flux convergence (or the vorticity flux) is positive and increasing with height.

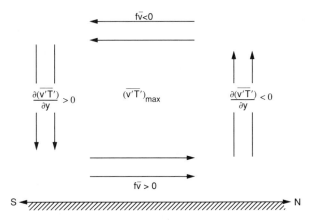

Fig. 10.4   Schematic Eulerian mean meridional circulation forced by poleward heat fluxes.

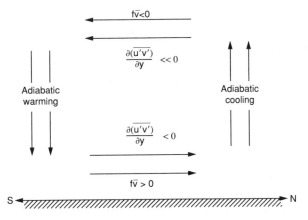

Fig. 10.5 Schematic Eulerian mean meridional circulation forced by a vertical gradient in the eddy momentum flux convergence.

This will be true in the Northern Hemisphere troposphere poleward of the core of the jet stream since $\overline{u'v'}$ tends to be poleward and to reach its maximum near the tropopause at about 30° (at the core of the mean jet stream) as shown in Fig. 10.6. For this configuration of momentum flux $\partial^2 \overline{u'v'}/\partial y\,\partial z < 0$ in the midlatitude troposphere, which again drives a mean meridional cell with $\bar\chi < 0$. From (10.11) it is clear that the Coriolis torque of this induced indirect meridional circulation is required to balance the

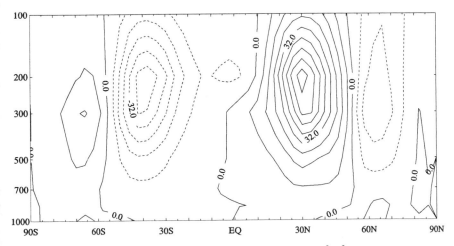

Fig. 10.6 Observed northward eddy momentum flux distribution ($m^2\,s^{-2}$) for Northern Hemisphere winter. (Adapted from Schubert *et al.*, 1990.)

acceleration owing to the momentum flux convergence, which would otherwise increase the vertical shear of the mean zonal wind and destroy the
thermal wind balance.

Thus, both the eddy heat flux and the eddy momentum flux distributions
tend to drive mean meridional cells in each hemisphere with rising motion
poleward of 45° and sinking motion equatorward of 45°. This eddy forcing
more than compensates the direct diabatic drive at midlatitudes and is
responsible for the observed thermally indirect *Ferrel cell.*

The resulting observed climatological Eulerian mean meridional circulation is shown in Fig. 10.7. It consists primarily of tropical Hadley cells
driven by diabatic heating and eddy-driven midlatitude Ferrel cells. There
are also minor thermally direct cells at polar latitudes. The meridional
circulation in the winter is much stronger than that in the summer, especially
in the Northern Hemisphere. This reflects the seasonal variation in both the
diabatic and eddy flux forcing terms in (10.15).

The zonal momentum balance in the upper troposphere in the tropical
and midlatitude cells is maintained by the balance between the Coriolis
force caused by the mean meridional drift and the eddy momentum flux
convergence. The heat balance is maintained by rising motion (adiabatic
cooling) balancing the diabatic heating in the tropics and the eddy heat
flux convergence at high latitudes and by subsidence (adiabatic warming)
balancing the eddy heat flux divergence in the subtropics.

Because of the appearance of eddy flux terms in both the mean momentum
and thermodynamic energy equations and the near cancellation of eddy

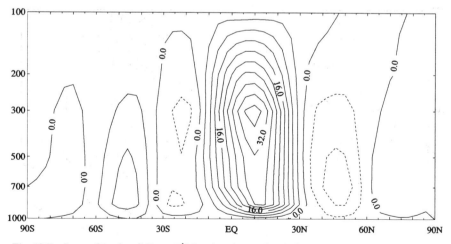

Fig. 10.7   Streamfunction (hPa m s^{-1}) for the observed Eulerian mean meridional circulation
for Northern Hemisphere winter, based on the data of Schubert *et al.* (1990).

and mean flow processes, it is rather inefficient to attempt to diagnose the net eddy forcing of the mean flow from the conventional Eulerian mean. It can be shown that similar eddy and mean flow compensation occurs in the Eulerian mean continuity equation for a long-lived tracer, so that tracer transport calculations are also inefficient in this formulation.

10.2.2 THE TRANSFORMED EULERIAN MEAN

An alternative approach to analysis of the zonal-mean circulation, which provides a clearer diagnosis of eddy forcing and also provides a more direct view of transport processes in the meridional plane, is the *transformed Eulerian mean* (TEM) formulation introduced by Andrews and McIntyre (1976). This transformation takes account of the fact that in (10.12) there tends to be a strong cancellation between the eddy heat flux convergence and adiabatic cooling, while the diabatic heating term is a small residual. Since in the mean an air parcel will rise to a higher equilibrium altitude only if its potential temperature is increased by diabatic heating, it is the *residual meridional circulation* associated with diabatic processes that determines the mean vertical circulation and hence the mean meridional mass flow.

The TEM equations can be obtained from (10.11) and (10.12) by defining the residual circulation $(\bar{v}^*, \bar{w}^*)$ as follows:

$$\bar{v}^* = \bar{v} - \rho_0^{-1} RH^{-1} \partial(\rho_0 \overline{v'T'}/N^2)/\partial z \qquad (10.16a)$$

$$\bar{w}^* = \bar{w} + RH^{-1} \partial(\overline{v'T'}/N^2)/\partial y \qquad (10.16b)$$

The residual vertical velocity defined in this manner clearly represents that part of the mean vertical velocity whose contribution to adiabatic temperature change is not canceled by the eddy heat flux divergence.

Substituting from (10.16) into (10.11) and (10.12) to eliminate $(\bar{v}, \bar{w})$ yields the TEM equations

$$\partial \bar{u}/\partial t - f_0 \bar{v}^* = +\rho_0^{-1} \nabla \cdot \mathbf{F} + \bar{X} \equiv \bar{G} \qquad (10.17)$$

$$\partial \bar{T}/\partial t + N^2 HR^{-1} \bar{w}^* = \bar{J}/c_p \qquad (10.18)$$

$$\partial \bar{v}^*/\partial y + \rho_0^{-1} \partial(\rho_0 \bar{w}^*)/\partial z = 0 \qquad (10.19)$$

where $\mathbf{F} \equiv \mathbf{j} F_y + \mathbf{k} F_z$, the *Eliassen–Palm flux* (EP flux), is a vector in the meridional (y, z) plane, which for large–scale quasi-geostrophic eddies has the components

$$F_y = -\rho_0 \overline{u'v'}, \qquad F_z = \rho_0 f_0 R \overline{v'T'}/(N^2 H) \qquad (10.20)$$

and $\bar{G}$ designates the total zonal force owing to both large-scale and small-scale eddies.

The TEM formulation clearly shows that the eddy heat and momentum fluxes do not act separately to drive changes in the zonal mean circulation but act only in the combination given by the divergence of the EP flux. The fundamental role of the eddies is thus to exert a zonal force. This eddy forcing of the zonal mean flow can be conveniently displayed by mapping the field of **F** and contouring the isolines of its divergence. When properly scaled by the basic state density, these contours give the zonal force per unit mass exerted by quasi-geostrophic eddies. The mean global EP flux divergence pattern for Northern Hemisphere winter is shown in Fig. 10.8. Note that in most of the extratropical troposphere the EP flux is convergent, so that the eddies exert an easterly zonal force on the atmosphere. On the seasonal time scale the zonal force owing to the EP flux divergence in (10.17) is nearly balanced by the Coriolis force of the residual mean meridional circulation. Conditions for this balance are discussed in the next subsection.

The structure of the residual mean meridional circulation can be determined by defining a residual streamfunction

$$\bar{\chi}^* \equiv \bar{\chi} + \rho_0 \frac{R}{H} \frac{\overline{v'T'}}{N^2}$$

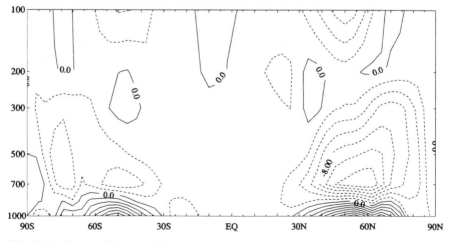

Fig. 10.8   Eliassen–Palm flux divergence divided by the standard density ρ_0 for Northern Hemisphere winter, based on the data of Schubert *et al.* (1990). (Units are m s^{-1} day^{-1}.)

It can then be shown by direct substitution into (10.14) and (10.15) that

$$\rho_0 \bar{v}^* = -\frac{\partial \bar{\chi}^*}{\partial z}, \qquad \rho_0 \bar{w}^* = \frac{\partial \bar{\chi}^*}{\partial y}$$

and

$$\frac{\partial^2 \bar{\chi}^*}{\partial y^2} + \rho_0 \frac{f_0^2}{N^2} \frac{\partial}{\partial z}\left(\frac{1}{\rho_0}\frac{\partial \bar{\chi}^*}{\partial z}\right) = \frac{\rho_0}{N^2}\left[\frac{\partial}{\partial y}\left(\frac{\kappa \bar{J}}{H}\right) + f_0 \frac{\partial \bar{G}}{\partial z}\right] \qquad (10.21)$$

In general, in the Northern (Southern) Hemisphere troposphere both the diabatic and EP flux contributions to the source term on the right in (10.21) are negative (positive), which implies that $\bar{\chi}^*$ itself is positive (negative) so that the residual meridional circulation consists of a single thermally direct overturning in each hemisphere, with the strongest cell in the winter hemisphere as shown in Fig. 10.9.

Unlike the conventional Eulerian mean, the residual mean vertical motion for time-averaged conditions is proportional to the rate of diabatic heating. It approximately represents the *diabatic circulation* in the meridional plane, that is, the circulation in which parcels that rise are diabatically heated and those that sink are diabatically cooled in order that their potential temperatures adjust to the local environment. The time-averaged residual mean meridional circulation thus approximates the mean motion of air parcels and hence, unlike the conventional Eulerian mean, provides an approximation to the mean advective transport of trace substances.

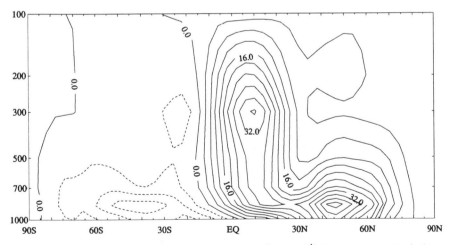

Fig. 10.9 Residual mean meridional streamfunction (hPa m s^{-1}) for Northern Hemisphere winter, based on the data of Schubert *et al.* (1990).

10.2.3   THE ZONAL MEAN POTENTIAL VORTICITY EQUATION

Further insight into the nature of the extratropical zonal mean circulation can be obtained by zonally averaging the quasi-geostrophic potential vorticity equation (6.25) to obtain

$$\partial \bar{q}/\partial t = -\partial(\overline{q'v'})/\partial y \tag{10.22}$$

where the zonal mean potential vorticity is

$$\bar{q} = f_0 + \beta y + \frac{1}{f_0}\frac{\partial^2 \bar{\Phi}}{\partial y^2} + \frac{f_0}{\rho_0}\frac{\partial}{\partial z}\left(\frac{\rho_0}{N^2}\frac{\partial \bar{\Phi}}{\partial z}\right) \tag{10.23}$$

and the eddy potential vorticity is

$$q' = \frac{1}{f_0}\left(\frac{\partial^2 \Phi'}{\partial x^2} + \frac{\partial^2 \Phi'}{\partial y^2}\right) + \frac{f_0}{\rho_0}\frac{\partial}{\partial z}\left(\frac{\rho_0}{N^2}\frac{\partial \Phi'}{\partial z}\right) \tag{10.24}$$

The quantity $\overline{q'v'}$ on the right-hand side in (10.22) is the divergence of the meridional flux of potential vorticity. According to (10.22), for adiabatic quasi-geostrophic flow the mean distribution of potential vorticity can be changed only if there is a nonzero flux of eddy potential vorticity. The zonal mean potential vorticity together with suitable boundary conditions completely determines the distribution of zonal mean geopotential and hence the zonal mean geostrophic wind and temperature distributions. Thus, eddy-driven mean flow accelerations require nonzero potential vorticity fluxes.

It can be shown that the potential vorticity flux is related to the eddy momentum and heat fluxes. We first note that

$$\overline{q'v'} = \overline{\frac{v'}{f_0}\frac{\partial^2 \Phi'}{\partial y^2}} + \frac{f_0}{\rho_0}\overline{v'\frac{\partial}{\partial z}\left(\frac{\rho_0}{N^2}\frac{\partial \Phi'}{\partial z}\right)}$$

Assuming that the eddy horizontal motion is geostrophic so that

$$f_0 v' = \partial\Phi'/\partial x \quad \text{and} \quad f_0 u' = -\partial\Phi'/\partial y$$

we can use the chain rule of differentiation and the fact that a perfect differential in x vanishes when zonally averaged to rewrite the terms on the right in the expression for potential vorticity flux as follows:

$$\overline{\frac{v'}{f_0}\frac{\partial^2 \Phi'}{\partial y^2}} = \frac{1}{f_0^2}\overline{\left(\frac{\partial\Phi'}{\partial x}\frac{\partial^2\Phi'}{\partial y^2}\right)}$$

$$= \frac{1}{f_0^2}\left[\frac{\partial}{\partial y}\overline{\left(\frac{\partial\Phi'}{\partial x}\frac{\partial\Phi'}{\partial y}\right)} - \frac{1}{2}\frac{\partial}{\partial x}\overline{\left(\frac{\partial\Phi'}{\partial y}\right)^2}\right] = -\frac{\partial}{\partial y}(\overline{u'v'})$$

and

$$\frac{f_0}{\rho_0}\overline{v'\frac{\partial}{\partial z}\left(\frac{\rho_0}{N^2}\frac{\partial\Phi'}{\partial z}\right)} = \frac{1}{\rho_0}\left[\frac{\partial}{\partial z}\left(\frac{\rho_0}{N^2}\overline{\frac{\partial\Phi'}{\partial x}\frac{\partial\Phi'}{\partial z}}\right) - \frac{\rho_0}{2N^2}\frac{\partial}{\partial x}\overline{\left(\frac{\partial\Phi'}{\partial z}\right)^2}\right]$$

$$= \frac{f_0}{\rho_0}\frac{\partial}{\partial z}\left(\frac{\rho_0}{N^2}\overline{v'\frac{\partial\Phi'}{\partial z}}\right)$$

Thus,

$$\overline{q'v'} = -\frac{\partial\overline{u'v'}}{\partial y} + \frac{f_0}{\rho_0}\frac{\partial}{\partial z}\left(\frac{\rho_0}{N^2}\overline{v'\frac{\partial\Phi'}{\partial z}}\right) \tag{10.25}$$

so that it is not the momentum flux $\overline{u'v'}$ or heat flux $\overline{v'\,\partial\Phi'/\partial z}$ that drives net changes in the mean flow distribution, but rather the combination given by the potential vorticity flux. Under some circumstances the eddy momentum flux and eddy heat flux may individually be large, but the combination (10.25) actually vanishes. This cancellation effect makes the traditional Eulerian mean formulation a poor framework for analysis of mean flow forcing by eddies.

Comparing (10.25) and (10.20) we see that the potential vorticity flux is proportional to the divergence of the EP flux vector:

$$\overline{q'v'} = \rho_0^{-1}\boldsymbol{\nabla}\cdot\mathbf{F} \tag{10.26}$$

Thus, the contribution of large-scale motions to the zonal force in (10.17) equals the meridional flux of quasi-geostrophic potential vorticity. If the motion is adiabatic and the potential vorticity flux is nonzero, (10.22) shows that the mean flow distribution must change in time. Thus, there cannot be complete compensation between the Coriolis torque and zonal force terms in (10.17).

10.3   The Angular Momentum Budget

In the previous section we used the quasi-geostrophic version of the zonal mean equations to show that large-scale eddies play an essential part in the maintenance of the zonal mean circulation in the extratropics. In particular, we contrasted the mean flow forcing as represented by the conventional Eulerian mean and TEM formulations. In this section we expand our consideration of the momentum budget by considering the overall balance of *angular* momentum for the atmosphere and the earth combined. Thus, rather than simply considering the balance of momentum for a given latitude and height in the atmosphere, we must consider the transfer of angular momentum between the earth and atmosphere and the flow of angular momentum in the atmosphere.

It would be possible to utilize the complete spherical coordinate version of the TEM equations for this analysis. But we are primarily concerned with the angular momentum balance for a zonal ring of air extending from the surface to the top of the atmosphere. In that case it proves simpler to use the conventional Eulerian mean formulation. It also proves convenient to use a special vertical coordinate, called the *sigma coordinate*, in which the surface of the earth is a coordinate surface.

Since the average rotation rate of the earth is itself observed to be very close to constant, the atmosphere must also on the average conserve its angular momentum. The atmosphere gains angular momentum from the earth in the tropics where the surface winds are easterly (i.e., where the angular momentum of the atmosphere is less than that of the earth) and gives up angular momentum to the earth in middle latitudes where the surface winds are westerly. Thus, there must be a net poleward transport of angular momentum within the atmosphere; otherwise the torque owing to surface friction would decelerate both the easterlies and westerlies. Furthermore, the angular momentum given by the earth to the atmosphere in the belt of easterlies must just balance the angular momentum given up to the earth in the westerlies if the global angular momentum of the atmosphere is to remain constant.

In the equatorial regions the poleward momentum transport is divided between the advection of absolute angular momentum by the poleward flow in the axially symmetric Hadley circulation and transport by eddies. In midlatitudes, however, it is primarily the eddy motions that transport momentum poleward, and the angular momentum budget of the atmosphere must be qualitatively as shown in Fig. 10.10.

As Fig. 10.10 suggests, in the winter season there is a maximum poleward flux of angular momentum at about 30° latitude and a maximum horizontal flux convergence at about 45°. This maximum in the flux convergence is a reflection of the strong energy conversion in the upper-level westerlies and is the mechanism whereby the atmosphere can maintain a positive zonal wind in the middle latitudes despite the momentum lost to the surface.

It is convenient to analyze the momentum budget in terms of absolute angular momentum since this is conserved for an air parcel in the absence of frictional or pressure torques. The absolute angular momentum per unit mass of atmosphere is

$$M = (\Omega a \cos \phi + u)a \cos \phi$$

where as before a is the radius of the earth.

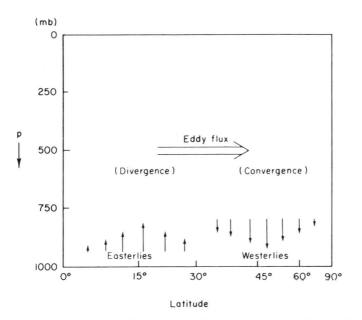

Fig. 10.10 Schematic mean angular momentum budget for the atmosphere-earth system.

The absolute angular momentum of an individual air parcel can be changed only by torques caused by the zonal pressure gradient and the eddy stresses. In isobaric coordinates Newton's second law in its angular momentum form is thus

$$\frac{DM}{Dt} = -a\cos\phi\left[\frac{\partial\Phi}{\partial x} + g\frac{\partial\tau_E^x}{\partial p}\right] \qquad (10.27)$$

where τ_E^x is the zonal component of the vertical eddy stress, and it is assumed that horizontal eddy stresses are negligible compared to the vertical eddy stress.

10.3.1 SIGMA COORDINATES

In neither the isobaric nor the log-pressure coordinate system does the lower boundary exactly coincide with a coordinate surface. In analytical studies it is usual to assume that the lower boundary can be approximated as a constant-pressure surface and to apply the approximate condition

$$\omega(p_s) = -\rho_0 gw(z_0)$$

as the lower boundary condition. Here we have assumed that the height of the ground z_0 is coincident with the pressure surface p_s (where p_s is usually set equal to 1000 mb). These assumptions are of course not strictly valid even when the ground is level. Pressure does change at the ground. But, more importantly, the height of the ground generally varies so that even if the pressure tendency were zero everywhere, the lower boundary condition should not be applied at a constant p_s. Rather, we should set $p_s = p_s(x, y)$. It is very inconvenient for mathematical analysis, however, to have a boundary condition that must be applied at a surface that is a function of the horizontal variables.

This problem can be overcome by defining a vertical coordinate that is proportional to pressure normalized by the surface pressure. The most common form of this coordinate is the *sigma coordinate*, defined as $\sigma \equiv p/p_s$, where $p_s(x, y, t)$ is the pressure at the surface. Thus, σ is a nondimensional independent vertical coordinate that decreases upward from a value of $\sigma = 1$ at the ground to $\sigma = 0$ at the top of the atmosphere. In sigma coordinates the lower boundary condition will always apply exactly at $\sigma = 1$. Furthermore, the vertical σ velocity defined by

$$\dot{\sigma} \equiv D\sigma/Dt$$

will always be zero at the ground even in the presence of sloping terrain. Thus, the lower boundary condition in the σ system is merely

$$\dot{\sigma} = 0 \qquad \text{at } \sigma = 1$$

To transform the dynamical equations from the isobaric system to the σ system we first transform the pressure gradient force in a manner analogous to that shown in Section 1.6.3. Applying (1.23) with p replaced by Φ, s replaced by σ, and z replaced by p, we find that

$$(\partial\Phi/\partial x)_\sigma = (\partial\Phi/\partial x)_p + \sigma(\partial \ln p_s/\partial x)(\partial\Phi/\partial\sigma) \tag{10.28}$$

Since any other variable will transform in an analogous way, we can write the general transformation as

$$\nabla_p(\ ) = \nabla_\sigma(\ ) - \sigma \nabla \ln p_s \, \partial(\ )/\partial\sigma \tag{10.29}$$

Applying the transformation (10.29) to the momentum equation (3.2) we get

$$\frac{D\mathbf{V}}{Dt} + f\,\mathbf{k}\times\mathbf{V} = -\nabla\Phi + \sigma\,\nabla \ln p_s\frac{\partial\Phi}{\partial\sigma} \tag{10.30}$$

where ∇ is now applied holding σ constant, and the total differential is

$$\frac{D}{Dt} = \frac{\partial}{\partial t} + \mathbf{V} \cdot \nabla + \dot{\sigma}\frac{\partial}{\partial \sigma} \tag{10.31}$$

The equation of continuity can be transformed to the σ system by first using (10.29) to express the divergence of the horizontal wind as

$$\nabla_p \cdot \mathbf{V} = \nabla_\sigma \cdot \mathbf{V} - \sigma(\nabla \ln p_s) \cdot \partial\mathbf{V}/\partial\sigma \tag{10.32}$$

To transform the term $\partial\omega/\partial p$ we first note that since p_s does not depend on σ

$$\frac{\partial}{\partial p} = \frac{\partial}{\partial(\sigma p_s)} = \frac{1}{p_s}\frac{\partial}{\partial\sigma}$$

Thus, the continuity equation (3.5) can be written as

$$p_s(\nabla_p \cdot \mathbf{V}) + \partial\omega/\partial\sigma = 0 \tag{10.33}$$

Now the sigma vertical velocity can be written as

$$\dot{\sigma} = \left(\frac{\partial\sigma}{\partial t} + \mathbf{V} \cdot \nabla\sigma\right)_p + \omega\frac{\partial\sigma}{\partial p} = -\frac{\sigma}{p_s}\left(\frac{\partial p_s}{\partial t} + \mathbf{V} \cdot \nabla p_s\right) + \frac{\omega}{p_s}$$

Differentiating the above with respect to σ and rearranging, we obtain the transformed continuity equation

$$\frac{\partial p_s}{\partial t} + \nabla \cdot (p_s\mathbf{V}) + p_s\frac{\partial\dot{\sigma}}{\partial\sigma} = 0 \tag{10.34}$$

With the aid of the equation of state and Poisson's equation (2.44) the hydrostatic approximation can be written in the sigma system as

$$\frac{\partial\Phi}{\partial\sigma} = -\frac{RT}{\sigma} = -\frac{R\theta}{\sigma}(p/p_0)^\kappa \tag{10.35}$$

where $p_0 = 10^3$ kPa (1000 mb).

Expanding the total derivative in (2.46) we may write the thermodynamic energy equation for sigma coordinates as

$$\frac{\partial\theta}{\partial t} + \mathbf{V} \cdot \nabla\theta + \dot{\sigma}\frac{\partial\theta}{\partial\sigma} = \frac{J}{c_p}\frac{\theta}{T} \tag{10.36}$$

10.3.2   THE ZONAL MEAN ANGULAR MOMENTUM

We now transform the angular momentum equation (10.27) into sigma coordinates to yield

$$\left(\frac{\partial}{\partial t}+\mathbf{V}\cdot\nabla+\dot{\sigma}\frac{\partial}{\partial\sigma}\right)M=-a\cos\phi\left(\frac{\partial\Phi}{\partial x}+\frac{RT}{p_s}\frac{\partial p_s}{\partial x}+\frac{g}{p_s}\frac{\partial\tau_E^x}{\partial\sigma}\right)\quad(10.37)$$

where we have used the hydrostatic approximation to transform the eddy stress term. Multiplying the continuity equation (10.34) by M and adding the result to (10.37) multiplied by p_s we obtain the flux form of the angular momentum equation:[1]

$$\frac{\partial(p_s M)}{\partial t}=-\nabla\cdot(p_s M\mathbf{V})-\frac{\partial(p_s M\dot{\sigma})}{\partial\sigma}$$

$$(10.38)$$

$$-a\cos\phi\left[p_s\frac{\partial\Phi}{\partial x}+RT\frac{\partial p_s}{\partial x}\right]-ga\cos\phi\frac{\partial\tau_E^x}{\partial\sigma}$$

To obtain the zonal mean angular momentum budget we must average (10.38) in longitude. Using the spherical coordinate expansion for the horizontal divergence as given in Appendix C, we have

$$\nabla\cdot(p_s M\mathbf{V})=\frac{1}{a\cos\phi}\left[\frac{\partial(p_s Mu)}{\partial\lambda}+\frac{\partial(p_s Mv\cos\phi)}{\partial\phi}\right]\quad(10.39)$$

We also observe that the bracketed term on the right in (10.38) can be rewritten as

$$\left[p_s\frac{\partial}{\partial x}(\Phi-RT)+\frac{\partial}{\partial x}(p_s RT)\right]\quad(10.40)$$

But with the aid of the hydrostatic equation (10.35) we can write

$$(\Phi-RT)=\Phi+\sigma\,\partial\Phi/\partial\sigma=\partial(\sigma\Phi)/\partial\sigma$$

Thus, recalling that p_s does not depend on σ we obtain

$$\left[p_s\frac{\partial\Phi}{\partial x}+RT\frac{\partial p_s}{\partial x}\right]=\left[\frac{\partial}{\partial\sigma}\left(p_s\sigma\frac{\partial\Phi}{\partial x}\right)+\frac{\partial}{\partial x}(p_s RT)\right]\quad(10.41)$$

---

[1] It may be shown (Problem 10.2) that in sigma coordinates the mass element $\rho_0\,dx\,dy\,dz$ takes the form $-g^{-1}p_s\,dx\,dy\,d\sigma$. Thus, p_s in sigma space plays a role similar to that of density in physical space.

Substituting from (10.39) and (10.41) into (10.38) and taking the zonal average gives

$$\frac{\partial(\overline{p_s M})}{\partial t} = -\frac{1}{\cos \phi} \frac{\partial}{\partial y} (\overline{p_s M v} \cos \phi)$$

$$-\frac{\partial}{\partial \sigma} \left[\overline{p_s M \dot{\sigma}} + ga \cos \phi (\overline{\tau_E^x}) + (a \cos \phi) \overline{\sigma p_s \frac{\partial \Phi}{\partial x}} \right] \tag{10.42}$$

The terms on the right in (10.42) represent the convergence of the horizontal flux of angular momentum and the convergence of the vertical flux of angular momentum, respectively. This differs from (10.11) not only because of the retention of nongeostrophic terms but also because the zonal averaging is carried out with σ rather than z held constant.

Integrating (10.42) vertically from the surface of the earth ($\sigma = 1$) to the top of the atmosphere ($\sigma = 0$) and recalling that $\dot{\sigma} = 0$ for $\sigma = 0, 1$ we have

$$\int_0^1 g^{-1} \frac{\partial}{\partial t} \overline{p_s M} \, d\sigma = -(g \cos \phi)^{-1} \int_0^1 \frac{\partial}{\partial y} (\overline{p_s M v} \cos \phi) \, d\sigma$$

$$-a \cos \phi [(\overline{\tau_E^x})_{\sigma=1} + \overline{p_s \, \partial h/\partial x}] \tag{10.43}$$

where $h(x, y) = g^{-1} \Phi(x, y, 1)$ is the height of the lower boundary ($\sigma = 1$), and we have assumed that the eddy stress vanishes at $\sigma = 0$.

Equation (10.43) expresses the angular momentum budget for a zonal ring of air of unit meridional width, extending from the ground to the top of the atmosphere. In the long-term mean the three terms on the right, representing the convergence of the meridional angular momentum flux, the torque due to small-scale turbulent fluxes at the surface, and the surface pressure torque, must balance. In the sigma coordinate system the surface pressure torque takes the particularly simple form $\overline{-p_s \, \partial h/\partial x}$. Thus, the pressure torque acts to transfer angular momentum from the atmosphere to the ground, provided that the surface pressure and the slope of the ground ($\partial h/\partial x$) are positively correlated. Observations indicate that this is generally the case in middle latitudes because there is a slight tendency for the surface pressure to be higher on the western sides of mountains than on the eastern sides (see Fig. 4.9). In midlatitudes of the Northern Hemisphere the surface pressure torque provides nearly half of the total atmosphere–surface momentum exchange, but in the tropics and the Southern Hemisphere the exchange is dominated by turbulent eddy stresses.

The role of eddy motions in providing the meridional angular momentum transport necessary to balance the surface angular momentum sinks can be elucidated best if we divide the flow into zonal mean and eddy components by letting

$$M = \bar{M} + M' = (\Omega a \cos \phi + \bar{u} + u')a \cos \phi$$

$$p_s v = \overline{(p_s v)} + (p_s v)'$$

where primes indicate deviations from the zonal mean. Thus, the meridional flux becomes

$$\overline{(p_s M v)} = [\Omega a \cos \phi \overline{p_s v} + \overline{\bar{u} p_s v} + \overline{u'(p_s v)'}]\Omega a \cos \phi \qquad (10.44)$$

The three terms on the right in this expression are called the Ω-momentum flux, the drift, and the eddy momentum flux, respectively.

The drift term is important in the tropics, but in midlatitudes it is small compared to the eddy flux and can be neglected in an approximate treatment. Furthermore, we can show that the Ω-momentum flux does not contribute to the vertically integrated flux. Averaging the continuity equation zonally and integrating vertically we obtain

$$\frac{\partial p_s}{\partial t} = -(\cos \phi)^{-1} \frac{\partial}{\partial y} \int_0^1 \overline{p_s v} \cos \phi \, d\sigma \qquad (10.45)$$

Thus, for time-averaged flow [where the left-hand side of (10.45) vanishes] there is no net mass flow across latitudes circles. The vertically integrated angular momentum flux is therefore given approximately by

$$\int_0^1 \overline{p_s M v} \, d\sigma \approx \int_0^1 \overline{u'(p_s v)'} a \cos \phi \, d\sigma \approx \int_0^1 a \cos \phi \, \bar{p}_s \overline{u'v'} \, d\sigma \qquad (10.46)$$

where we have assumed that the fractional change in p_s is small compared to the change in v' so that $(p_s v)' \approx \bar{p}_s v'$. The angular momentum flux is thus proportional to the negative of the meridional EP flux.

In the Northern Hemisphere, as shown in Fig. 10.10, the eddy momentum flux is positive and decreasing with latitude in the belt of westerlies. For quasi-geostrophic flow positive eddy momentum flux requires that the eddies be asymmetric in the horizontal plane with the trough and ridge axes tilting as indicated in Fig. 10.11. When the troughs and ridges on the average have southwest-to-northeast phase tilt, the zonal flow will be larger than average $(u' > 0)$ where the meridional flow is poleward $(v' > 0)$ and less than average $(u' < 0)$ where the flow is equatorward. Thus, $\overline{u'v'} > 0$ and the eddies will systematically transport positive zonal momentum poleward.

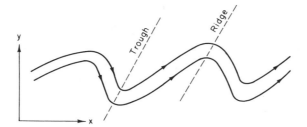

Fig. 10.11 Schematic streamlines for a positive eddy momentum flux.

As shown in (10.42) the total vertical momentum flux consists of the flux owing to large-scale motions $\overline{p_s M \dot{\sigma}}$, the flux owing to the pressure torque $a \cos \phi \overline{\sigma p_s \partial \Phi / \partial x}$, and the flux owing to small-scale turbulent stresses $ga \cos \phi \overline{\tau_E^x}$. As was previously mentioned, the last two are responsible for the transfer of momentum from the earth to the atmosphere in the tropics and from the atmosphere to the earth in midlatitudes. Outside the planetary boundary layer, however, the vertical momentum transport in the troposphere is primarily due to the Ω-momentum flux, $\Omega a \cos \phi \, \overline{p_s \dot{\sigma}}$.

Estimates of the torques owing to turbulent transfer from the surface and the pressure torque owing to large-scale topography have been attempted by several investigators. One such estimate of the latitudinal variation of the eastward torque exerted by the earth on the atmosphere is shown in Fig. 10.12a. The northward flux of angular momentum required to balance the estimated total surface torque is indicated in Fig.10.12b. This flux can also be directly estimated from wind data using (10.46). In Fig. 10.12b an estimate based on the observed winds is compared with the northward angular momentum flux required to balance the torques shown in Fig. 10.12a. Considering the many uncertainties in these measurements, the agreement is remarkable. It should also be mentioned here that except for the belt within 10° of the equator, almost all of the northward flux is due to the eddy flux term $\overline{u'v'}$. Thus, the momentum budget and the energy cycle both depend critically on the transports by the eddies.

10.4 The Lorenz Energy Cycle

In the previous section we discussed the interaction between the zonally averaged flow and longitudinally varying eddy motions in terms of the angular momentum balance. It is also useful to examine the exchange of energy between the eddies and the mean flow. As in Section 10.2, we limit the analysis to quasi-geostrophic flow on a midlatitude β plane. The

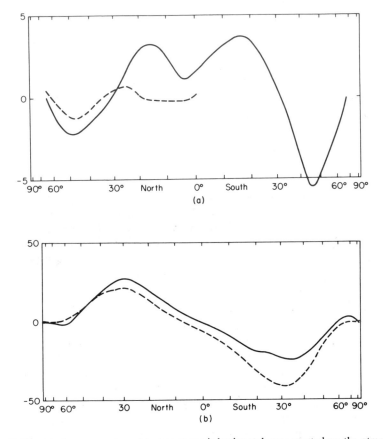

Fig. 10.12 (a) Average eastward torque per unit horizontal area exerted on the atmosphere by surface friction (solid curve) and by mountains in the Northern Hemisphere (dashed curve) in units of 10^5 kg s^{-2} (b) The observed transport of angular momentum in units of 10^{18} kg m^2 s^{-2} (solid curve) and the required transport (dashed curve) as given by the observed surface torques. (After Lorenz, 1967.)

Eulerian mean equations can then be written as

$$\partial \bar{u} / \partial t - f_0 \bar{v} = -\partial(\overline{u'v'})/\partial y + \bar{X} \qquad (10.47)$$

$$f_0 \bar{u} = -\partial \bar{\Phi}/\partial y \qquad (10.48)$$

$$\frac{\partial}{\partial t}\left(\frac{\partial \bar{\Phi}}{\partial z}\right) + \bar{w} N^2 = \frac{\kappa}{H}\bar{J} - \frac{\partial}{\partial y}\left(\overline{v'\frac{\partial \Phi'}{\partial z}}\right) \qquad (10.49)$$

$$\partial \bar{v}/\partial y + \rho_0^{-1}\, \partial(\rho_0 \bar{w})/\partial z = 0 \qquad (10.50)$$

Here we have used the hydrostatic approximation (10.3) to express the temperature in (10.49) in terms of the differential geopotential thickness, and we have again neglected vertical eddy fluxes and advection by the mean meridional circulation in (10.47) and (10.49). We have, however, included a turbulent drag force $\bar{X}$ in (10.47), since dissipation by unresolved turbulent eddies is an essential element in the energy balance.

In order to analyze the exchange of energy between the mean flow and the eddies, we require a similar set of dynamical equations for the eddy motion. For simplicity we assume that the eddies satisfy the following linearized set of equations:[2]

$$\left(\frac{\partial}{\partial t}+\bar{u}\frac{\partial}{\partial x}\right)u'-\left(f_0-\frac{\partial\bar{u}}{\partial y}\right)v'=-\frac{\partial\Phi'}{\partial x}+X' \qquad (10.51)$$

$$\left(\frac{\partial}{\partial t}+\bar{u}\frac{\partial}{\partial x}\right)v'+f_0u'=-\frac{\partial\Phi'}{\partial y}+Y' \qquad (10.52)$$

$$\left(\frac{\partial}{\partial t}+\bar{u}\frac{\partial}{\partial x}\right)\frac{\partial\Phi'}{\partial z}+v'\frac{\partial}{\partial y}\left(\frac{\partial\bar{\Phi}}{\partial z}\right)+N^2w'=\frac{\kappa J'}{H} \qquad (10.53)$$

$$\frac{\partial u'}{\partial x}+\frac{\partial v'}{\partial y}+\frac{1}{\rho_0}\frac{\partial(\rho_0w')}{\partial z}=0 \qquad (10.54)$$

where X' and Y' are the zonally varying components of drag owing to unresolved turbulent motions.

We now define a global average

$$\langle\ \rangle\equiv A^{-1}\int_0^\infty\int_0^D\int_0^L(\ )\,dx\,dy\,dz$$

where L is the distance around a latitude circle, D is the meridional extent of the midlatitude β plane, and A designates the total horizontal area of the β plane. Then for any quantity Ψ

$\langle\partial\Psi/\partial x\rangle=0$

$\langle\partial\Psi/\partial y\rangle=0$ if Ψ vanishes at $y=\pm D$

$\langle\partial\Psi/\partial z\rangle=0$ if Ψ vanishes at $z=0$ and $z\to\infty$

[2] A similar analysis can be carried out for the nonlinear case.

An equation for the evolution of the mean flow kinetic energy can then be obtained by multiplying (10.47) by $\rho_0 \bar{u}$ and (10.48) by $\rho_0 \bar{v}$ and adding the results to get

$$\rho_0 \frac{\partial}{\partial t}\left(\frac{\bar{u}^2}{2}\right) = -\rho_0 \bar{v}\frac{\partial \bar{\Phi}}{\partial y} - \rho_0 \bar{u}\frac{\partial}{\partial y}(\overline{u'v'}) + \rho_0 \bar{u}\bar{X}$$

Integrating over the entire volume we get

$$\frac{d}{dt}\left\langle \frac{\rho_0 \bar{u}^2}{2}\right\rangle = +\left\langle \rho_0 \bar{\Phi}\frac{\partial \bar{v}}{\partial y}\right\rangle + \left\langle \rho_0 \overline{u'v'}\frac{\partial \bar{u}}{\partial y}\right\rangle + \left\langle \rho_0 \bar{u}\bar{X}\right\rangle \tag{10.55}$$

where we have assumed that $\bar{v} = 0$ and $\overline{u'v'} = 0$ for $y = \pm D$. The terms on the right-hand side in (10.55) can be interpreted as the work done by the zonal mean pressure force, the conversion of eddy kinetic energy to zonal mean kinetic energy, and dissipation by the zonal mean eddy stress. Alternatively, the first term on the right can be rewritten with the aid of the continuity equation to yield

$$\left\langle \bar{\Phi}\frac{\partial \rho_0 \bar{v}}{\partial y}\right\rangle = -\left\langle \bar{\Phi}\frac{\partial \rho_0 \bar{w}}{\partial z}\right\rangle = \left\langle \rho_0 \bar{w}\frac{\partial \bar{\Phi}}{\partial z}\right\rangle$$

where we have assumed that $\rho_0 \bar{w} = 0$ at $z = 0, \infty$. Thus, averaged over the whole domain the pressure work term is proportional to the correlation between the zonal mean vertical mass flux $\rho_0 \bar{w}$ and the zonal mean temperature (or thickness). This term will be positive if on the average warm air is rising and cold air sinking, that is, if there is a conversion from potential to kinetic energy.

In Section 8.3.1 we showed that in the quasi-geostrophic system the available potential energy is proportional to the square of the deviation of temperature from a standard atmosphere profile divided by the static stability. In terms of differential thickness, the zonal-mean available potential energy is defined as

$$\bar{P} \equiv \frac{1}{2}\left\langle \frac{\rho_0}{N^2}\left(\frac{\partial \bar{\Phi}}{\partial z}\right)^2\right\rangle$$

Multiplying (10.49) through by $\rho_0(\partial\bar{\Phi}/\partial z)/N^2$ and averaging over space gives

$$
\frac{d}{dt}\left\langle\frac{\rho_0}{2N^2}\left(\frac{\partial\bar{\Phi}}{\partial z}\right)^2\right\rangle = -\left\langle\rho_0\bar{w}\frac{\partial\bar{\Phi}}{\partial z}\right\rangle + \left\langle\frac{\rho_0\kappa\bar{J}}{N^2H}\left(\frac{\partial\bar{\Phi}}{\partial z}\right)\right\rangle
$$

$$
-\left\langle\frac{\rho_0}{N^2}\frac{\partial\bar{\Phi}}{\partial z}\frac{\partial}{\partial y}\overline{\left(v'\frac{\partial\Phi'}{\partial z}\right)}\right\rangle
$$

(10.56)

The first term on the right is just equal and opposite to the first term on the right in (10.55), which confirms that this term represents a conversion between zonal mean kinetic and potential energies. The second term involves the correlation between temperature and diabatic heating; it expresses the generation of zonal mean potential energy by diabatic processes. The final term, which involves the meridional eddy heat flux, expresses the conversion between zonal mean and eddy potential energy.

That the second term on the right in (10.55) and the final term in (10.56) represent conversion between zonal mean and eddy energies can be confirmed by performing analogous operations on the eddy equations (10.51)–(10.53) to obtain equations for the eddy kinetic and available potential energies:

$$
\frac{d}{dt}\left\langle\rho_0\frac{\overline{u'^2+v'^2}}{2}\right\rangle = +\left\langle\rho_0\overline{\Phi'\left(\frac{\partial u'}{\partial x}+\frac{\partial v'}{\partial y}\right)}\right\rangle
$$

$$
-\left\langle\rho_0\overline{u'v'}\frac{\partial\bar{u}}{\partial y}\right\rangle + \langle\rho_0(\overline{u'X'}+\overline{v'Y'})\rangle
$$

(10.57)

$$
\frac{d}{dt}\left\langle\frac{\rho_0}{N^2}\overline{\left(\frac{\partial\Phi'}{\partial z}\right)^2}\right\rangle = -\left\langle\rho_0\overline{w'\frac{\partial\Phi'}{\partial z}}\right\rangle + \left\langle\frac{\rho_0\kappa\overline{J'\,\partial\Phi'/\partial z}}{N^2H}\right\rangle
$$

$$
-\left\langle\frac{\rho_0}{N^2}\left(\frac{\partial^2\bar{\Phi}}{\partial z\,\partial y}\right)\overline{\left(v'\frac{\partial\Phi'}{\partial z}\right)}\right\rangle
$$

(10.58)

The first term on the right in (10.57) can be rewritten using the continuity equation (10.54) as

$$
\left\langle\rho_0\overline{\Phi'\left(\frac{\partial u'}{\partial x}+\frac{\partial v'}{\partial y}\right)}\right\rangle = -\left\langle\overline{\Phi'\frac{\partial(\rho_0w')}{\partial z}}\right\rangle = \left\langle\rho_0\overline{w'\left(\frac{\partial\Phi'}{\partial z}\right)}\right\rangle
$$

which is equal to minus the first term on the right in (10.58). Thus, this term expresses the conversion between eddy kinetic and eddy potential energy for the Eulerian mean formulation. Similarly, the last term in (10.58) is equal to minus the last term in (10.56) and thus represents conversion between eddy and zonal mean available potential energy.

The Lorenz energy cycle can be expressed compactly by defining zonal mean and eddy kinetic and available potential energies:

$$\bar{K} \equiv \left\langle \rho_0 \frac{\bar{u}^2}{2} \right\rangle, \qquad K' \equiv \left\langle \rho_0 \frac{\overline{u'^2 + v'^2}}{2} \right\rangle$$

$$\bar{P} \equiv \frac{1}{2} \left\langle \frac{\rho_0}{N^2} \left(\frac{\partial \bar{\Phi}}{\partial z} \right)^2 \right\rangle, \qquad P' \equiv \frac{1}{2} \left\langle \frac{\rho_0}{N^2} \overline{\left(\frac{\partial \Phi'}{\partial z} \right)^2} \right\rangle$$

the energy transformations

$$[\bar{P} \cdot \bar{K}] \equiv \left\langle \rho_0 \bar{w} \frac{\partial \bar{\Phi}}{\partial z} \right\rangle, \qquad [P' \cdot K'] \equiv \left\langle \rho_0 \overline{w' \frac{\partial \Phi'}{\partial z}} \right\rangle$$

$$[K' \cdot \bar{K}] \equiv \left\langle \rho_0 \overline{u'v'} \frac{\partial \bar{u}}{\partial y} \right\rangle, \qquad [P' \cdot \bar{P}] \equiv \left\langle \frac{\rho_0}{N^2} \overline{v' \frac{\partial \Phi'}{\partial z}} \frac{\partial^2 \bar{\Phi}}{\partial y \, \partial z} \right\rangle$$

and the sources and sinks

$$\bar{R} \equiv \left\langle \frac{\rho_0}{N^2} \frac{\kappa \bar{J}}{H} \frac{\partial \bar{\Phi}}{\partial z} \right\rangle, \qquad R' \equiv \left\langle \frac{\rho_0}{N^2} \overline{\frac{\kappa J'}{H} \frac{\partial \Phi'}{\partial z}} \right\rangle$$

$$\bar{\varepsilon} \equiv \langle \rho_0 \bar{u} \bar{X} \rangle, \qquad \varepsilon' \equiv \langle \rho_0 (\overline{u'X' + v'Y'}) \rangle$$

Equations (10.55)-(10.58) can then be expressed in the simple form

$$d\bar{K}/dt = [\bar{P} \cdot \bar{K}] + [K' \cdot \bar{K}] + \bar{\varepsilon} \qquad (10.59)$$

$$d\bar{P}/dt = -[\bar{P} \cdot \bar{K}] + [P' \cdot \bar{P}] + \bar{R} \qquad (10.60)$$

$$dK'/dt = [P' \cdot K'] - [K' \cdot \bar{K}] + \varepsilon' \qquad (10.61)$$

$$dP'/dt = -[P' \cdot K'] - [P' \cdot \bar{P}] + R' \qquad (10.62)$$

Here $[A \cdot B]$ designates conversion from energy form A to form B.

Adding (10.59)-(10.62) we obtain an equation for the rate of change of total energy (kinetic plus available potential):

$$d(\bar{K} + K' + \bar{P} + P')/dt = \bar{R} + R' + \bar{\varepsilon} + \varepsilon' \qquad (10.63)$$

For adiabatic inviscid flow the right side vanishes and the total energy $\bar{K} + K' + \bar{P} + P'$ is conserved. In this system the zonal mean kinetic energy does not include a contribution from the mean meridional flow because the zonally averaged meridional momentum equation was replaced by the geostrophic approximation. (Likewise, use of the hydrostatic approximation means that neither the mean nor the eddy vertical motion is included in the total kinetic energy.) Thus, the quantities that are included in the total energy depend on the particular model used. For any model the definitions of energy must be consistent with the approximations employed.

In the long-term mean the left side of (10.63) must vanish. Thus, the production of available potential energy by zonal mean and eddy diabatic processes must balance the mean plus eddy kinetic energy dissipation:

$$\bar{R} + R' = -\bar{\varepsilon} - \varepsilon' \qquad (10.64)$$

Since solar radiative heating is a maximum in the tropics, where the temperatures are high, it is clear that $\bar{R}$, the generation of zonal mean potential energy by the zonal mean heating, will be positive. For a *dry* atmosphere in which eddy diabatic processes are limited to radiation and diffusion R', the diabatic production of eddy available potential energy should be negative because the thermal radiation emitted to space from the atmosphere increases with increasing temperature and thus tends to reduce horizontal temperature contrasts in the atmosphere. For the earth's atmosphere, however, the presence of clouds and precipitation greatly alters the distribution of R'. Present estimates (see Fig. 10.13) suggest that in the Northern

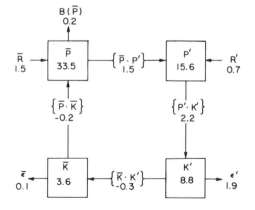

Fig. 10.13 The observed mean energy cycle for the Northern Hemisphere. Numbers in the squares are energy amounts in units of 10^5 J m^{-2}. Numbers next to the arrows are energy transformation rates in units of W m^{-2}. $B(\bar{P})$ represents a net energy flux into the Southern Hemisphere. Other symbols are defined in the text. (Adapted from Oort and Peixoto, 1974.)

Hemisphere R' is positive and nearly half as large as $\bar{R}$. Thus, diabatic heating generates both zonal mean and eddy available potential energy.

The equations (10.59)–(10.62) together provide a complete description of the quasi-geostrophic energy cycle from the conventional Eulerian mean point of view. The content of these equations is summarized by means of the four-box diagram of Fig. 10.13. In this diagram the squares represent reservoirs of energy and the arrows indicate sources, sinks, and conversions. The observed direction of the conversion terms in the troposphere for the Northern Hemisphere annual mean is indicated by the arrows. It should be emphasized that the direction of the various conversions cannot be theoretically deduced by reference to the energy equations alone. It also should be emphasized that the conversion terms given here are a result of the particular type of zonal average model used. The analogous energy equations for the TEM equations have rather different conversions. Thus, the energy transformations given in the present analysis should be regarded not as fundamental properties of the atmosphere but rather as properties of the Eulerian mean system.

Nevertheless, since the conventional Eulerian mean model is generally used as a basis for the study of baroclinic waves, the four-box energy diagram presented here does provide a useful framework for considering the role of weather disturbances in maintenance of the general circulation. The observed energy cycle as summarized in Fig. 10.13 suggests the following qualitative picture:

1.   The zonal mean diabatic heating generates mean zonal available potential energy through a net heating of the tropics and cooling of the polar regions.

2.   Baroclinic eddies transport warm air poleward and cold air equatorward and transform the mean available potential energy to eddy available potential energy.

3.   At the same time eddy available potential energy is transformed into eddy kinetic energy by the vertical motions in the eddies.

4.   The zonal kinetic energy is maintained primarily by the conversions from eddy kinetic energy owing to the correlation $\overline{u'v'}$. This will be discussed further in the next section.

5.   The energy is dissipated by surface and internal friction in the eddies and mean flow.

In summary, the observed atmospheric energy cycle as given by the Eulerian mean formulation is consistent with the notion that baroclinically unstable eddies are the primary disturbances responsible for the energy exchange in midlatitudes. It is through the eddy motions that the kinetic energy lost through turbulent stresses is replaced, and it is the eddies that are primarily responsible for the poleward heat transport to balance the radiation deficit in the polar regions. In addition to the transient baroclinic eddies, forced stationary orographic waves and free Rossby waves may also contribute substantially to the poleward heat flux. The direct conversion of mean available potential energy to mean kinetic energy by symmetric overturning is, on the other hand, small and negative in middle latitudes but positive in the tropics, where it plays an important role in the maintenance of the mean Hadley circulation.

10.5 Longitudinally Dependent Time-Averaged Flow

So far in this chapter we have concentrated on the zonally averaged component of the general circulation. For a planet with a longitudinally uniform surface the flow averaged over a season should be completely characterized by the zonally symmetric component of the circulation, since for such a hypothetical planet the statistics of zonally asymmetric transient eddies (i.e., weather disturbances) should be independent of longitude. On the earth, however, large-scale topography and continent–ocean heating contrasts provide strong forcing for longitudinally asymmetric planetary scale time-mean motions. Such motions, usually referred to as *stationary waves*, are especially strong in the Northern Hemisphere during the winter season.

Observations indicate that tropospheric stationary waves generally tend to have an equivalent barotropic structure; that is, wave amplitude generally increases with height, but phase lines tend to be vertical. Although nonlinear processes may be significant in the formation and maintenance of stationary waves, the climatological stationary wave pattern can to a first approximation be described in terms of forced barotropic Rossby waves. When superposed on the zonal mean circulation, such waves produce local regions of enhanced and diminished time mean westerly winds, which strongly influence the development and propagation of transient weather disturbances. They thus represent essential features of the climatological flow.

10.5.1 STATIONARY ROSSBY WAVES

The most significant of the time-mean zonally asymmetric circulation features is the pattern of stationary planetary waves excited in the Northern Hemisphere by the flow over the Himalayas and the Rockies. It was shown in Section 7.7.2 that the quasi-stationary wave pattern along 45° latitude could be accounted for to a first approximation as the forced wave response when mean westerlies impinge on the large-scale topography. More detailed analysis shows that zonally asymmetric heat sources also contribute to the forcing of the climatological stationary wave pattern. Some controversy remains, however, concerning the relative importance of heating and orography in forcing the observed stationary wave pattern.

The discussion of topographic Rossby waves in 7.7.2 used a β plane channel model in which it was assumed that wave propagation was parallel to latitude circles. In reality, however, large-scale topographic features and heat sources are confined in latitude as well as longitude, and the stationary waves excited by such forcing may propagate energy meridionally as well as zonally. For a quantitatively accurate analysis of the barotropic Rossby wave response to a local source it is necessary to utilize the barotropic vorticity equation in spherical coordinates and to include the latitudinal dependence of the mean zonal wind. The mathematical analysis for such a situation is beyond the scope of this book. It is possible, however, to obtain a qualitative notion of the nature of the wave propagation for this case by generalizing the β plane analysis of Section 7.7. Thus, rather than assuming that propagation is limited to a channel of specified width, we assume that the β plane extends to plus and minus infinity in the meridional direction and that Rossby waves can propagate both northward and southward without reflection from artificial walls.

The free barotropic Rossby wave solution then has the form (7.90) and satisfies the dispersion relation (7.91) where l is the meridional wave number, which is now allowed to vary. From (7.93) it is clear that for a specified zonal wave number, k, the free solution is stationary for l given by

$$l^2 = \beta/\bar{u} - k^2 \tag{10.65}$$

For example, westerly flow over an isolated mountain that primarily excites a response at a given k will produce stationary waves with both positive and negative l satisfying (10.65). As remarked in Section 7.7.1, although Rossby wave phase propagation relative to the mean wind is always westward, this is not true of the group velocity. From (7.91) we readily find that the x and y components of group velocity are

$$c_{gx} = \frac{\partial \nu}{\partial k} = \bar{u} + \beta \frac{k^2 - l^2}{(k^2 + l^2)^2} \tag{10.66}$$

$$c_{gy} = \frac{\partial \nu}{\partial l} = \frac{2\beta kl}{(k^2 + l^2)^2} \tag{10.67}$$

For stationary waves these may be expressed alternatively as

$$c_{gx} = \frac{2\bar{u}k^2}{k^2 + l^2}, \qquad c_{gy} = \frac{2\bar{u}kl}{k^2 + l^2} \tag{10.68}$$

The group velocity vector for stationary Rossby waves is perpendicular to the wave crests. It always has an eastward zonal component and has a northward or southward meridional component depending on whether l is positive or negative. The magnitude is given by

$$|c_g| = 2\bar{u} \cos \alpha \tag{10.69}$$

(see Problem 10.9). Here, as shown in Fig. 10.14 for the case of positive l, α is the angle between lines of constant phase and the y axis.

Since energy propagates at the group velocity, (10.68) indicates that the stationary wave response to a localized topographic feature should consist of two wave trains, one with $l > 0$ extending eastward and northward and the other with $l < 0$ extending eastward and southward. An example computed using spherical geometry is given in Fig. 10.15. Although the positions of individual troughs and ridges remain fixed for stationary waves, the wave trains in this example do not decay in time since effects of dissipation are counteracted by energy propagation from the source at the Rossby wave group velocity.

For the climatological stationary wave distribution in the atmosphere the excitation comes from a number of sources, both topographic and thermal, distributed around the globe. Thus, it is not easy to trace out distinct paths of wave propagation. Nevertheless, detailed calculations using spherical

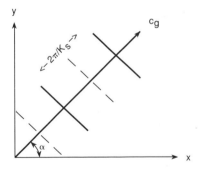

Fig. 10.14   Stationary plane Rossby wave in a westerly flow. Ridges (solid lines) and troughs (dashed lines) are oriented at an angle α to the y-axis, and the group velocity relative to the ground, c_g, is oriented at an angle α to the x axis. The wavelength is $2\pi/K_s$. (Adapted from Hoskins, 1983.)

Fig. 10.15   Vorticity pattern generated on a sphere when a constant angular velocity westerly flow impinges on a circular forcing centered at 30°N and 45°W of the central point. The figures from left to right show the response at 2, 4, and 6 days after switch on of the forcing. Five contour intervals correspond to the maximum vorticity response that would occur in 1 day if there were no wave propagation. Heavy lines correspond to zero contours. The pattern is drawn on a projection in which the sphere is viewed from infinity. (After Hoskins, 1983.)

geometry suggest that two-dimensional barotropic Rossby wave propagation provides a reasonable first approximation for the observed departure of the extratropical time-mean flow from zonal symmetry.

10.5.2.   JET STREAMS AND STORM TRACKS

When the longitudinally asymmetric geopotential anomalies associated with stationary waves are added to the zonal mean geopotential distribution the resulting time-mean field includes local regions of enhanced meridional geopotential gradient that are manifested in the wind field of the Northern Hemisphere by the Asian and North American jet streams. The existence of these two jets can be inferred from the January mean 500-mb geopotential height field shown in Fig. 6.3. Note the strong meridional gradients in height associated with the troughs centered just east of the Asian and North American continents (the same features can be seen in annual mean charts, although with somewhat reduced intensity). The zonal flow associated with these semipermanent troughs is illustrated in Fig. 6.2. In addition to the two intense jet cores in the western Pacific and western Atlantic there is a weaker jet centered over North Africa and the Middle East. Figure 6.2 shows dramatically the large deviations from zonal symmetry in the jet stream structure. In midlatitudes the zonal wind speed varies by nearly a factor of 3 between the core of the Asian jet and the low wind speed area in western North America. Although, as was argued above, the climatological stationary wave distribution on which the Asian and North American jets are superposed is apparently forced primarily by orography, the structure of the jets also appears to be influenced by continent–ocean heating contrasts. Thus, the strong vertical shear in the Asian and North American jets reflects a thermal wind balance consistent with the very strong meridional temperature gradients that occur in winter near the eastern edges of the Asian and North American continents owing to the contrast between warm water to the southeast and cold land to the northwest. A satisfactory description of the jet streams must account, however, not only for their thermal structure but also for the westerly acceleration that air parcels must experience as they enter the jet and the deceleration as they leave the jet core.

To understand the momentum budget in the jet streams and its relationship to the observed distribution of weather we consider the zonal component of the momentum equation, which (if we neglect the β effect) may be written in the form (6.30)

$$\frac{Du_g}{Dt} = f_0(v - v_g) \equiv f_0 v_a \qquad (10.70)$$

where v_a is the meridional component of the ageostrophic wind. This equation indicates that the westerly acceleration $(Du_g/Dt > 0)$ that air parcels experience as they enter the jet can be provided only by a poleward ageostrophic wind component $(v_a > 0)$, and, conversely, the easterly acceleration that air parcels experience as they leave the jet requires an equatorward ageostrophic motion. This meridional flow together with its accompanying vertical circulation is illustrated in Fig. 10.16. Note that this secondary circulation is thermally direct upstream of the jet core. A magnitude of $v_a \sim 2\text{-}3 \text{ m s}^{-1}$ is required to account for the observed zonal wind acceleration. This is an order of magnitude stronger than the zonal mean indirect cell (Ferrel cell) that prevails in midlatitudes. Downstream of the jet core, on the other hand, the secondary circulation is thermally indirect but much stronger than the zonally averaged Ferrel cell. It is interesting to note that the vertical motion pattern on the poleward (cyclonic shear) side of the jet is similar to that associated with deep transient baroclinic eddies in the sense that subsidence occurs to the west of the stationary trough associated with the jet and ascent occurs east of the trough (see, for example, Fig. 6.11).

Since the growth rate of baroclinically unstable synoptic-scale disturbances is proportional to the strength of the basic state thermal wind, it is not surprising that the Pacific and Atlantic jet streams are important source regions for storm development. Typically, transient baroclinic waves develop in the jet entrance region, grow as they are advected downstream, and decay in the jet exit region. The role of these transient eddies in maintenance of the jet stream structure is rather complex. Transient eddy heat fluxes, which are strong and poleward in the storm tracks, appear to act to weaken the climatological jets. The transient eddy vorticity flux in the upper troposphere, on the other hand, appears to act to maintain the jet structure. In both cases the secondary ageostrophic circulation associated with the jet

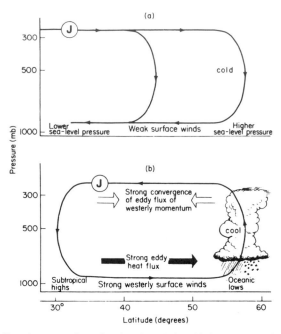

Fig. 10.16   Meridional cross sections showing the relationship between the time-mean secondary meridional circulation (continuous thin lines with arrows) and the jet stream
(denoted by J) at locations (a) upstream and (b) downstream from the jet stream
cores. (After Blackmon *et al.*, 1977. Reproduced with permission of the American
Meteorological Society.)

tends to partly balance the influence of the transient eddy fluxes in order
to maintain the mean thermal and momentum balances.

10.6   Low-Frequency Variability

An understanding of the general circulation requires consideration not
only of the zonal mean and stationary wave components and their variations
with the annual cycle but also of irregular variability on time scales longer
than that of individual transient eddies. The term *low-frequency variability*
is generally used to describe such components of the general circulation.
The observed spectrum of low-frequency variability ranges from weather
anomalies lasting only 7–10 days to interannual variability on the scale of
several years (see Section 11.1.6).

One possible cause of atmospheric low-frequency variability is forcing
owing to anomalies in sea surface temperature (SST), which themselves

arise from coherent air–sea interaction. Because of the large thermal inertia of the oceanic surface mixed layer, such anomalies tend to have time scales much longer than those associated with subseasonal variations in the atmosphere; they are thought to be of greatest significance on the seasonal and interannual time scales.

It is believed, however, that substantial variability on subseasonal time scales can arise in midlatitudes in the absence of anomalous SST forcing, as a result of internal nonlinear atmospheric dynamics, although SST anomalies may tend to favor the occurrence of certain types of variations. One example of internally generated low-frequency variability is the forcing of large-scale anomalies by potential vorticity fluxes of high-frequency transient waves. This process appears to be important in the maintenance of the high-amplitude quasi-stationary wave disturbances called *blocking* patterns. Some types of blocking may also be related to special nonlinear wave patterns called *solitary waves*, in which damping by Rossby wave dispersion is balanced by intensification owing to nonlinear advection. Although most internal mechanisms involve nonlinearity, there is some evidence that the longitudinally dependent time-mean flow may be unstable to linear barotropic normal modes that are stationary in space but oscillate in time at low frequencies. Such modes, which are global in scale, may be responsible for some observed teleconnection patterns.

10.6.1 CLIMATE REGIMES

It has long been observed that the extratropical circulation appears to alternate between a so-called high-index state, corresponding to a circulation with strong zonal flow and weak waves, and a low-index state with weak zonal flow and high-amplitude waves. This behavior suggests that there may exist more than one climate regime consistent with a given external forcing and that the observed climate may switch back and forth between regimes in a chaotic fashion. Whether the high-index and low-index states actually correspond to distinct quasi-stable atmospheric climate regimes is a matter of controversy. The general notion of vacillation between two quasi-stable flow regimes can, however, be demonstrated quite convincingly in laboratory experiments. Such experiments will be described briefly in Section 10.7.

The concept of climate regimes can also be demonstrated in a highly simplified model of the atmosphere developed by Charney and DeVore (1979). They examined the equilibrium mean states that can result when a damped topographic Rossby wave interacts with the zonal mean flow. Their model is an extension of the topographic Rossby wave analysis given in Section 7.7.2. In this model the wave disturbance is governed by (7.99), which is the linearized form of the barotropic vorticity equation (7.94) with weak damping added. The zonal mean flow is governed by the barotropic momentum equation

$$\frac{\partial \bar{u}}{\partial t} = -D(\bar{u}) - \kappa(\bar{u} - U_e) \tag{10.71}$$

where the first term on the right designates forcing by interaction between the waves and the mean flow, and the second term represents a linear relaxation toward an externally determined basic state flow, U_e.

The zonal mean equation (10.71) can be obtained from (7.94) by dividing the flow into zonal mean and eddy parts and taking the zonal average to get

$$\frac{\partial}{\partial t}\left(-\frac{\partial \bar{u}}{\partial y}\right) = -\frac{\partial}{\partial y}\left(\overline{v_g' \zeta_g'}\right) - \frac{f_0}{H}\frac{\partial}{\partial y}\left(\overline{v_g' h_T}\right)$$

which after integrating in y and adding the external forcing term yields (10.71) with

$$D(\bar{u}) = -\overline{v_g'\zeta_g'} - (f_0/H)\overline{v_g' h_T} \tag{10.72}$$

As shown in Problem 10.5, the eddy vorticity flux [the first term on the right in (10.72)] is proportional to the divergence of the eddy momentum flux. The second term, which is sometimes referred to as the *form drag*, is the equivalent in the barotropic model of the surface pressure torque term in the angular momentum balance equation (10.43).

If h_T and the eddy geostrophic streamfunction are assumed to consist of single harmonic wave components in x and y, as given by (7.96) and (7.97), respectively, the vorticity flux vanishes, and with the aid of (7.100) the form drag can be expressed as

$$D(\bar{u}) = -\left(\frac{f_0}{H}\right)\overline{v_g' h_T} = \left(\frac{rK^2 f_0^2}{2\bar{u}H^2}\right)\frac{h_0^2 \cos^2 ly}{(K^2 - K_s^2)^2 + \varepsilon^2} \tag{10.73}$$

where K_S is the resonant stationary Rossby wave number defined in (7.93).

It is clear from (10.73) that the form drag will have a strong maximum when $\bar{u} = \beta/K^2$, as shown schematically in Fig. 10.17. The last term on the right in (10.71), on the other hand, will decrease linearly as $\bar{u}$ increases. Thus, for suitable parameters there will be three values of $\bar{u}$ (labeled A, B, and C in Fig. 10.17) for which the wave drag just balances the external forcing so that steady-state solutions may exist. By perturbing the solution about the points A, B, and C it is easily shown (Problem 10.12) that point B corresponds to an unstable equilibrium, while the equilibria at points

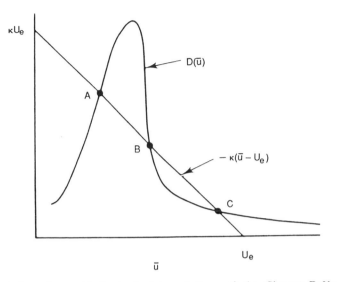

Fig. 10.17   Schematic graphical steady-state solutions of the Charney–DeVore model. (Adapted from Held, 1983.)

A and C are stable. Solution A corresponds to a low-index equilibrium, with high-amplitude waves analogous to a blocking regime. Solution C corresponds to a high-index equilibrium with strong zonal flow and weak waves. Thus, for this very simple model there are two possible "climates" associated with the same forcing.

The Charney–DeVore model is a highly oversimplified model of the atmosphere. Models that contain more degrees of freedom do not generally have multiple steady solutions. Rather, there is a tendency for the (unsteady) solutions to cluster about certain climate regimes and to shift between regimes in an unpredictable fashion. Such behavior is characteristic of a wide range of nonlinear dynamical systems and is popularly known as *chaos* (see Section 13.8).

10.6.2   Sea Surface Temperature Anomalies

Sea surface temperature (SST) anomalies influence the atmosphere by altering the flux of latent and sensible heat from the ocean and thus providing anomalous heating patterns. The efficacy of such anomalies in exciting global scale responses depends on their ability to generate Rossby waves. A thermal anomaly can generate a Rossby wave response only by perturbing the vorticity field. This requires that the thermal anomaly produce an anomalous vertical motion field, which in turn produces anomalous vortex tube stretching.

For low-frequency disturbances the thermodynamic energy equation (10.5) may be approximated as

$$\mathbf{V} \cdot \mathbf{\nabla} T + wN^2HR^{-1} \approx J/c_p \qquad (10.74)$$

Thus, diabatic heating can be balanced by horizontal temperature advection or by adiabatic cooling owing to vertical motion. The ability of diabatic heating produced by a sea surface temperature anomaly to generate Rossby waves depends on which of these processes dominates. In the extratropics SST anomalies primarily generate low-level heating, and this is balanced mainly by horizontal temperature advection. In the tropics positive SST anomalies are associated with enhanced convection, and the resulting diabatic heating is balanced by adiabatic cooling. Tropical anomalies have their greatest effect in the western Pacific, where the average sea surface temperature is very high, so that even a small positive anomaly can generate large increases in evaporation owing to the exponential increase of saturation vapor pressure with temperature. By continuity of mass the upward motion in cumulonimbus convection requires convergence at low levels and divergence in the upper troposphere. The low-level convergence acts to sustain the convection by moistening and destabilizing the environment,

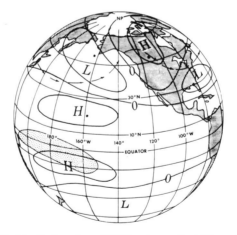

Fig. 10.18   The Pacific North American (PNA) pattern of middle and upper tropospheric height anomalies for Northern Hemisphere winter during an ENSO event in the tropical Pacific. Region of enhanced tropical precipitation is shown by shading. Arrows depict a 500-mb streamline for the anomaly conditions. H and L designate anomaly highs and lows, respectively. The anomaly pattern propagates along a great circle path with an eastward component of group velocity, as predicted by stationary Rossby wave theory. (After Horel and Wallace, 1981. Reproduced with permission of the American Meteorological Society.)

while the upper-level divergence generates a vorticity anomaly. If the mean flow is westerly in the region of upper-level divergence, the forced vorticity anomaly will form a stationary Rossby wave train. The observed upper tropospheric height anomalies during Northern Hemisphere winter associated with such an anomaly are shown schematically in Fig. 10.18. The pattern strongly suggests a train of stationary Rossby waves that emanates from the equatorial source region and follows a great circle path, as predicted by barotropic Rossby wave theory (Section 10.5.1). In this manner tropical SST anomalies may generate low-frequency variability in the extratropics. It is possible that the effects of SST anomalies and of internal variability are not completely independent. In particular, it is more likely that the atmosphere will tend to reside preferentially in regimes whose anomalous flow is correlated with the pattern in Fig. 10.18 than would be the case in the absence of SST anomalies.

10.7   Laboratory Simulation of the General Circulation

We have shown that the gross features of the general circulation outside the tropics can be understood within the framework of the quasi-geostrophic system; the observed zonal mean circulation and the atmospheric energy cycle are both rather well described by quasi-geostrophic theory. The fact that the quasi-geostrophic model, in which the spherical earth is replaced by the β plane, can successfully model many of the essential features of the general circulation suggests that the fundamental properties of the general circulation are not dependent on parameters unique to a spherical planet but may be common to all rotating differentially heated fluids. That the conjecture is in fact correct can be demonstrated in the laboratory with a rather simple apparatus.

In one group of experiments the apparatus consists of a cyclindrical vessel that is rotated about its vertical axis. The vessel is heated at its rim and cooled at its center. These experiments are often referred to as *dishpan* experiments, since some of the first experiments of this type utilized an ordinary dishpan. The fluid in the dishpan experiments crudely represents one hemisphere in the atmosphere, with the rim of the dishpan corresponding to the equator and the center to the pole. Because the geometry is cylindrical rather than spherical the β effect is not modeled for baroclinic motions in this type of system.[3] Thus the dishpan experiments omit dynamical effects of the atmospheric meridional vorticity gradient as well as the geometrical curvature terms, which were neglected in the quasi-geostrophic model.

[3] For barotropic motions the radial height gradient of the rotating fluid in the cylinder creates an "equivalent" β effect (see Section 4.3).

For certain combinations of rotation and heating rates, the flow in the dishpan appears to be axially symmetric with a steady azimuthal flow that is in thermal wind equilibrium with the radial temperature gradient and a superposed direct meridional circulation with rising motion near the rim and sinking near the center. This symmetric flow is usually called the *Hadley regime* since the flow is essentially that of a Hadley cell.

For other combinations of rotation and heating rates, however, the observed flow is not symmetric. It consists, rather, of irregular wavelike fluctuations and meandering zonal jets. In such experiments velocity tracers on the surface of the fluid reveal patterns very similar to those on midlatitude upper-air weather charts, while tracers near the lower boundary reveal structures similar to atmospheric fronts. This type of flow is usually called the *Rossby regime*, although it should be noted that the observed waves are not Rossby waves since there is no β effect in the tank.

Experiments have also been carried out using an apparatus in which the fluid is contained in the annular region between two coaxial cylinders of different radii. In these *annulus* experiments, the walls of the inner and outer cylinders are held at constant temperatures so that a precisely controlled temperature difference can be maintained across the annular region. Very regular wave patterns can be obtained for certain combinations of rotation and heating. In some cases the wave patterns are steady, while in other cases they undergo regular periodic fluctuations called *vacillation cycles* (see Fig. 10.19). As mentioned in the last section, vacillation is an example of internally generated low-frequency variability, which may be analogous to some atmospheric climate variability.

In order to determine whether the annulus experiments are really valid analogs of the atmospheric circulation or merely bear an accidental resemblance to atmospheric flow, it is necessary to analyze the experiments quantitatively. Mathematical analysis of the experiments can proceed from essentially the same equations that are applied to the atmosphere, except that cylindrical geometry replaces spherical geometry and temperature replaces potential temperature in the heat equation. (Since water is nearly incompressible, adiabatic temperature changes are negligible following the motion.) In addition, the equation of state must be replaced by an appropriate measure of the relationship between temperature and density:

$$\rho = \rho_0[1 - \varepsilon(T - T_0)] \qquad (10.75)$$

where ε is the thermal expansion coefficient ($\varepsilon \approx 2 \times 10^{-4}$ K^{-1} for water) and ρ_0 is the density at the mean temperature T_0.

In Chapter 1 we indicated that the character of the motion in a fluid is crucially dependent on the characteristic scales of parameters such as the

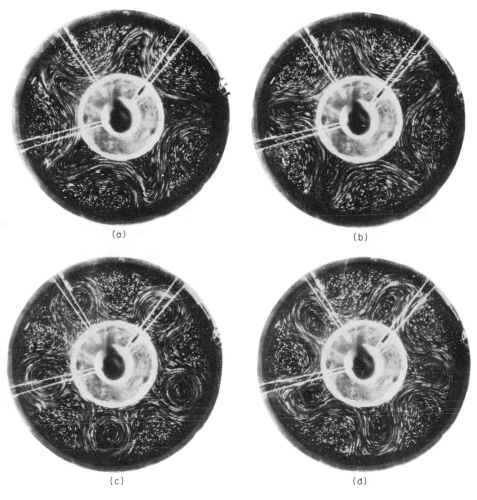

(a) (b)

(c) (d)

Fig. 10.19 Time exposures showing the motion of surface tracer particles in a rotating annulus. The four photographs illustrate various stages of a five-wave tilted trough vacillation cycle. The period of the vacillation cycle is 16.25 revolutions and the photographs are at intervals of 4 revolutions. (Photographs by Dave Fultz.)

velocity, pressure perturbation, length, and time. In the laboratory the scales of these parameters are generally many orders of magnitude different from their scales in nature. It is still possible, however, to produce quasi-geostrophic motions in the laboratory provided that the motions are slow enough so that the flow is approximately hydrostatic and geostrophic. As we saw in Section 2.4.2, the geostrophy of a flow does not depend on the absolute value of the scaling parameters, but rather on a nondimensional ratio of these parameters called the Rossby number. For the annulus experiments, the maximum horizontal scale is set by the dimensions of the tank so that it is convenient to define the Rossby number as

$$\text{Ro} \equiv U/[\Omega(b-a)]$$

where Ω is the angular velocity of the tank, U is a typical relative velocity of the fluid, and $b-a$ is the difference between the radius b of the outer wall and the radius a of the inner wall of the annular region.

Using the hydrostatic approximation,

$$\partial p/\partial z = -\rho g$$

Together with the geostrophic relationship

$$2\Omega \mathbf{V}_g = \rho_0^{-1} \mathbf{k} \times \nabla p$$

and the equation of state (10.75) we can obtain a *thermal wind* relationship in the form

$$\frac{\partial \mathbf{V}_g}{\partial z} = \frac{\varepsilon g}{2\Omega} \mathbf{k} \times \nabla T \tag{10.76}$$

Letting U denote the scale of the geostrophic velocity, H the mean depth of fluid, and δT the radial temperature difference across the width of the annulus, we obtain from (10.76)

$$U \sim \varepsilon g H \, \delta T/[2\Omega(b-a)] \tag{10.77}$$

Substituting the value of U in (10.77) into the formula for the Rossby number yields the *thermal Rossby number*

$$\text{Ro}_T \equiv \varepsilon g H \, \delta T/[2\Omega^2(b-a)^2]$$

This nondimensional number is the best measure of the range of validity of quasi-geostrophic dynamics in the annulus experiments. Provided that

$\text{Ro}_T \ll 1$ the quasi-geostrophic theory should be valid for motions in the annulus. For example, in a typical experiment

$$H \approx b - a \approx 10 \text{ cm}, \qquad \delta T \approx 10 \ K, \qquad \Omega \approx 1 \ s^{-1}$$

Thus, $\text{Ro}_T \approx 10^{-1}$.

Annulus experiments carried out over a wide range of rotation rates and temperature contrasts produce results that can be classified consistently on a log–log plot of the thermal Rossby number versus a nondimensional measure of the rotation rate:

$$(G^*)^{-1} \equiv (b - a)\Omega^2/g \tag{10.78}$$

Some results are summarized in Fig. 10.20. The heavy solid line separates the axially symmetric Hadley regime from the wavy Rossby regime. These results can best be understood qualitatively by considering an experiment

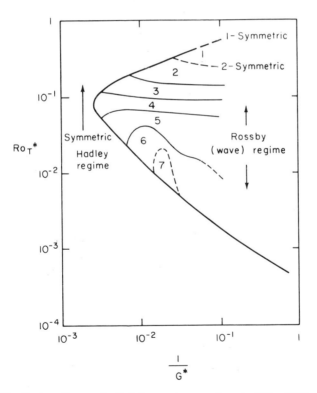

Fig. 10.20   Regime diagram for Fultz's annulus experiments. (After Phillips, 1963.)

in which the thermal Rossby number is slowly increased from zero by gradually imposing a temperature difference $T_b - T_a$. For small $T_b - T_a$ the motion is of the Hadley type, with weak horizontal and vertical temperature gradients in the fluid. As the horizontal temperature contrast is increased, however, the mean thermal wind must also increase, until at some critical value of Ro_T the flow becomes baroclinically unstable. According to the theory presented in Chapter 8, the wavelength of maximum instability is proportional to the ratio of the static stability to the square of the rotation rate. Thus, as can be seen in Fig. 10.20, the wave number observed when the flow becomes unstable decreases (that is, wavelength increases) as the rotation rate is reduced. Furthermore, since baroclinic waves transport heat vertically as well as laterally, they will tend to increase the static stability of the fluid. Therefore, as the thermal Rossby number is increased within the Rossby regime the increased heat transport by the waves will raise the static stability and hence increase the wavelength of the wave of maximum instability. The flow then undergoes transitions in which the observed wave number decreases until finally the static stability becomes so large that the flow is stable to even the largest wave that can fit the tank. The flow then returns to a symmetric Hadley circulation stabilized by a high static stability, which is maintained by a vigorous direct meridional circulation. This regime is usually called the *upper* symmetric regime to distinguish it from the *lower* symmetric regime which occurs for very weak heating.

The laboratory studies, despite their many idealizations, can model the most important of those features of the general circulation that are not dependent on the topography of the earth or continent–ocean heating contrasts. Specifically, we find, perhaps surprisingly, that the β effect (i.e., the planetary vorticity gradient) is not essential for the development of circulations that look very much like tropospheric synoptic systems. Thus, observed midlatitude waves should be regarded essentially as baroclinic waves modified by the β effect, not as Rossby waves in a baroclinic current.

In addition to demonstrating the primacy of baroclinic instability, the experiments also confirm that internal diabatic heating owing to condensation or radiative processes is not an essential mechanism for simulation of large-scale midlatitude circulations. The laboratory experiments thus enable us to separate the essential mechanisms from second-order effects in a manner not easily accomplished by observation of the atmosphere itself. Also, since laboratory simulation experiments have typical rotation rates of 10 rpm, it is possible to model many years of "atmospheric" flow in a short time, so that it is feasible to accumulate accurate statistics even for very low frequency variability. In addition, since temperatures and velocities can be measured at uniform intervals, the experiments can provide excellent sets of data for testing numerical weather prediction models.

10.8 Numerical Simulation of the General Circulation

In the previous section we discussed the role of laboratory experiments in contributing to a qualitative understanding of the general circulation of the atmosphere. Although laboratory experiments can elucidate most of the gross features of the general circulation, there are many details that cannot possibly be duplicated in the laboratory. For example, the possible long-term climatic effects of aerosols and trace gases that are added to the atmosphere as a result of human activities could not possibly be predicted on the basis of laboratory experiments. As another example, the influence of an ice-free Arctic Ocean on the global climate would also be extremely difficult to simulate in the laboratory. Since the conditions of the atmosphere cannot all be duplicated in the laboratory, the only practical manner in which to quantitatively simulate the present climate, or to predict possible climate modifications resulting from intentional or unintentional human intervention, is by numerical simulation with the aid of supercomputers.

Atmospheric general circulation models (GCMs) are similar to large-scale numerical weather prediction models (see Chapter 13) in that they attempt to simulate explicitly synoptic-scale weather disturbances. But, whereas weather prediction is an *initial value* problem, which requires that the evolution of the flow be computed from a specified initial state, general circulation modeling is a *boundary value* problem in which the average circulation is computed for specified external forcing conditions.

Although the dynamical equations used in general circulation models are the same as those used in short-range numerical prediction models, general circulation models are generally more complex since in simulations of time scales beyond a few days physical processes that are unimportant for the short-term evolution may become crucial. Thus, parameterizations are needed for physical processes such as surface heat and moisture fluxes, moist convection, turbulent mixing, and radiation. On the other hand, since a forecast of the flow evolution from a specific initial state is not needed, the problems associated with specifying initial data in forecast models do not arise in GCMs. In recent years, as numerical prediction has been extended into the middle range of 1–2 weeks, more physical processes have been included in the forecast models, and the differences between global weather forecast models and general circulation models have become smaller.

10.8.1 THE DEVELOPMENT OF GCMS

The success of the quasi-geostrophic model in short-range prediction suggests that for simulating the gross features of the general circulation such a model might be adequate provided that diabatic heating and frictional dissipation were included in a suitable manner. Phillips (1956) made the first attempt to model the general circulation numerically. His experiment employed a two-level quasi-geostrophic forecast model modified so that it included boundary layer friction and a latitudinally dependent radiative heating. The heating rates chosen by Phillips were based on estimates of the net diabatic heating rates necessary to balance the poleward heat transport at 45°N computed from observational data. Despite the severe limitations of this model, the simulated circulation in many respects resembled the observed.

Phillips' experiment, although it was an extremely important advance in dynamic meteorology, suffered from a number of shortcomings as a general circulation model. Perhaps the gravest shortcoming in his model was the specification of diabatic heating as a fixed function of latitude only. In reality, the atmosphere must to some degree determine the distribution of its own heat sources. This is true not only for condensation heating, which obviously depends on the distribution of vertical motion and water vapor, but also for radiative heating. Both the net solar heating and net infrared heating are sensitive to the distribution of clouds, and infrared heating depends on the atmospheric temperature as well.

Another important limitation of Phillips' two-level model was that static stability could not be predicted but had to be specified as an external parameter. This limitation is a serious one because the static stability of the atmosphere is obviously controlled by the motions.

It would be possible to design a quasi-geostrophic model in which the diabatic heating and static stability were motion dependent. Indeed such models have been used to some extent, especially in theoretical studies of the annulus experiments. To model the global circulation completely, however, requires a dynamical framework which is valid in the equatorial zone. Thus, it is desirable to base a GCM on the global primitive equations. Because of its enormous complexity and many important applications, general circulation modeling has become a highly specialized activity, which cannot possibly be covered adequately in a short space. Here we can only give a summary of the primary physical processes represented and present an example of an application in climate modeling. A brief discussion of the technical aspects of the formulation of numerical prediction models will be given in Chapter 13.

10.8.2 DYNAMICAL FORMULATION

Most general circulation models are based on the primitive equations in the σ-coordinate form introduced in Section 10.3.1. As was pointed out in that section, σ coordinates make it possible to retain the dynamical advantages of pressure coordinates but simplify the specification of boundary conditions at the surface.

The prediction equations for a σ-coordinate GCM are the horizontal momentum equation (10.30), the mass continuity equation (10.34), the thermodynamic energy equation (10.36), and a moisture continuity equation, which can be expressed as

$$\frac{D}{Dt}(q_v) = P_v \qquad (10.79)$$

where q_v is the water vapor mixing ratio and P_v is the sum of all sources and sinks. In addition, we require the hydrostatic equation (10.35) to provide a diagnostic relationship between the pressure and temperature fields. Finally, a relationship is needed to determine the evolution of the surface pressure $p_s(x, y, t)$. This is given by integrating (10.34) vertically and using the boundary conditions $\dot{\sigma} = 0$ at $\sigma = 0$ and 1 to obtain

$$\frac{\partial p_s}{\partial t} = -\int_0^1 \nabla \cdot (p_s \mathbf{V}) \, d\sigma \qquad (10.80)$$

Vertical variations are generally represented by dividing the atmosphere into a number of levels and utilizing a finite difference grid. Most GCMs have 10 to 20 levels at 1–3-km intervals extending from the surface to about 30-km altitude. Some models, however, have many more levels extending nearly to the mesopause. Horizontal resolution of global models varies widely, from an effective grid size of as much as 1000-km to as little as about 100 km.

10.8.3   PHYSICAL PROCESSES AND PARAMETERIZATIONS

The various types of surface and atmospheric processes represented in a typical GCM and the interactions among these processes are shown schematically in Fig. 10.21. The most important classes of physical processes are (i) radiative, (ii) cloud and precipitation, and (iii) turbulent mixing and exchange.

As was pointed out in Section 10.1, the fundamental process that drives the circulation of the atmosphere is the differential radiative heating that results from the fact that solar absorption at the surface is greater than long-wave emission to space at low latitudes, while long-wave emission

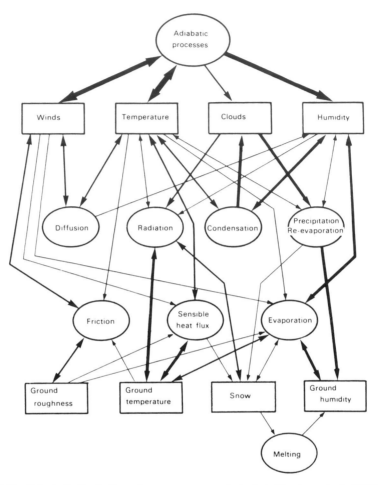

Fig. 10.21 Schematic illustration of processes commonly included in atmospheric GCMs and their interactions. The thickness of each arrow gives a rough indication of the importance of the interaction that the particular arrow represents. (After Simmons and Bengtsson, 1984.)

dominates over solar absorption at high latitudes, at least in the winter hemisphere. The general circulation of the atmosphere and the oceans provides the meridional and vertical heat transfer required for balance.

Most of the solar radiation absorbed at the surface is used to evaporate water and hence to moisten the atmosphere. Solar heating is realized in the atmosphere primarily in the form of latent heat release associated with convective clouds. The global distribution of evaporation clearly depends on the sea surface temperature, which is itself dependent on the general circulation of the oceans as well as on interaction with the atmosphere. Although some efforts are under way to run coupled atmosphere–ocean GCMs in which the sea surface temperature is predicted, at present most atmospheric GCMs use externally specified monthly or seasonal mean sea surface temperatures. Over land, on the other hand, the surface temperature adjusts very quickly to changes in the fluxes of solar and infrared radiation and is determined from a surface energy-balance equation.

Atmospheric radiative heating by solar radiation and heating and cooling by long-wave thermal radiation are computed using radiative transfer models of varying sophistication. Zonally averaged distributions of radiatively important constituents such as carbon dioxide, ozone, and even cloudiness are generally employed. But the more complete models do utilize model predicted zonally and time-varying cloudiness in their radiation codes.

Boundary layer fluxes of momentum, heat, and moisture are parameterized in most GCMs by use of bulk aerodynamic formulas (see Section 5.3.1). Typically the fluxes are specified to be proportional to the magnitude of the horizontal velocity at the lowest atmospheric level times the difference between the field variable at the boundary and its value at the lowest atmospheric level. In some models the boundary layer is explicitly resolved by locating several prediction levels within the lowest 2 km and utilizing the model-predicted boundary layer static stability in the parameterization of turbulent fluxes.

The hydrological cycle is usually represented by a combination of parameterization and explicit prediction. Water vapor mixing ratio is generally one of the explicitly predicted fields. The distributions of layer clouds and large-scale precipitation are then determined from the predicted distribution of humidity by requiring that when the predicted humidity exceeds 100% enough vapor is condensed to reduce the mixing ratio to saturation or less. Parameterizations in terms of the mean state thermal and humidity structure must be used to represent the distributions of convective clouds and precipitation.

10.8.4 THE GFDL "SKYHI" MODEL

The SKYHI model developed at the Geophysical Fluid Dynamics Laboratory (GFDL) of the National Oceanic and Atmospheric Administration (NOAA) is a very high resolution general circulation model that includes both the lower and the middle atmosphere. The GFDL model has 40 vertical levels from the surface of the earth to the mesopause. Several different horizontal resolutions of the model have been employed in order to test the effects of horizontal resolution on the simulated climatology.

In common with other models, the coarse-resolution versions of this model tend to produce a simulated climate in which the polar winter stratosphere is much too cold. The resulting (thermal wind balanced) polar night jet is much too strong and is not separated from the tropospheric jet stream. This cold polar bias apparently is caused by lack of sufficient zonal mean eddy drag to balance the zonal momentum budget.

The reason that a zonal drag force is required to maintain the observed mean wind speeds is as follows: The observed temperature in the high-latitude upper troposphere and lower stratosphere is far above radiative equilibrium. Strong adiabatic warming through subsidence in the polar stratosphere must exist to maintain this disequilibrium in the presence of radiative cooling. Continuity of mass then requires a strong poleward drift to balance the subsidence. The Coriolis force associated with this poleward drift must be balanced by a zonal drag force owing to either the EP flux

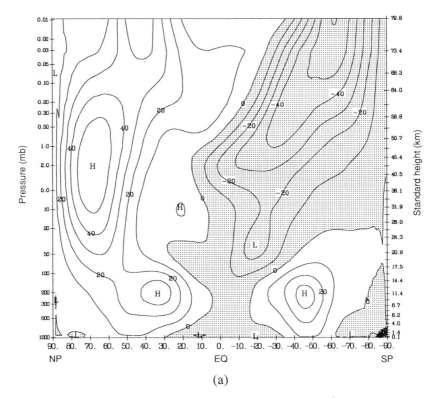

(a)

Fig. 10.22   Meridional cross sections of (a) mean zonal wind (m s^{-1}) and (b) temperature (K) for Northern Hemisphere winter, as simulated by the GFDL "SKYHI" model. (Courtesy of Dr. J. Mahlman.) (*Figure continues.*)

divergence of resolved eddies or the turbulent drag of small-scale eddies [see Eq. (10.17)]; otherwise there would be strong mean flow acceleration and the polar regions would cool toward radiative equilibrium.

There is evidence that most GCMs fail to simulate correctly the observed polar night jet structure because they underpredict the intensity of the EP fluxes owing to resolved eddies, and they lack sufficient resolution to produce the small-scale vertically propagating gravity waves that appear to be an essential component of the wave drag.

The highest-resolution version (1° latitude by 1.2° longitude) of the SKYHI model appears to overcome both these problems. It correctly simulates the planetary and synoptic-scale eddy EP fluxes and also produces a broad spectrum of inertio-gravity waves. The net result, as shown in Fig. 10.22, agrees quite well with observations (Figs. 6.1, 12.2, and 12.3) and appears to be adequate to account for the observed weak zonal winds and strong departure of temperature from radiative equilibrium in the Northern Hemisphere polar winter stratosphere.

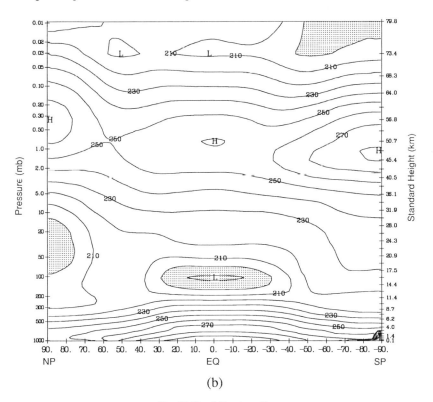

(b)

Fig. 10.22   (*Continued*)

Problems

10.1. Starting with the thermodynamic energy equation (2.42), derive the log-pressure version (10.5).

10.2. Show that in the σ-coordinate system a mass element $\rho_0\, dx\, dy\, dz$ takes the form $-g^{-1} p_s\, dx\, dy\, d\sigma$.

10.3. Compute the mean zonal wind $\bar{u}$ at the 200-mb level at 30°N under the assumptions that $\bar{u} = 0$ at the equator and that the absolute angular momentum is independent of latitude. What is the implication of this result for the role of eddy motions?

10.4. Show by scale analysis that advection by the mean meridional circulation can be neglected in the zonally averaged equations (10.11) and (10.12) for quasi-geostrophic motions.

10.5. Show that for quasi-geostrophic eddies the next to last term in square brackets on the right-hand side in (10.15) is proportional to the vertical derivative of the eddy meridional relative vorticity flux.

10.6. Starting from Eqs. (10.16)–(10.19), derive the governing equation for the residual streamfunction (10.21).

10.7. Using the observed data given in Fig.10.13, compute the time required for each possible energy transformation or loss to restore or deplete the observed energy stores. (A watt equals $1\,\mathrm{J\,s^{-1}}$.)

10.8. Compute the surface torque per unit horizontal area exerted on the atmosphere by topography for the following distribution of surface pressure and surface height:

$$p_s = p_0 + \hat{p}\sin kx, \qquad h = \hat{h}\sin(kx - \gamma)$$

where $p_0 = 10^2\,\mathrm{kPa}$, $\hat{p} = 1\,\mathrm{kPa}$, $\hat{h} = 2.5\times 10^3\,\mathrm{m}$, $\gamma = \pi/6\,\mathrm{rad}$, and $k = 1/(a\cos\phi)$, where $\phi = \pi/4\,\mathrm{rad}$ is the latitude and a is the radius of the earth. Express the answer in $\mathrm{kg\,s^{-2}}$.

10.9. Starting from (10.66) and (10.67), show that the group velocity relative to the ground for stationary Rossby waves is perpendicular to the wave crests and has a magnitude given by (10.68).

10.10. Derive the expression (10.76) for the thermal wind in the dishpan experiments.

10.11. Consider a thermally stratified liquid contained in a rotating annulus of inner radius 0.8 m, outer radius 1.0 m, and depth 0.1 m. The temperature at the bottom boundary is held constant at T_0. The fluid is assumed to satisfy the equation of state (10.75) with $\rho_0 = 10^3$ kg m^{-3} and $\varepsilon = 2 \times 10^{-4}$ K^{-1}. If the temperature increases linearly with height along the outer radial boundary at a rate of 1°C cm^{-1} and is constant with height along the inner radial boundary, determine the geostrophic velocity at the upper boundary for a rotation rate of $\Omega =$ 1 rad s^{-1}. (Assume that the temperature depends linearly on radius at each level.)

10.12. Show by considering $\partial \bar{u}/\partial t$ for small perturbations about the equilibrium points in Fig. 10.17 that point B is an unstable equilibrium point while points A and C are stable.

Suggested References

Lorenz, *The Nature and Theory of the General Circulation of the Atmosphere*, although somewhat out of date, contains an excellent survey of the subject including observational and theoretical aspects.

Hide (1966) reviews a number of types of laboratory experiments with rotating fluids including the rotating-annulus experiments.

Washington and Parkinson, *An Introduction to Three-Dimensional Climate Modeling*, is an excellent text, which covers the physical basis and computational aspects of general circulation modeling.

Simmons and Bengtsson (1984) provide a brief survey of the development and use of general circulation models for climate studies.

Fels *et al.* (1980) describe the GFDL "SKYHI" model and give some results of climate sensitivity studies.

Chapter

11 | Tropical Dynamics

Throughout the previous chapters of this book, we have emphasized circulation systems of the extratropical regions (that is, the regions poleward of about 30° latitude). This emphasis should not be regarded as an indication of a lack of interesting motion systems in the tropics; it is a result, rather, of the relative complexity of the dynamics of tropical circulations. There is no simple theoretical framework, analogous to quasi-geostrophic theory, which can be used to provide an overall understanding of large-scale tropical motions.

Outside the tropics, the primary energy source for synoptic-scale disturbances is the zonal available potential energy associated with the latitudinal temperature gradient. Observations indicate that latent heat release and radiative heating are usually secondary contributors to the energetics of extratropical synoptic-scale systems. In the tropics, on the other hand, the storage of available potential energy is small owing to the very small temperature gradients in the tropical atmosphere. Latent heat release appears to be the primary energy source, at least for those disturbances that originate within the equatorial zone. Most latent heat release in the tropics occurs in association with convective cloud systems, although much of the actual precipitation falls from mesoscale regions of stratiform clouds within such systems; the cloud systems are themselves generally embedded in large-scale circulations. The diabatic heating associated with tropical precipitation not only drives a local response in the atmospheric circulation but may also, through excitation of equatorial waves, induce a remote response. Thus, there is a strong interaction among cumulus convection, mesoscale, and large-scale circulations, which is of primary importance for understanding tropical motion systems. Furthermore, the distribution of diabatic heating in the tropics is strongly influenced by sea surface temperature (SST) variations, and these in turn are strongly influenced by the motion of the atmosphere.

An understanding of tropical circulations thus requires consideration of equatorial wave dynamics, interactions of cumulus convection and meso-scale circulations with large-scale motions, and air–sea interactions. Detailed treatment of all these topics is beyond the scope of an introductory text. Nevertheless, because the tropics play a fundamental role in the general circulation of the atmosphere and coupling between tropical and middle latitudes is an important consideration in extratropical extended-range forecasting, some discussion of the tropics is required even in a text with an extratropical emphasis.

Of course, it is not always possible to distinguish clearly between tropical and extratropical systems. In the subtropical regions (~30° latitude) circula-tion systems characteristic of both tropical and extratropical regions may be observed depending on the season and geographic location. To keep the discussion in this chapter as simple as possible, we focus primarily on the zone well equatorward of 30° latitude, where the influence of middle-latitude systems should be a minimum.

11.1 The Observed Structure of Large-Scale Tropical Circulations

Owing to the characteristics of the heat source, as well as the smallness of the Coriolis parameter, large-scale equatorial motion systems have several distinctive characteristic structural features that are quite different from those of midlatitude systems. Many of these can be understood in terms of the equatorial wave modes to be discussed in Section 11.3. Before discussing equatorial wave theory, however, it is useful to review some of the major observed circulation features of the tropical atmosphere.

11.1.1 THE INTERTROPICAL CONVERGENCE ZONE

Traditionally, the tropical general circulation was thought to consist of a thermally direct Hadley circulation in which air in the lower troposphere in both hemispheres moved equatorward toward the *intertropical convergence zone* (ITCZ) where by continuity considerations it was forced to rise uniformly and move poleward, thus transporting heat away from the equator in the upper troposphere in both hemispheres. This simple model of large-scale overturning is not, however, consistent with the observed vertical profile of equivalent potential temperature (θ_e). As indicated in Fig. 11.1 the mean tropical atmosphere is conditionally stable above about 600 mb. Thus, a large-scale upward mass flow, were it to exist, would be up the gradient of θ_e in the upper troposphere and would actually *cool* the upper troposphere in the region of the ITCZ. Such a circulation could not generate potential energy and would not, therefore, satisfy the heat balance in the equatorial zone.

It appears that the only way in which heat can effectively be brought from the surface to the upper troposphere in the ITCZ is through pseudoadiabatic ascent in the cores of large cumulonimbus clouds (often referred to as "hot towers"). For such motions, cloud parcels approximately conserve θ_e. They can, therefore, arrive in the upper troposphere with moderate temperature excesses. Thus, the heat balance of the equatorial

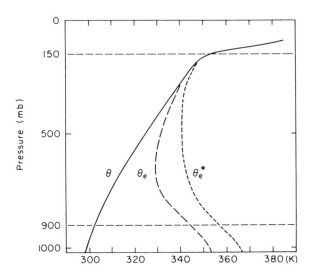

Fig. 11.1   Typical sounding in the tropical atmosphere showing the vertical profiles of potential temperature θ, equivalent potential temperature θ_e, and the equivalent potential temperature θ_e^* of a hypothetically saturated atmosphere with the same temperature at each level. This figure should be compared with Fig. 9.10, which shows similar profiles for a midlatitude squall line sounding. (After Ooyama, 1969. Reproduced with permission of the American Meteorological Society.)

zone can be accounted for, at least qualitatively, provided that the vertical motion in the ITCZ is confined primarily to the updrafts of individual convective cells. Riehl and Malkus (1958) estimated that only 1500–5000 individual hot towers would need to exist simultaneously around the globe to account for the required vertical heat transport in the ITCZ.

This view of the ITCZ as a narrow zonal band of vigorous cumulus convection has been confirmed beyond doubt by observations, particularly satellite cloud photos. An example is given in Fig. 11.2, which shows infrared cloud brightness temperatures in the tropics for the period August 14 to

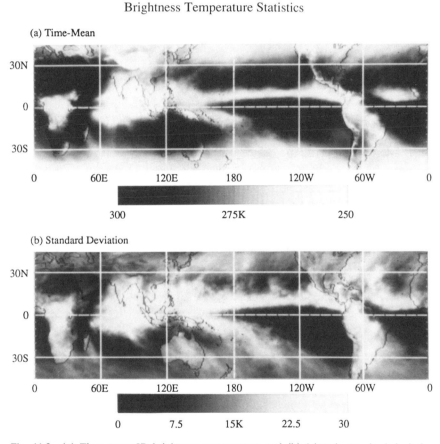

Brightness Temperature Statistics

(a) Time-Mean

(b) Standard Deviation

Fig. 11.2 (a) Time-mean IR brightness temperature and (b) 3-hourly standard deviations about the time-mean IR brightness temperature for 14 August to 17 December 1983. Low values of brightness temperature indicate the presence of high cold cirrus anvil clouds. (After Salby *et al.*, 1991.)

December 17, 1983. Low brightness temperatures signal the presence of deep anvil clouds characteristic of convective storms. The ITCZ appears as a line of deep convective cloud extending across the Atlantic and Pacific oceans between about 5° and 10°N.

Observations indicate that within the ITCZ precipitation greatly exceeds the moisture supplied by evaporation from the ocean surface below. Thus, much of the vapor necessary to maintain the convection in the ITCZ must be supplied by the converging trade wind flow in the lower troposphere. In this manner the large-scale flow provides the latent heat supply required to sustain the convection. The convective heating in turn produces large-scale midtropospheric temperature perturbations and (through hydrostatic adjustment) surface and upper-level pressure perturbations, which maintain the low-level inflow.

The above description of the ITCZ is actually oversimplified. In reality, the ITCZ over the oceans rarely appears as a long unbroken band of heavy convective cloudiness, and it almost never is found right at the equator. Rather, it is usually made up of a number of distinct cloud clusters, with scales of the order of a few hundred kilometers, which are separated by regions of relatively clear skies. The strength of the ITCZ is also quite variable in both space and time. It is particularly persistent and well defined over the Pacific and Atlantic between about 5° and 10°N latitude (as in Fig. 11.2) and occasionally appears in the western Pacific between 5° and 10°S.

Figure 11.2 shows that not only is the mean deep convection associated with the ITCZ found in the 5°–10°N latitude belt but also during the period shown in the figure, the standard deviation of deep convection is a maximum there. This is consistent with the idea that the ITCZ is the locus of transient cloud clusters, rather than simply a region of steady-state precipitation and mean uplift. The dry zones along the equator in the oceanic regions are a particularly striking feature in Fig. 11.2.

As the above discussion suggests, the vertical mass flux associated with the ITCZ has important regional variations. Nevertheless, there is a significant zonal mean component, which constitutes the upward mass flux of the mean Hadley circulation. This Hadley circulation consists of overturning thermally direct cells in the low latitudes of both hemispheres, as shown in Fig. 10.7. The center of the Hadley circulation is located at the mean latitude of the ITCZ. As shown in Fig. 10.7, the winter hemisphere branch of the Hadley cell is much stronger than the summer hemisphere branch. Observations indicate that two Hadley cells, symmetric about the equator, are rarely observed even in the equinoctial seasons. Rather, the northern cell dominates from November to March, the southern cell dominates from May to September, and rapid transitions occur in April and October (see Oort, 1983).

11.1.2 EQUATORIAL WAVE DISTURBANCES

The variance observed in the cloudiness associated with the ITCZ, as illustrated in Fig. 11.2b, is generally caused by transient precipitation zones associated with weak equatorial wave disturbances that propagate westward along the ITCZ. That such westward-propagating disturbances exist and are responsible for a large part of the cloudiness in the ITCZ can easily be seen by viewing time–longitude sections constructed from daily satellite pictures cut into thin zonal strips. An example is shown in Fig. 11.3. The well-defined bands of cloudiness that slope from right to left down the page define the locations of the cloud clusters as functions of longitude and time. Clearly, much of the cloudiness in the 5–10°N latitude zone of the Pacific is associated with westward-moving disturbances. The slope of the cloud lines in Fig. 11.3 implies a westward propagation speed of about 8–10 m s^{-1}. The longitudinal separation of the cloud bands is about 3000–4000 km, corresponding to a period range of about 4–5 d for this type of disturbance.

Diagnostic studies indicate that these westward-propagating disturbances are driven by the release of latent heat in convective precipitation areas accompanying the waves. The vertical structure of this type of disturbance is shown in schematic form in Fig. 11.4. Since the approximate thermodynamic energy equation (11.10) requires that the vertical motion be proportional to the diabatic heating, the maximum large-scale vertical velocities occur in the convective zone. By continuity there must thus be convergence at low levels in the convective zone and divergence in the upper troposphere. Hence, provided that the absolute vorticity is positive (negative) for Northern (Southern) Hemisphere disturbances, there will be cyclonic vorticity generation in the lower troposphere and anticyclonic vorticity generation in the upper troposphere owing to the divergence term in the vorticity equation. The process of adjustment between the mass and velocity fields will then generate a pressure minimum (trough) at low levels.[1] Thus, the thickness (or layer mean temperature) in the convective zone must be greater than in the surrounding environment.

The strongest convective activity in these waves occurs where the midtropospheric temperatures are warmer than average (although generally by less than 1°C). This is also the location of maximum upward motion. The correlations between temperature and vertical motion and temperature and diabatic heating are thus both positive, and the potential energy generated by the diabatic heating is immediately converted to kinetic energy [i.e., the

[1] The terms "trough" and "ridge" as used by tropical meteorologists designate pressure minima and maxima, respectively, just as in midlatitudes. In the easterlies of the Northern Hemisphere tropics, however, the zonal mean pressure increases with latitude so that the pattern of isobars depicting a tropical trough will resemble the pattern associated with a ridge in middle latitudes (i.e., there is a poleward deflection of the isobars).

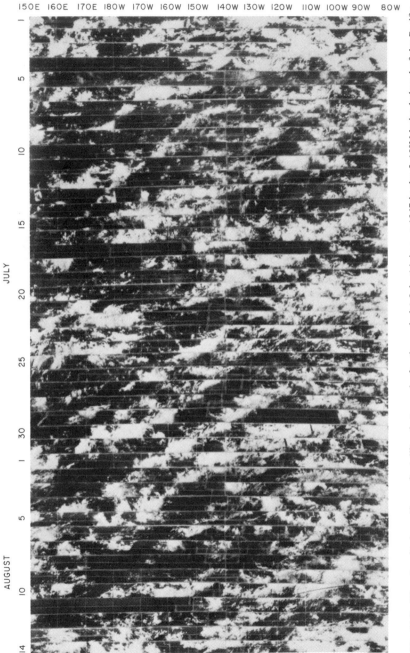

Fig. 11.3 Time-longitude sections of satellite photographs for the period 1 July–14 August 1967 in the 5–10°N latitude band of the Pacific. The westward progression of the cloud clusters is indicated by the bands of cloudiness sloping down the page from right to left. (After Chang, 1970. Reproduced with permission of the American Meteorological Society.)

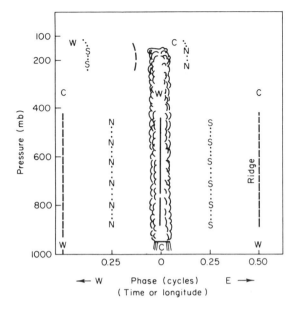

Fig. 11.4 Schematic model for equatorial wave disturbances showing trough axis (solid line), ridge axis (dashed lines), and axes of northerly and southerly wind components (dotted lines). Regions of warm and cold air are designated by W and C, respectively. (After Wallace, 1971.)

$[P' \cdot K']$ conversion balances R' in (10.62)]. There is, in this approximation, no storage in the form of available potential energy. The energy cycle of these disturbances, therefore, differs remarkably from that of midlatitude baroclinic systems, in which the available potential energy greatly exceeds the kinetic energy.

For latent heat release by cumulonimbus clouds to be an effective energy source for large-scale disturbances there must be an interaction between the convective scale and the large scale, as mentioned in Section 9.7.2. In such interaction large-scale convergence at low levels moistens and destabilizes the environment so that small-scale thermals can easily reach the level of free convection and produce deep cumulus convection. The cumulus cells, in turn, act cooperatively to provide the large-scale heat source that drives the secondary circulation responsible for the low-level convergence.

A typical vertical profile of divergence in the precipitation zone of a synoptic-scale equatorial wave disturbance in the western Pacific is shown in Fig. 11.5. Convergence is not limited to low-level frictional inflow in the planetary boundary layer but extends up to nearly 400 mb, which is the height where the hot towers achieve their maximum buoyancy. The deep

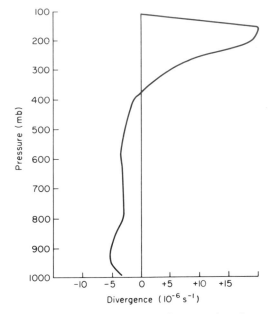

Fig. 11.5 Vertical profile of 4°-square area average divergence based on composites of many equatorial disturbances. (Adapted from Williams, 1971.)

convergence implies that there must be substantial entrainment of mid-tropospheric air into the convective cells. Because the midtropospheric air is relatively dry, this entrainment will require considerable evaporation of liquid water to bring the mixture of cloud and environment air to saturation. It will thus reduce the buoyancy of the cloud air and may in fact produce negatively buoyant convective downdrafts if there is sufficient evaporative cooling. However, in the large cumulonimbus clouds present in equatorial waves the central core updrafts are protected from entrainment by the surrounding cloud air so that they can penetrate nearly to the tropopause without dilution by environmental air. It is these undiluted cores that constitute the "hot towers" referred to in Section 11.1.1. Since the hot towers are responsible for much of the vertical heat and mass transport above the boundary layer in the ITCZ, and the wave disturbances contain most of the active convective precipitation areas along the ITCZ, it is obvious that equatorial waves play an essential role in the general circulation of the atmosphere.

11.1.3 AFRICAN WAVE DISTURBANCES

The considerations of the previous subsection are valid for ITCZ disturbances over most regions of the tropical oceans. In the region of the North

African continent, however, local effects owing to surface conditions create a unique situation that requires separate discussion. During the Northern Hemisphere summer intense surface heating over the Sahara generates a strong positive temperature gradient in the lower troposphere between the equator and about 25°N. The resulting easterly thermal wind is responsible for the existence of a strong easterly jet core near 650 mb centered near 16°N as shown in Fig. 11.6. Synoptic-scale disturbances are observed to form and propagate westward in the cyclonic shear zone to the south of this jet core. Occasionally such disturbances are progenitors of tropical storms and hurricanes in the western Atlantic. The average wavelength of observed African wave disturbances is about 2500 km and the westward propagation speed is about 8 m s^{-1}, implying a period of about 3.5 d. The disturbances have horizontal velocity perturbations that reach maximum amplitude at the 650-mb level, as indicated in Fig. 11.7. Although there is considerable organized convection associated with these waves, they do not appear to be primarily driven by latent heat release but depend, rather, on barotropic and baroclinic conversions of energy from the easterly jet.

In Fig. 11.8 the absolute vorticity profile for the African easterly jet shown in Fig. 11.6 is plotted. The shaded region indicates the area in which the vorticity gradient is negative. Thus, it is clear that the African jet satisfies the necessary condition for barotropic instability discussed in Section 8.4.2.[2]

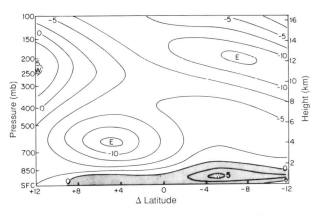

Fig. 11.6 Mean zonal wind distribution in the North African region (30°W to 10°E longitude) for the period 23 August 1974 to 19 September 1974. Latitude is shown relative to latitude of maximum disturbance amplitude at 700 mb (about 12°N). The contour interval is 2.5 m s^{-1}. (After Reed et al., 1977. Reproduced with permission of the American Meteorological Society.)

[2] It should be noted here that the profile shown in Fig. 11.6 is not a zonal mean. Rather, it is a time mean for a limited longitudinal domain. Provided that the longitudinal scale of variation of this time mean zonal flow is large compared to the scale of the disturbance, the time mean flow may be regarded as a locally valid basic state for linear stability calculations.

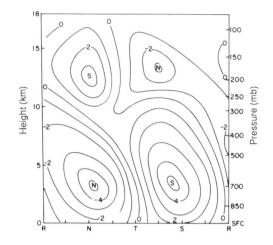

Fig. 11.7   Vertical cross section along the reference latitude of Fig. 11.6 showing perturbation meridional velocities in m s^{-1}. R, N, T, S refer to ridge, northwind, trough, and southwind sectors of the wave, respectively. (After Reed *et al.*, 1977. Reproduced with permission of the American Meteorological Society.)

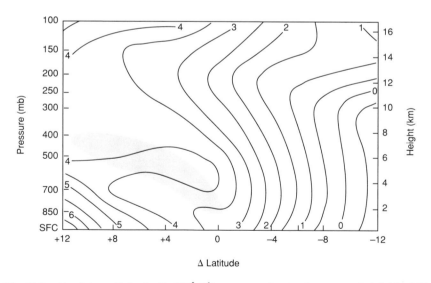

Fig. 11.8   Absolute vorticity (units 10^{-5} s^{-1}) corresponding to the mean wind field of Fig. 11.6. Shading shows region where $\beta - \partial^2 \bar{u}/\partial y^2$ is negative. (After Reed *et al.*, 1977. Reproduced with permission of the American Meteorological Society.)

Baroclinic instability owing to the strong easterly shear in the lower troposphere also appears to play a role in these disturbances. Thus, both barotropic and baroclinic conversions from the mean flow energy appear to be important for the generation of African wave disturbances.

Since such disturbances continue to exist in the absence of strong mean wind shears after they have propagated westward into the Atlantic, it is unlikely that either baroclinic or barotropic instability continues to be the primary energy source for their maintenance. Rather, diabatic heating through precipitating convective systems appears to be the main energy source for such waves over the ocean.

11.1.4 TROPICAL MONSOONS

The term "monsoon" is commonly used in a rather general sense to designate any seasonally reversing circulation system. Much of the tropics is influenced to some extent by monsoons. The most extensive monsoon circulation by far, however, is the complex circulation associated with the tropical region of Asia. This monsoon completely dominates the climate of the Indian subcontinent, producing warm wet summers and cool dry winters. An idealized model of the structure of the summer monsoons is indicated in Fig. 11.9. The basic drive for this monsoon circulation is provided by the contrast in the thermal properties of the land and sea surfaces. Since

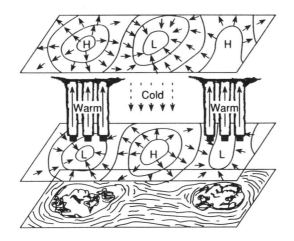

Fig. 11.9 Schematic representation of a summer monsoon circulation. Solid lines represent geopotential contours at 1000 mb (lower plane) and 200 mb (upper plane). Short solid arrows indicate direction of vertical motion in the midtroposphere. (After Wallace and Hobbs, 1977.)

the thin layer of soil that responds to the seasonal changes in surface temperature has a small heat capacity compared to the heat capacity of the upper layer of the oceans that responds on a similar time scale, the absorption of solar radiation raises the surface temperature over the land much more rapidly than over the oceans. This surface warming leads to enhanced cumulus convection and hence to latent heat release, which produces warm temperatures throughout the troposphere. Thus, as indicated in Fig. 11.9, the 100–20 kPa (1000–200 mb) layer thickness becomes larger over the land than over the ocean. As a result there is a pressure gradient force at the upper levels directed from the land to the ocean. The divergent wind that develops in response to this pressure gradient (shown by the small arrows in Fig. 11.9) causes a net mass transport out of the air column above the continent and thereby generates a surface low over the continent (sometimes called a *thermal low*). A compensating convergent wind then develops at low levels. This low-level flow produces a convergence of moisture, which by increasing the equivalent potential temperature in the boundary layer makes the environment more favorable for development of the cumulus convection that is the primary energy source for the monsoon circulation.

The low-level convergence and upper-level divergence over the continent constitute a secondary circulation that generates cyclonic vorticity at the lower levels and anticyclonic vorticity at the upper levels. Thus, the vorticity adjusts towards geostrophic balance. From Fig. 11.9 it is clear that there is a positive correlation between the vertical motion and temperature field. Therefore, the monsoon circulation converts eddy potential energy to eddy kinetic energy, just as midlatitude baroclinic eddies do.

Unlike the case of baroclinic eddies, however, the primary energy cycle of the monsoons does not involve the zonal mean potential or kinetic energy. Rather, eddy potential energy is generated directly by diabatic heating (latent and radiative heating), the eddy potential energy is converted to eddy kinetic energy by a thermally direct secondary circulation, and the eddy kinetic energy is frictionally dissipated. (A portion of the eddy kinetic energy may be converted to zonal kinetic energy.) In a dry atmosphere monsoon circulations would still exist; however, since the diabatic heating would then be confined to a shallow layer near the surface, they would be much weaker than the observed monsoons. The presence of cumulus convection and its concomitant latent heat release greatly amplifies the eddy potential energy generation and makes the summer monsoons among the most important features of the global circulation.

In the winter season the thermal contrast between the land and sea reverses so that the circulation is just opposite to that shown in Fig. 11.9. As a result the continents are cool and dry and the precipitation is found over the relatively warm oceans.

11.1.5 THE WALKER CIRCULATION

The pattern of diabatic heating in the equatorial regions exhibits strong departures from zonal symmetry. These are caused by longitudinal variations in sea surface temperature due mainly to the effects of wind-driven ocean currents. Such SST variations produce zonally asymmetric atmospheric circulations, which in some regions dominate over the Hadley circulation. Of particular significance is the east–west overturning along the equator which is shown schematically in Fig. 11.10. Several overturning cells are indicated, which are associated with diabatic heating over equatorial Africa, Central and South America, and the Maritime Continent (i.e., the Indonesian region). The dominant cell both in zonal scale and amplitude, however, is that in the equatorial Pacific. This cell is referred to as the *Walker circulation*, after G. T. Walker, who first documented the surface pressure pattern associated with it.

As suggested by Fig. 11.10, this pressure pattern consists of low surface pressure in the western Pacific and high surface pressure in the eastern Pacific. The resulting westward-directed pressure gradient force drives mean surface easterlies in the equatorial Pacific, which are much stronger than the zonal mean surface easterlies and by horizontal vapor transport provide a moisture source for convection in the western Pacific in addition to that

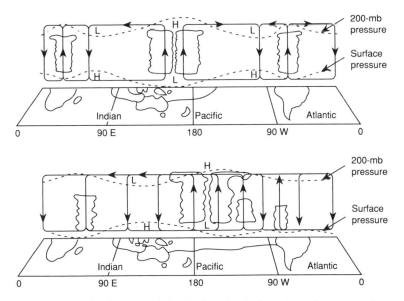

Fig. 11.10 Schematic diagrams of the Walker circulations along the equator for normal conditions (upper panel) and El Niño conditions (lower panel). (After Webster, 1983 and Webster and Chang, 1988.)

provided by the high evaporation rates caused by the high sea surface temperatures in that region.

The wind stress owing to the time mean equatorial easterly surface winds over the Pacific has a strong influence on the heat balance of the ocean surface layer. It advects warm surface waters into the western Pacific and produces poleward drifts in the oceanic Ekman layer, which by continuity drive an equatorial upwelling. It is this upwelling that accounts for the cold tongue of water along the equator, which in turn is a major reason for the equatorial dry zone exhibited in Fig. 11.2.

11.1.6 EL NIÑO AND THE SOUTHERN OSCILLATION

The east–west pressure gradient associated with the Walker circulation undergoes an irregular interannual variation. This global scale "see-saw" in pressure, and its associated changes in patterns of wind, temperature, and precipitation, was named the *Southern Oscillation* by Walker. This oscillation can be clearly seen by comparing time series of surface pressure anomalies (i.e., departures from the long-term mean) for locations on the western and eastern sides of the equatorial Pacific. As shown in Fig. 11.11, surface pressure anomalies at Darwin, Australia, and Tahiti are negatively correlated and have strong variations in the period range of 2–5 years. During periods of large pressure difference between these stations the Walker circulation is unusually strong and has the same general structure as its time mean pattern. In periods of weak pressure difference the Walker

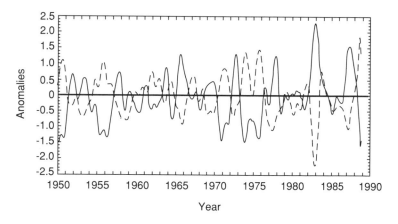

Fig. 11.11 Time series of anomalies in sea-level pressures at Darwin (solid) and Tahiti (dashed) from 1950 to 1988 smoothed to eliminate fluctuations with periods less than a year. (After Trenberth, 1984.)

circulation shifts in position and the region of maximum rainfall shifts eastward into the Pacific (see Fig. 11.10, lower panel).

The oscillations in wind stress owing to the Southern Oscillation are associated with changes in the circulation of the ocean and the sea surface temperatures that are referred to as *El Niño* (Spanish for "the child"). This term was originally applied to a warming of the coastal waters of Peru and Ecuador, which occurs annually near Christmastime (hence, El Niño refers to the Christ child). It is now used in a more general sense for the large-scale oceanic anomalies associated with the Southern Oscillation, which may begin near the coast but over the course of several months spread westward along the equator to produce large-scale SST anomalies over much of the equatorial Pacific.

This entire complex of atmospheric and oceanic variations is now referred to as ENSO (for "El Niño–Southern Oscillation"). It is a dramatic example of atmosphere–ocean coupling. At present, a number of competing theories exist to account for the interannual variability associated with ENSO. All of these theories depend in one manner or another on the dynamical and thermodynamical coupling between the atmosphere and the ocean. ENSO appears to be associated with a wide range of interannual climate anomalies in the extratropical regions as well as in the tropics. Thus, a model that could reliably predict the occurrence of ENSO events would be of great practical utility.

11.2 Scale Analysis of Large-Scale Tropical Motions

Despite the uncertainties involved in the interaction between the convective and synoptic scales, some information on the character of synoptic-scale motions in the tropics can be obtained through the methods of scale analysis. The scaling arguments can be carried out most conveniently if the governing equations are written in the log-pressure coordinate system introduced in Chapter 8:

$$\left(\frac{\partial}{\partial t} + \mathbf{V} \cdot \nabla + w^* \frac{\partial}{\partial z^*}\right) \mathbf{V} + f\,\mathbf{k} \times \mathbf{V} = -\nabla \Phi \tag{11.1}$$

$$\partial \Phi / \partial z^* = RT/H \tag{11.2}$$

$$\partial u/\partial x + \partial v/\partial y + \partial w^*/\partial z^* - w^*/H = 0 \tag{11.3}$$

$$\left(\frac{\partial}{\partial t} + \mathbf{V} \cdot \nabla\right) T + \frac{w^* N^2 H}{R} = \frac{J}{c_p} \tag{11.4}$$

We wish to compare the magnitudes of the various terms in (11.1)–(11.4) for synoptic-scale motions in the tropics. We first note that an upper limit on the vertical velocity scale W is imposed by the continuity equation (11.3). Thus, following the discussion of Section 4.5,

$$\partial u/\partial x + \partial v/\partial y \leq U/L$$

But for motions with vertical scales comparable to the density scale height H,

$$\partial w^*/\partial z^* - w^*/H \sim W/H$$

so that the vertical velocity scale must satisfy the constraint $W \leq HU/L$ if the horizontal divergence and vertical stretching terms are to balance in the continuity equation. We next define characteristic scales for the various field variables as follows:

$H \sim 10^4$ m	vertical length scale
$L \sim 10^6$ m	horizontal length scale
$U \sim 10$ m s^{-1}	horizontal velocity scale
$W \leq HU/L$	vertical velocity scale
$\delta\Phi$	geopotential fluctuation scale
$L/U \sim 10^5$ s	time scale for advection

Here we have assumed magnitudes for the horizontal length and velocity scales that are typical for observed values in synoptic-scale systems both in the tropics and at midlatitudes. We now wish to show how the corresponding characteristic scales for vertical velocity and geopotential fluctuations are limited by the dynamic constraints imposed by conservation of mass, momentum, and thermodynamic energy.

We can estimate the magnitude of the geopotential fluctuation $\delta\Phi$ by scaling the terms in the momentum equation. For this purpose it is convenient to compare the magnitude of the horizontal inertial acceleration,

$$(\mathbf{V} \cdot \nabla)\mathbf{V} \sim U^2/L$$

with each of the other terms in (11.1) as follows:

$$|\partial \mathbf{V}/\partial t|/|(\mathbf{V} \cdot \nabla)\mathbf{V}| \sim 1 \tag{11.5}$$

$$|w^* \,\partial \mathbf{V}/\partial z^*|/|(\mathbf{V} \cdot \nabla)\mathbf{V}| \sim WL/UH \leq 1 \tag{11.6}$$

$$|f\,\mathbf{k} \times \mathbf{V}|/|\mathbf{V} \cdot \nabla)\mathbf{V}| \sim fL/U = \mathrm{Ro}^{-1} \leq 1 \tag{11.7}$$

$$|\nabla \Phi|/|(\mathbf{V} \cdot \nabla)\mathbf{V}| \sim \delta\Phi/U^2 \qquad (11.8)$$

We have previously shown that in middle latitudes where $f \sim 10^{-4}\,\mathrm{s}^{-1}$, the Rossby number Ro is small so that to a first approximation the Coriolis force and pressure gradient force terms balance. In that case $\delta\Phi \sim fUL$. In the equatorial region, however, $f \le 10^{-5}\,\mathrm{s}^{-1}$ and the Rossby number is of order unity or greater. Therefore, it is not appropriate to assume that the Coriolis force balances the pressure gradient force. In fact, (11.5)–(11.8) show that if the pressure gradient force is to be balanced in (11.1), the geopotential perturbation must scale as $\delta\Phi \sim U^2 \sim 100\ \mathrm{m}^2\,\mathrm{s}^{-2}$, and geopotential perturbations associated with equatorial synoptic-scale disturbances will be an order of magnitude smaller than those for midlatitude systems of similar scale.

This constraint on the amplitude of the geopotential fluctuations in the tropics has profound consequences for the structure of synoptic-scale tropical motion systems. These consequences can be easily understood by applying scaling arguments to the thermodynamic energy equation. It is first necessary to obtain an estimate of the temperature fluctuations. The hydrostatic approximation (11.2) implies that for systems whose vertical scale is comparable to the scale height

$$T = (H/R)\,\partial\Phi/\partial z^* \sim (\delta\Phi/R) \sim U^2/R \sim 0.3\ \mathrm{K} \qquad (11.9)$$

Therefore, deep tropical systems are characterized by practically negligible synoptic-scale temperature fluctuations. Referring to the thermodynamic energy equation, we find that for such systems

$$\left(\frac{\partial}{\partial t} + \mathbf{V} \cdot \nabla\right) T \sim 0.3\ \mathrm{K\,d}^{-1}$$

In the absence of precipitation the diabatic heating is primarily caused by emission of long-wave radiation, which tends to cool the troposphere at a rate of $J/c_p \sim 1\ \mathrm{K\,d}^{-1}$. Since the actual temperature fluctuations are small, this radiative cooling must be approximately balanced by adiabatic warming owing to subsidence. Thus, to a first approximation (11.4) becomes a diagnostic relationship for w^*:

$$w^*(N^2 H/R) \approx J/c_p \qquad (11.10)$$

For the tropical troposphere, $N^2 H/R \sim 3\ \mathrm{K\,km}^{-1}$; thus the vertical motion scale must satisfy $W \sim 0.3\ \mathrm{cm\,s}^{-1}$, and $WL/UH \sim 0.03$ in (11.6). Therefore, in the absence of precipitation the vertical motion is constrained to be even smaller than in extratropical synoptic systems of a similar scale. Not only can the vertical advection term be neglected in (11.1) but also, from the continuity equation (11.3), the divergence of the horizontal wind is $\sim 3 \times 10^{-7}\,\text{s}^{-1}$. Thus, the flow is nearly nondivergent.

The quasi-nondivergent nature of the flow in the absence of convective disturbances in the tropics makes it possible to simplify the governing equations for that situation. A theorem of Helmholtz[3] states that any velocity field can be divided into a *nondivergent* part $\mathbf{V}_\psi$ plus an *irrotational* part $\mathbf{V}_e$ such that

$$\mathbf{V} = \mathbf{V}_\psi + \mathbf{V}_e$$

where $\nabla \cdot \mathbf{V}_\psi = 0$ and $\nabla \times \mathbf{V}_e = 0$.

For a two-dimensional velocity field the nondivergent part can be expressed in terms of a *streamfunction* ψ defined by letting

$$\mathbf{V}_\psi = \mathbf{k} \times \nabla \psi \tag{11.11}$$

or in Cartesian components:

$$u_\psi = -\partial \psi / \partial y, \qquad v_\psi = \partial \psi / \partial x$$

from which it is easily verified that $\nabla \cdot \mathbf{V}_\psi = 0$ and $\zeta = \mathbf{k} \cdot \nabla \times \mathbf{V}_\psi = \nabla^2 \psi$. Because the isolines of ψ correspond to streamlines for the nondivergent velocity, and the distance separating the isolines of ψ is inversely proportional to the magnitude of the nondivergent velocity, the spatial distribution of $\mathbf{V}_\psi$ can be easily pictured by plotting lines of constant ψ on a synoptic chart.

We now approximate $\mathbf{V}$ by its nondivergent part $\mathbf{V}_\psi$ in (11.1) and neglect the small vertical advection term to obtain an approximate momentum equation valid for synoptic-scale motions in the tropics outside regions of precipitation:

$$\frac{\partial \mathbf{V}_\psi}{\partial t} + (\mathbf{V}_\psi \cdot \nabla)\mathbf{V}_\psi + f\,\mathbf{k} \times \mathbf{V}_\psi = -\nabla \Phi \tag{11.12}$$

Using the vector identity

$$(\mathbf{V} \cdot \nabla)\mathbf{V} = \nabla\left(\frac{\mathbf{V} \cdot \mathbf{V}}{2}\right) + \mathbf{k} \times \nabla \zeta$$

[3] See, for example, Bourne and Kendall (1968, p. 190).

we can rewrite (11.12) as

$$\frac{\partial \mathbf{V}_\psi}{\partial t} = -\nabla \left(\Phi + \frac{\mathbf{V}_\psi \cdot \mathbf{V}_\psi}{2} \right) - \mathbf{k} \times \mathbf{V}_\psi (\zeta + f) \qquad (11.13)$$

We next take $\mathbf{k} \cdot \nabla \times (11.13)$ to obtain the vorticity equation

$$\left(\frac{\partial}{\partial t} + \mathbf{V}_\psi \cdot \nabla \right)(\zeta + f) = 0 \qquad (11.14)$$

valid for nondivergent flow. This equation shows that in the absence of condensation heating, synoptic-scale circulations in the tropics in which the vertical scale is comparable to the scale height of the atmosphere must be barotropic; absolute vorticity is conserved following the nondivergent horizontal wind. Such disturbances cannot convert potential energy to kinetic energy. They must be driven by barotropic conversion of mean flow kinetic energy or by lateral coupling either to midlatitude systems or to precipitating tropical disturbances.

Because both the nondivergent velocity and the vorticity can be expressed in terms of the streamfunction, (11.14) requires only the field of ψ at any level in order to make a prediction. The pressure distribution is neither required nor predicted. Rather, it must be determined diagnostically. The relationship of the pressure and streamfunction fields can be obtained by taking $\nabla \cdot (11.13)$. This yields a diagnostic relationship between the geopotential and streamfunction fields, which is usually referred to as the *nonlinear balance equation*:

$$\nabla^2[\Phi + \tfrac{1}{2}(\nabla\psi)^2] = \nabla \cdot [(f + \nabla^2\psi)\nabla\psi] \qquad (11.15)$$

For the special case of stationary circularly symmetric flow (11.15) is equivalent to the gradient wind approximation. Unlike the gradient wind, however, the balance in (11.15) does not require information on trajectory curvature and thus can be solved for Φ from knowledge of the instantaneous distribution of ψ on an isobaric surface. Alternatively, if the Φ distribution is known (11.15) can be solved for ψ. In this case the equation is quadratic so that there are generally two possible solutions, which correspond to the normal and anomalous gradient wind cases.

Such a balance condition is valid only when the above scaling arguments apply. These, however, have been based on the assumptions that the depth scale is comparable to the scale height of the atmosphere and that the horizontal scale is of order 1000 km. There is a special class of planetary-scale motions for which the divergence term in the vorticity equation is important even outside regions of active precipitation (see Section 11.4). For such motions the pressure field cannot be diagnosed from a balance relationship. Rather, the pressure distribution must be predicted from the primitive equation form of the dynamics equations.

For precipitating synoptic-scale systems in the tropics, the above scaling considerations require considerable modification. Precipitation rates in such systems are typically of order 2 cm d^{-1}. This implies condensation of $m_w = 20$ kg water for an atmospheric column of 1 m^2 cross section. Since the latent heat of condensation is $L_c \approx 2.5 \times 10^6$ J kg^{-1}, this precipitation rate implies an addition of heat energy to the atmospheric column of

$$m_w L_c \sim 5 \times 10^7 \text{ J m}^{-2} \text{ d}^{-1}$$

If this heat is distributed uniformly over the entire atmospheric column of mass $p_0/g \approx 10^4$ kg m^{-2}, then the average heating rate per unit mass of air is

$$J/c_p \approx [L_c m_w / c_p (p_0/g)] \sim 5 \text{ K d}^{-1}$$

In reality, the condensation heating owing to deep convective clouds is not distributed evenly over the entire vertical column but is a maximum between 300 and 400 mb, where the heating rate can be as high as 10 K d^{-1}. In this case the approximate thermodynamic energy equation (11.10) implies that the vertical motion on the synoptic scale in precipitating systems must have a magnitude of order $W \sim 3$ cm s^{-1} in order that the adiabatic cooling can balance the condensation heating in the 300–400-mb layer. Therefore, the average vertical motion in precipitating disturbances in the tropics is an order of magnitude larger than the vertical motion outside the disturbances. As a result, the flow in these disturbances has a relatively large divergent component so that the barotropic vorticity equation (11.14) is no longer a reasonable approximation, and the full primitive equations must be used to analyze the flow.

11.3   Condensation Heating

The manner in which the atmosphere is heated by condensation of water vapor depends crucially on the nature of the condensation process. In particular, it is necessary to differentiate between latent heat release through large-scale vertical motion (that is, synoptic-scale forced uplift) and the latent heat release owing to deep cumulus convection. The former process, which is generally associated with midlatitude synoptic systems, can be easily incorporated into the thermodynamic energy equation in terms of the synoptic-scale field variables. The large-scale heating field resulting from the cooperative action of many cumulonimbus cells, on the other hand, involves representation of this type of latent heating in terms of the synoptic-scale field variables, which is much more difficult.

Before considering the problem of condensation heating by cumulus convection, it is worth indicating briefly how the condensation heating by large-scale forced uplift can be included in a prediction model. The approximate thermodynamic energy equation for a pseudoadiabatic process (9.38) states that

$$\frac{D \ln \theta}{Dt} \approx -\left(\frac{L_c}{c_p T}\right)\frac{Dq_s}{Dt} \tag{11.16}$$

The change in q_s following the motion is caused primarily by ascent, so that

$$\frac{Dq_s}{Dt} \approx \begin{cases} w \, \partial q_s/\partial z & \text{for } w>0 \\ 0 & \text{for } w<0 \end{cases} \tag{11.17}$$

and (11.16) can be written in the form

$$\left(\frac{\partial}{\partial t}+\mathbf{V}\cdot\nabla\right)\ln\theta + w\left(\frac{\partial \ln \theta}{\partial z}+\frac{L_c}{c_p T}\frac{\partial q_s}{\partial z}\right)\approx 0 \tag{11.18}$$

for regions where $w>0$. But from (9.40)

$$\frac{\partial \ln \theta_e}{\partial z} \approx \frac{\partial \ln \theta}{\partial z}+\frac{L_c}{c_p T}\frac{\partial q_s}{\partial z}$$

so that (11.18) can be written in a form valid for both positive and negative vertical motion as

$$\left(\frac{\partial}{\partial t}+\mathbf{V}\cdot\nabla\right)\theta + w\Gamma_e\approx 0 \tag{11.19}$$

where Γ_e is an *equivalent static stability* defined by

$$\Gamma_e \approx \begin{cases} \theta \, \partial \ln \theta_e/\partial z & \text{for } q \geq q_s \quad \text{and} \quad w>0 \\ \partial \theta/\partial z & \text{for } q < q_s \quad \text{or} \quad w<0 \end{cases}$$

Thus, in the case of condensation owing to large-scale forced ascent ($\Gamma_e > 0$) the thermodynamic energy equation has essentially the same form as for adiabatic motions except that the static stability is replaced by the equivalent static stability. As a consequence, the local temperature changes induced by the forced ascent will be smaller than for the case of forced dry ascent with the same lapse rate.

If, on the other hand, $\Gamma_e < 0$, the atmosphere is conditionally unstable and condensation will occur primarily through cumulus convection. In that case (11.17) is still valid but the vertical velocity must be that of the individual cumulus updrafts, not the synoptic-scale w. Thus, a simple formulation of the thermodynamic energy equation in terms of only the synoptic-scale variables is not possible. We can still, however, simplify the thermodynamic energy equation to some extent. We recall from Section 11.2 [see Eq. (11.10)] that owing to the smallness of temperature fluctuations in the tropics the adiabatic cooling and diabatic heating terms must approximately balance. Thus, (11.16) becomes approximately

$$w \frac{\partial \ln \theta}{\partial z} \approx -\frac{L_c}{c_p T} \frac{Dq_s}{Dt} \tag{11.20}$$

The synoptic-scale vertical velocity w that appears in (11.20) is the average of very large vertical motions in the active convection cells and small vertical motions in the environment. Thus, if we let w' be the vertical velocity in the convective cells and $\bar{w}$ the vertical velocity in the environment, we have

$$w = aw' + (1 - a)\bar{w} \tag{11.21}$$

where a is the fractional area occupied by the convection. With the aid of (11.17) we can then write (11.20) in the form

$$w \frac{\partial \ln \theta}{\partial z} \approx -\frac{L_c}{c_p T} aw' \frac{\partial q_s}{\partial z} \tag{11.22}$$

The problem is then to express the condensation heating term on the right in (11.22) in terms of the synoptic-scale field variables.

This problem of *parameterizing* the cumulus convective heating is one of the most challenging areas in tropical meteorology. A simple approach that has been used successfully in some theoretical studies[4] is based on the fact that, since the storage of water in the clouds is rather small, the total

---

[4] See for example, Stevens and Lindzen (1978).

vertically integrated heating rate owing to condensation must be approximately proportional to the net precipitation rate:

$$-\int_{z_c}^{z_T} (\rho a w' \partial q_s/\partial z)\, dz = P \qquad (11.23)$$

where z_c and z_T represent the cloud base and cloud top heights, respectively, and P is the precipitation rate (kg m^{-2} s^{-1}).

Since relatively little moisture goes into changing the atmospheric vapor mixing ratio, the net precipitation rate must approximately equal the moisture convergence into an atmospheric column plus surface evaporation

$$P = -\int_{0}^{z_m} \boldsymbol{\nabla} \cdot (\rho q \mathbf{V})\, dz + E \qquad (11.24)$$

where E is the evaporation rate (kg m^{-2} s^{-1}) and z_m is the top of the moist layer ($z_m \approx 2$ km over much of the equatorial oceans). Substituting into (11.24) from the approximate continuity equation for q,

$$\boldsymbol{\nabla} \cdot (\rho q \mathbf{V}) + \partial(\rho q w)/\partial z \approx 0 \qquad (11.25)$$

we obtain

$$P = (\rho w q)_{z_m} + E \qquad (11.26)$$

Using (11.26) we can relate the vertically averaged heating rate to the synoptic-scale variables $w(z_m)$ and $q(z_m)$.

We still, however, need to determine the distribution of the heating in the vertical. The most common approach is to use an empirically determined vertical distribution based on observations. In that case (11.18) can be written as

$$\left(\frac{\partial}{\partial t} + \mathbf{V} \cdot \boldsymbol{\nabla}\right) \ln \theta + w \frac{\partial \ln \theta}{\partial z} = \frac{L_c}{\rho c_p T} \eta(z)[(\rho w q)_{z_m} + E] \qquad (11.27)$$

where $\eta(z) = 0$ for $z < z_c$ and $z > z_T$ and $\eta(z)$ for $z_c \le z \le z_T$ is a weighting function that must satisfy

$$\int_{z_c}^{z_T} \eta(z)\, dz = 1$$

Recalling that the diabatic heating must be approximately balanced by adiabatic cooling as indicated in (11.22), we see from (11.27) that $\eta(z)$ will have a vertical structure similar to that of the large-scale vertical mass flux, ρw. Observations indicate that for many tropical synoptic-scale disturbances $\eta(z)$ reaches its maximum at about the 40-kPa (400-mb) level, consistent with the divergence pattern shown in Fig. 11.5.

The above formulation is designed to model average tropical conditions. In reality, the vertical distribution of diabatic heating is determined by the local distribution of cloud heights. Thus, the cloud height distribution is apparently a key parameter in cumulus parameterization. A cumulus parameterization scheme in which this distribution is determined in terms of the large-scale variables was developed by Arakawa and Schubert (1974). Discussion of this scheme is beyond the scope of this text.

11.4 Equatorial Wave Theory

Equatorial waves are an important class of eastward and westward-propagating disturbances that are trapped about the equator (that is, they decay away from the equatorial region). Diabatic heating by organized tropical convection may excite equatorial wave motions. Through such waves the dynamical effects of convective storms can be communicated over large longitudinal distances in the tropics. Such waves thus can produce remote responses to localized heat sources. Furthermore, by influencing the pattern of low-level moisture convergence they can partly control the spatial and temporal distribution of convective heating. A complete development of equatorial wave theory would be rather complicated. In order to introduce equatorial waves in the simplest possible context, we here use a shallow-water model and concentrate on the *horizontal* structure. Vertical propagation in a stratified atmosphere will be discussed in Chapter 12.

11.4.1 EQUATORIAL ROSSBY AND ROSSBY–GRAVITY MODES

It turns out that the horizontal structures of various equatorial wave modes can be elucidated clearly by considering free oscillations within the context of a shallow-water model analogous to that introduced in Section 7.3.2. For simplicity we consider the linearized momentum and continuity equations for a fluid system of mean depth h_e in a motionless basic state. Since we are interested only in the tropics, we utilize Cartesian geometry on an *equatorial β plane*. In this approximation terms proportional to cos ϕ are replaced by unity and terms proportional to sin ϕ are replaced by y/a, where y is the distance from the equator and a is the radius of the earth.

The Coriolis parameter in this approximation is given by

$$f \approx \beta y \tag{11.28}$$

where $\beta \equiv 2\Omega/a$, and Ω is the angular velocity of the earth. The resulting linearized shallow-water equations for perturbations on a motionless basic state of mean depth h_e may be written as

$$\partial u'/\partial t - \beta y v' = -\partial \Phi'/\partial x \tag{11.29}$$

$$\partial v'/\partial t + \beta y u' = -\partial \Phi'/\partial y \tag{11.30}$$

$$\partial \Phi'/\partial t + g h_e (\partial u'/\partial x + \partial v'/\partial y) = 0 \tag{11.31}$$

where $\Phi' = gh'$ is the geopotential disturbance and primed variables designate perturbation fields.

The x and t dependences may be separated by specifying solutions in the form of zonally propagating waves:

$$\begin{pmatrix} u' \\ v' \\ \Phi' \end{pmatrix} = \begin{bmatrix} \hat{u}(y) \\ \hat{v}(y) \\ \hat{\Phi}(y) \end{bmatrix} \exp[i(kx - \nu t)] \tag{11.32}$$

Substitution of (11.32) into (11.29)–(11.31) then yields a set of ordinary differential equations in y for the meridional structure functions $\hat{u}$, $\hat{v}$, $\hat{\Phi}$:

$$-i\nu\hat{u} - \beta y\hat{v} = -ik\hat{\Phi} \tag{11.33}$$

$$-i\nu\hat{v} + \beta y\hat{u} = -\partial\hat{\Phi}/\partial y \tag{11.34}$$

$$-i\nu\hat{\Phi} + gh_e(ik\hat{u} + \partial\hat{v}/\partial y) = 0 \tag{11.35}$$

If (11.33) is solved for $\hat{u}$ and the result substituted into (11.34) and (11.35) we obtain

$$(\beta^2 y^2 - \nu^2)\hat{v} = ik\beta y\hat{\Phi} + i\nu\,\partial\hat{\Phi}/\partial y \tag{11.36}$$

$$(\nu^2 - gh_e k^2)\hat{\Phi} + i\nu gh_e\left(\frac{\partial\hat{v}}{\partial y} - \frac{k}{\nu}\beta y\hat{v}\right) = 0 \tag{11.37}$$

Finally, (11.37) can be substituted into (11.36) to eliminate $\hat{\Phi}$, yielding a second-order differential equation in the single unknown, $\hat{v}$:

$$\frac{\partial^2\hat{v}}{\partial y^2} + \left[\left(\frac{\nu^2}{gh_e} - k^2 - \frac{k}{\nu}\beta\right) - \frac{\beta^2 y^2}{gh_e}\right]\hat{v} = 0 \tag{11.38}$$

Since (11.38) is homogeneous, we expect that nontrivial solutions satisfying the requirement of decay at large $|y|$ will exist only for certain values of ν, corresponding to frequencies of the normal mode disturbances.

Before discussing this equation in detail it is worth considering the asymptotic limits that occur when either $h_e \to \infty$ or $\beta = 0$. In the former case, which is equivalent to assuming that the motion is nondivergent, (11.38) reduces to

$$\frac{\partial^2 \hat{v}}{\partial y^2} + \left[-k^2 - \frac{k}{\nu}\beta \right] \hat{v} = 0$$

Solutions exist of the form $\hat{v} \sim \exp(ily)$, provided that ν satisfies the Rossby wave dispersion relationship, $\nu = -\beta k / (k^2 + l^2)$. This illustrates that for nondivergent barotropic flow equatorial dynamics is in no way special. The rotation of the earth enters only in the form of the β effect; it is not dependent on f. On the other hand, if $\beta = 0$ all influence of rotation is eliminated and (11.38) reduces to the shallow-water gravity wave model, which has nontrivial solutions for

$$\nu = \pm \left[gh_e(k^2 + l^2) \right]^{1/2}$$

Returning to (11.38), we seek solutions for the meridional distribution of $\hat{v}$, subject to the boundary condition that the disturbance fields vanish for $|y| \to \infty$. This boundary condition is necessary since the approximation $f \approx \beta y$ is not valid for latitudes much beyond $\pm 30°$, so that solutions must be equatorially trapped if they are to be good approximations to the exact solutions on the sphere. Equation (11.38) differs from the classic equation for a harmonic oscillator in y because the coefficient in square brackets is not a constant but is a function of y. For sufficiently small y this coefficient is positive and solutions oscillate in y, while for large y, solutions either grow or decay in y. Only the decaying solutions, however, can satisfy the boundary conditions.

It turns out[5] that solutions to (11.38) which satisfy the condition of decay far from the equator exist only when the constant part of the coefficient in square brackets satisfies the relationship

$$\frac{\sqrt{gh_e}}{\beta}\left(-\frac{k}{\nu}\beta - k^2 + \frac{\nu^2}{gh_e} \right) = 2n + 1; \qquad n = 0, 1, 2, \ldots \qquad (11.39)$$

[5] See Matsuno (1966).

which is a cubic dispersion equation determining the frequencies of permitted equatorially trapped free oscillations for zonal wave number k and meridional mode number n. These solutions can be expressed most conveniently if y is replaced by the nondimensional meridional coordinate

$$\xi \equiv (\beta/\sqrt{gh_e})^{1/2}y$$

Then the solution has the form

$$\hat{v}(\xi) = H_n(\xi) \exp(-\xi^2/2) \tag{11.40}$$

where $H_n(\xi)$ designates the nth *Hermite polynomial*. The first few of these polynomials have the values

$$H_0 = 1, \quad H_1(\xi) = 2\xi, \quad H_2(\xi) = 4\xi^2 - 2$$

Thus, the index n corresponds to the number of nodes in the meridional velocity profile in the domain $|y| < \infty$.

In general, the three solutions of (11.39) can be interpreted as eastward- and westward-moving equatorially trapped gravity waves and westward-moving equatorial Rossby waves. The case $n = 0$ (for which the meridional velocity perturbation has a Gaussian distribution centered at the equator) must be treated separately. In this case the dispersion relationship (11.39) factors as

$$\left(\frac{\nu}{\sqrt{gh_e}} - \frac{\beta}{\nu} - k\right)\left(\frac{\nu}{\sqrt{gh_e}} + k\right) = 0 \tag{11.41}$$

The root $\nu/k = -\sqrt{gh_e}$, corresponding to a westward-propagating gravity wave, is not permitted since the second term in parentheses in (11.41) was implicitly assumed not to vanish when (11.36) and (11.37) were combined to eliminate $\hat{\Phi}$. The roots given by the first term in parentheses in (11.41) are

$$\nu = k\sqrt{gh_e}\left[\frac{1}{2} \pm \frac{1}{2}\left(1 + \frac{4\beta}{k^2\sqrt{gh_e}}\right)^{1/2}\right] \tag{11.42}$$

The positive root corresponds to an eastward-propagating equatorial inertio-gravity wave, while the negative root corresponds to a westward-propagating wave, which resembles an inertio-gravity wave for long zonal scale ($k \to 0$) and resembles a Rossby wave for zonal scales characteristic of synoptic-scale disturbances. This mode is generally referred to as a *Rossby–gravity* wave. The horizontal structure of the westward-propagating $n = 0$ solution is shown in Fig. 11.12, while the relationship between frequency and zonal wave number for this and several other equatorial wave modes is shown in Fig. 11.13.

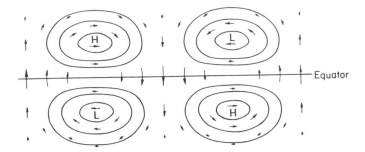

Fig. 11.12   Plan view of horizontal velocity and height perturbations associated with an equatorial Rossby-gravity wave. (Adapted from Matsuno, 1966.)

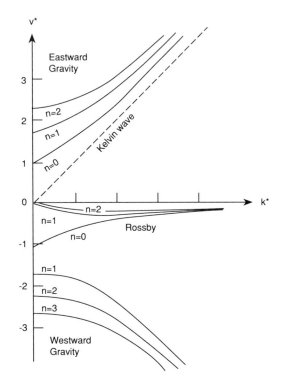

Fig. 11.13   Dispersion diagram for free equatorial waves. Frequency and zonal wave numbers have been nondimensionalized by defining

$$\nu^* \equiv \nu/(\beta\sqrt{gh_e})^{1/2}, \qquad k^* \equiv k(\sqrt{gh_e}/\beta)^{1/2}$$

Curves show dependence of frequency on zonal wave number for eastward and westward gravity modes and for Rossby and Kelvin modes.

11.4.2 EQUATORIAL KELVIN WAVES

In addition to the modes discussed in the previous section, there is another equatorial wave that is of great practical importance. For this mode, which is called the equatorial *Kelvin wave*, the meridional velocity perturbation vanishes and (11.33)–(11.35) are reduced to the simpler set

$$-iv\hat{u} = -ik\hat{\Phi} \tag{11.43}$$

$$\beta y\hat{u} = -\partial\hat{\Phi}/\partial y \tag{11.44}$$

$$-iv\hat{\Phi} + gh_e(ik\hat{u}) = 0 \tag{11.45}$$

Combining (11.43) and (11.45) to eliminate $\hat{\Phi}$, we see that the Kelvin wave dispersion equation is identical to that for ordinary shallow-water gravity waves:

$$c^2 \equiv (v/k)^2 = gh_e \tag{11.46}$$

According to (11.46) the phase speed c can be either positive or negative. But, if (11.43) and (11.44) are combined to eliminate $\hat{\Phi}$ we obtain a first-order equation for determining the meridional structure:

$$\beta y\hat{u} = -c\,\partial\hat{u}/\partial y \tag{11.47}$$

which may be integrated immediately to yield

$$\hat{u} = u_0\exp(-\beta y^2/2c) \tag{11.48}$$

where u_0 is the amplitude of the perturbation zonal velocity at the equator. Equation (11.48) shows that if solutions decaying away from the equator are to exist, the phase speed must be positive ($c > 0$). Thus, Kelvin waves are eastward propagating and have zonal velocity and geopotential perturbations that vary in latitude as Gaussian functions centered on the equator. The *e*-folding decay width is given by

$$Y_K = |2c/\beta|^{1/2}$$

which for a phase speed $c = 30$ m s^{-1} gives $Y_K \approx 1600$ km.

The perturbation wind and geopotential structure for the Kelvin wave are shown in plan view in Fig. 11.14. In the zonal direction the force balance is exactly that of an eastward-propagating shallow-water gravity wave. A vertical section along the equator would thus be the same as that shown in Fig. 7.6. The meridional force balance for the Kelvin mode is an exact geostrophic balance between the zonal velocity and the meridional pressure gradient. It is the change in sign of the Coriolis parameter at the equator that permits this special type of equatorial mode to exist.

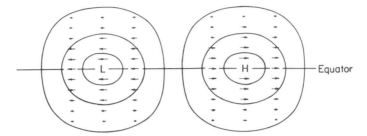

Fig. 11.14   Plan view of horizontal velocity and height perturbations associated with an
equatorial Kelvin wave. (Adapted from Matsuno, 1966.)

11.5   Steady Forced Equatorial Motions

Not all zonally asymmetric circulations in the tropics can be explained
on the basis of inviscid equatorial wave theory. For quasi-steady circulations
the zonal pressure gradient force must be balanced by turbulent drag rather
than by inertia. The Walker circulation may be regarded as a quasi-steady
circulation that is generated by diabatic heating. The simplest models of
such circulations specify the diabatic heating and use the equations of
equatorial wave theory to compute the atmospheric response. These models,
however, neglect the fact that the distribution of diabatic heating is highly
dependent on the mean wind and equivalent potential temperature distributions in the boundary layer. These in turn depend on the surface pressure
and the moisture distribution, which are themselves dependent on the
motion field. Thus, in a consistent model the diabatic heating cannot be
regarded as an externally specified quantity but must be obtained as part
of the solution by, for example, using the cumulus parameterization scheme
illustrated in Section 11.3.

As indicated in (11.27), this scheme requires information on the vertical
distribution of convective heating in order to solve for the temperature
perturbation throughout the troposphere. There is, however, evidence that
the essential features of stationary equatorial circulations can be partly
explained on the basis of a model that involves only the boundary layer.
This is perhaps not surprising since the dynamics of convective systems
depends crucially on evaporation and moisture convergence in the boundary
layer. Over the tropical oceans the boundary layer can be approximated as
a mixed layer of about 2 km depth (see Fig. 5.2), which is capped by an
inversion, across which there is a density discontinuity $\delta\rho$. The virtual
temperature in the mixed layer is strongly correlated with the sea surface
temperature. If we assume that the pressure field is uniform at the top of
the mixed layer, the surface pressure will be determined by hydrostatic

mass adjustment within the mixed layer. Positive sea surface temperature anomalies and negative boundary layer height anomalies will produce low surface pressures, and vice versa. If the boundary layer depth does not vary too much, the surface pressure gradient will thus tend to be proportional to the sea surface temperature gradient. The dynamics of steady circulations in such a mixed layer can be approximated by assuming that the vertically averaged pressure gradient force is proportional to the surface pressure gradient determined hydrostatically as indicated above.

We furthermore assume that the surface eddy stress can be parameterized as proportional to the mean velocity in the mixed layer, with a rate coefficient α. The x and y components of the vertically averaged momentum equation for steady motions in the mixed layer can then be written as follows:

$$\alpha u - \beta y v + \partial \Phi / \partial x = 0 \tag{11.49}$$

$$\alpha v + \beta y u + \partial \Phi / \partial y = 0 \tag{11.50}$$

Here the pressure gradient force is represented in terms of the perturbation geopotential Φ. According to the above arguments Φ is dependent on the virtual temperature in the boundary layer and the boundary layer height. If the latter is assumed to be proportional to the mean convergence in the boundary layer, the model can be closed by an equation that is analogous to the continuity equation for the shallow-water system:

$$\alpha \Phi + c_B^2 (1 - \varepsilon)(\partial u / \partial x + \partial v / \partial y) = -\Gamma \theta_v \tag{11.51}$$

where ε is a coefficient that is zero in the absence of convection and of order $3/4$ when convection occurs. This parameter allows for ventilation of the boundary layer by convection, which will tend to limit the total surface pressure perturbation. (See Battisti et al., 1992, for a more complete discussion.)

Equations (11.49)–(11.51) are a closed set for prediction of the boundary layer variables u, v, Φ for a specified boundary layer perturbation virtual temperature θ_v. But, since the parameter ε depends on the presence or absence of convection, the system can be solved only by iteration through using (11.24) to test for the presence of convection. This model can be used to compute the steady surface circulation. A sample calculation of the anomalous circulation for temperature anomalies corresponding to a typical ENSO event is shown in Fig. 11.15. Note that the region of convergence is narrower than the region of warm SST anomalies due to the convective feedback in the boundary layer model.

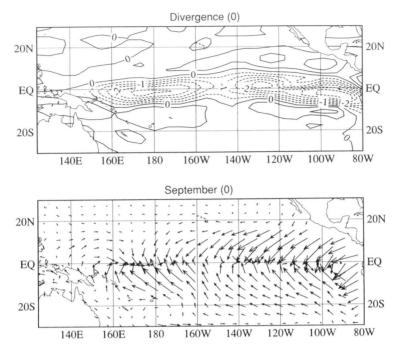

Fig. 11.15 Steady surface circulation forced by sea surface temperature gradients in the equatorial Pacific. (After Battisti *et al.*, 1992.)

Problems

11.1. Suppose that the relative vorticity at the top of an Ekman layer at 15°N is $\zeta = 2 \times 10^{-5}\,\mathrm{s}^{-1}$. Let the eddy viscosity coefficient be $K_m = 10\,\mathrm{m}^2\,\mathrm{s}^{-1}$, and the water vapor mixing ratio at the top of the Ekman layer be $12\,\mathrm{g\,kg}^{-1}$. Use the method of Section 11.3 to estimate the precipitation rate owing to moisture convergence in the Ekman layer.

11.2. As mentioned in Section 11.1.3, barotropic instability is a possible energy source for some equatorial disturbances. Consider the following profile for an easterly jet near the equator:

$$\bar{u}(y) = -u_0 \sin^2[l(y - y_0)]$$

where u_0, y_0, and l are constants and y is the distance from the equator. Determine the necessary conditions for this profile to be barotropically unstable.

11.3. Show that the nonlinear terms in the balance equation (11.13)

$$G(x, y) \equiv -\nabla^2(\tfrac{1}{2}\nabla\psi \cdot \nabla\psi) + \nabla \cdot (\nabla\psi\nabla^2\psi)$$

may be written in Cartesian coordinates as

$$G(x, y) = 2[(\partial^2\psi/\partial x^2)(\partial^2\psi/\partial y^2) - (\partial^2\psi/\partial x\partial y)^2]$$

11.4. With the aid of the results of Problem 11.3, show that if f is assumed to be constant the balance equation (11.15) is equivalent to the gradient wind equation (3.15) for a circularly symmetric geopotential perturbation given by

$$\Phi = \Phi_0(x^2 + y^2)/L^2$$

where Φ_0 is a constant geopotential and L a constant length scale. Hint: Assume that $\Psi(x, y)$ has the same functional dependence on (x, y) as does Φ.

11.5. Starting from the perturbation equations (11.29)–(11.31) show that the sum of kinetic plus available potential energy is conserved for equatorial waves. Hence, show that for the Kelvin wave there is an equipartition of energy between kinetic and available potential energy.

11.6. Solve for the meridional dependence of the zonal wind and geopotential perturbations for a Rossby–gravity mode in terms of the meridional velocity distribution (11.40).

11.7. Use the linearized model (11.49)–(11.50) to compute the meridional distribution of divergence in the mixed layer for a situation in which the geostrophic wind is given by

$$u_g = u_0 \exp(-\beta y^2/2c), \qquad v_g = 0$$

where u_0 and c are constants.

Suggested References

Philander, *El Niño, La Niña, and the Southern Oscillation*, is a well-written text that covers the dynamics of both the atmospheric and oceanographic aspects of ENSO at an advanced level.

Trenberth (1991) is an excellent review article on ENSO.

Wallace (1971) is a detailed review of the structure of synoptic-scale tropospheric wave disturbances in the equatorial Pacific.

Webster (1983) presents an excellent review of the observed structure and dynamics of large-scale tropical circulations.

Chapter

12 | Middle Atmosphere Dynamics

The *middle atmosphere* is generally regarded as the region extending from the tropopause (about 10–16 km altitude depending on latitude) to about 100 km. The bulk of the middle atmosphere consists of two main layers, the *stratosphere* and the *mesosphere*, which are distinguished on the basis of temperature stratification (Fig. 12.1). The stratosphere, which has large static stability associated with an overall increase of temperature with height, extends from the tropopause to the *stratopause* at about 50 km. The mesosphere, which has a lapse rate similar to that in the troposphere, extends from the stratopause to the *mesopause* at about 80 km.

Previous chapters of this text have focused almost exclusively on the dynamics of the troposphere. The troposphere accounts for about 85% of the total mass of the atmosphere and virtually all atmospheric water. There can be little doubt that processes occurring in the troposphere are primarily responsible for weather disturbances and climate variability. Nevertheless, the middle atmosphere cannot be neglected. The troposphere and the middle atmosphere are linked through radiative and dynamical processes, which must be represented in global forecast and climate models. They are also linked through the exchange of trace substances that are important in the

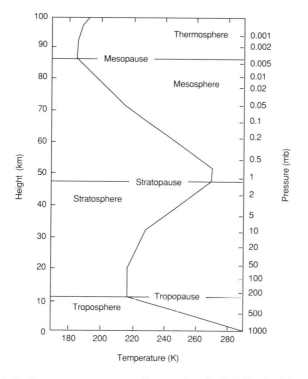

Fig. 12.1   Midlatitude mean temperature profile. Based on the *U.S. Standard Atmosphere*, 1976.

photochemistry of the ozone layer. In this chapter we focus primarily on dynamical processes in the lower part of the middle atmosphere and their links to the troposphere.

12.1   Structure and Circulation of the Middle Atmosphere

Zonal mean temperature cross sections for January and July in the lower and middle atmosphere are shown in Fig. 12.2. Because very little solar radiation is absorbed in the troposphere, the thermal structure of the troposphere is maintained by an approximate balance among infrared radiative cooling, vertical transport of sensible and latent heat away from the surface by small-scale eddies, and large-scale heat transport by synoptic-scale eddies. The net result is a mean temperature structure in which the surface temperature has its maximum in the equatorial region and decreases toward both the winter and summer poles. There is also a rapid decrease in altitude with a lapse rate of about 6°C km^{-1}.

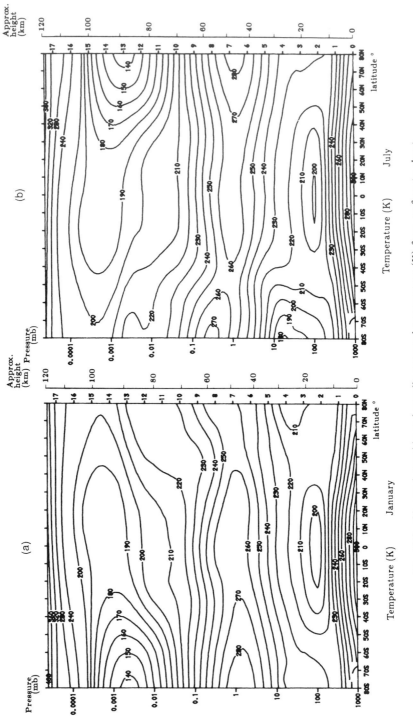

Fig. 12.2 Observed monthly and zonally averaged temperature (K) from surface to about 120 km for (a) January and (b) July. (After Fleming et al., 1990.)

In the stratosphere, on the other hand, infrared radiative cooling is in the mean balanced primarily by radiative heating owing to the absorption of solar ultraviolet radiation by ozone. As a result of the solar heating in the ozone layer, the mean temperature in the stratosphere increases with height to a maximum at the stratopause near 50 km. Above the stratopause temperature decreases with height owing to the reduced solar heating of ozone.

The meridional temperature structure in the middle atmosphere is also quite different from that in the troposphere. In the lower stratosphere, where the temperature is strongly influenced by upper tropospheric processes, there are a temperature minimum at the equator and maxima at the summer pole and in midlatitudes of the winter hemisphere. Above about 30 mb the temperature decreases uniformly from summer pole to winter pole, in qualitative accord with radiative equilibrium conditions.

Mean zonal wind climatologies for the middle atmosphere are usually derived from the satellite-observed temperature field. This is done by using the geostrophic winds on an isobaric surface in the lower stratosphere (obtained from conventional meteorological analyses) as a lower boundary condition and integrating the thermal wind equation vertically. January and July mean zonal wind cross sections are shown in Fig. 12.3. The main features are an easterly jet in the summer hemisphere and a westerly jet in the winter hemisphere, with maxima in the wind speeds occurring near the 60-km level. Of particular significance are the high-latitude westerly jets in the winter hemispheres. These polar night jets provide wave guides for the vertical propagation of quasi-stationary planetary waves. In the Northern Hemisphere the EP flux convergence owing to such waves occasionally leads to rapid deceleration of the mean zonal flow and an accompanying *sudden stratospheric warming* in the polar region, as will be discussed in Section 12.4.

The zonal mean flow in the equatorial middle atmosphere is strongly influenced by vertically propagating equatorial wave modes, especially the Kelvin and Rossby–gravity modes. These waves interact with the mean flow to produce a long-period oscillation called the *quasi-biennial oscillation.* This oscillation produces large year-to-year variability in the mean zonal wind in the equatorial middle atmosphere, which is not shown in the long-term means of Fig. 12.3.

12.2 The Zonal Mean Circulation of the Middle Atmosphere

As discussed in Section 10.1, the general circulation of the atmosphere considered as a whole can be regarded to a first approximation as the atmospheric response to the diabatic heating caused by absorption of solar radiation at the surface. Thus, it is reasonable to say that the atmosphere

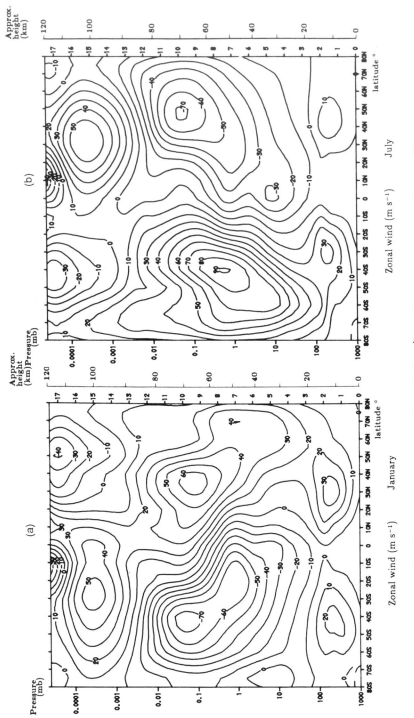

Fig. 12.3 Zonal mean geostrophic winds (m s⁻¹) based on the temperatures shown in Fig. 12.2 for (a) January and (b) July.

is driven by differential diabatic heating. For an open subregion of the atmosphere, such as the middle atmosphere, it is not correct, however, to assume that the circulation is driven by diabatic heating. It is rather, necessary to consider the transfer of momentum and energy between that subregion and the rest of the atmosphere.

In the absence of eddy motions the zonal mean temperature of the middle atmosphere would relax to a radiatively determined state in which, except for a small lag owing to thermal inertia, the temperature would correspond to an annually varying radiative equilibrium following the annual cycle in solar heating. The circulation in such a situation would consist only of a zonal-mean zonal flow in thermal wind balance with the meridional temperature gradient. Neglecting small effects of the annual cycle, there would be no meridional or vertical circulation and no stratosphere–troposphere exchange for such a hypothetical state.

Departures from this radiatively determined state must be maintained by eddy transports. Thus, rather than *causing* the mean circulation, the radiative heating and cooling patterns observed in the middle atmosphere are a *result* of the eddies driving the flow away from a state of radiative balance. This eddy-driven circulation has meridional and vertical wind components that induce substantial local departures from radiative equilibrium, especially in the winter stratosphere and near the mesopause in both winter and summer. The observed latitudinally dependent temperature distribution in the middle atmosphere arises from a balance between the net radiative drive and the heat transport plus local temperature change produced by these motions.

12.2.1   THE TRANSFORMED EULERIAN MEAN

A cross section of the radiatively determined temperature during Northern Hemisphere winter is shown in Fig. 12.4. It should be compared to the observed temperature profile for the same season that was shown in Fig. 12.2. Although the rather uniform increase of temperature from winter pole to summer pole in the region 30–60 km altitude is qualitatively consistent with the radiatively determined distribution, the actual temperature difference between the two poles is much smaller than in the radiatively determined state. Above 60 km even the sign of gradient is opposite to that in the radiative solution; the summer polar mesopause is much colder than the winter polar mesopause.

The dynamical role of eddies in maintaining the observed zonal mean temperature distribution can be elucidated by utilizing the transformed Eulerian mean (TEM) equations introduced in Section 10.2.2.[1] Recall that

---

[1] As in Chapter 10 we express the log-pressure coordinate simply as z rather than z^*.

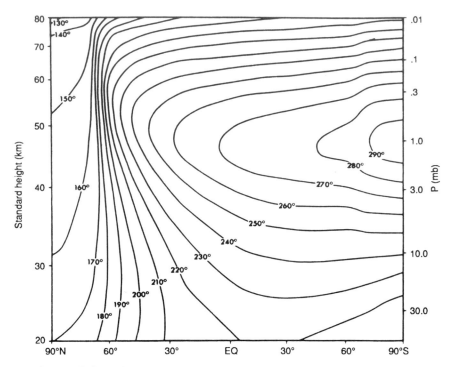

Fig. 12.4 Radiatively determined middle atmosphere temperature distribution for January 15 from a radiative model that is time-marched through an annual cycle. Realistic tropospheric temperatures and cloudiness are used to determine the upward radiative flux at the tropopause. (After Fels, 1985.)

in this formulation the zonal mean momentum and thermodynamic energy equations have the form

$$\partial \bar{u}/\partial t - f_0 \bar{v}^* = \rho_0^{-1} \mathbf{\nabla} \cdot \mathbf{F} + \bar{X} \equiv \bar{G} \qquad (12.1)$$

$$\partial \bar{T}/\partial t + N^2 H R^{-1} \bar{w}^* = \bar{J}/c_p \qquad (12.2)$$

where $\mathbf{F}$ designates the Eliassen–Palm (EP) flux owing to large-scale eddies, $\bar{X}$ is the zonal force owing to small-scale eddies (e.g., gravity wave drag), and $\bar{G}$ designates the total *zonal force*.

   To understand how eddies can lead to departures of the zonal-mean temperature distribution in the middle atmosphere from the radiatively determined state, we use the TEM system of equations to consider a hierarchy of simple models. The first of these is a steady-state model in which there is no seasonal cycle. The time derivatives in (12.1) and (12.2) are then zero so that we obtain simple diagnostic relations for the residual circulation:

$$-f_0 \bar{v}^* = \bar{G}, \qquad N^2 H R^{-1} \bar{w}^* = \bar{J}/c_p \qquad (12.3)$$

The Coriolis force owing to the residual meridional drift is balanced by the zonal force caused by the large-scale and small-scale eddies; the adiabatic cooling caused by the residual vertical motion is balanced by radiative heating (and vice versa). The relationships in (12.3) are not independent but are connected by the continuity equation (10.19), which implies that

$$-\partial \bar{G}/\partial y + f_0 \rho_0^{-1} \, \partial(\rho_0 \bar{J} \kappa / N^2 H)/\partial z = 0 \qquad (12.4)$$

Equation (12.4) shows how the diabatic heating rate must be related to the eddy forcing in this simple steady-state model. If the eddy forcing vanishes ($\bar{G} = 0$) the residual meridional drift must vanish. The continuity equation then shows that if $\bar{w}^*$ is zero at the lower boundary it will be zero everywhere and so the diabatic heating must also be zero everywhere. Thus, the simplest possible model for the circulation of the middle atmosphere is one in which there is no seasonal cycle, the eddy forcing vanishes, and the zonal mean temperature is in a steady-state radiative balance.

We next include time dependence in the form of an annual cycle of solar heating, but again assume that the eddy forcing vanishes. In this case it turns out that the actual temperature is quite close to radiative equilibrium so that we can parameterize the diabatic heating as proportional to the departure of the zonal mean temperature $\bar{T}(y, z, t)$ from its radiative equilibrium value $T_r(y, z, t)$:

$$\bar{J}/c_p = -\alpha_r [\bar{T} - T_r(y, z, t)] \qquad (12.5)$$

where α_r is the Newtonian cooling rate. Substituting from (12.5) into (12.2) and using the continuity equation (10.19) and the thermal wind equation (10.13) to combine (12.1) and (12.2), we obtain the temperature tendency equation

$$\left[\frac{\partial^2}{\partial y^2} + \frac{\partial}{\partial z} \left(\rho_0^{-1} \frac{\partial}{\partial z} \rho_0 \varepsilon \right) \right] \frac{\partial \bar{T}}{\partial t} + \frac{\partial}{\partial z} \left\{ \rho_0^{-1} \frac{\partial}{\partial z} [\alpha_r \rho_0 \varepsilon (\bar{T} - T_r)] \right\} = 0 \qquad (12.6)$$

where $\varepsilon(z) \equiv f_0^2 / N^2$. Here the forcing of the zonal mean temperature tendency is approximately proportional to the departure of temperature from the radiative equilibrium value. The temperature response will, however, lag the forcing (owing to the finite relaxation time) and the response will be smoothed in space owing to the elliptic operator in the tendency term.

In this annually varying model the diabatic heating is caused by the thermal inertia of the atmosphere; it is not imposed by an external radiative drive. The long-wave contribution to the diabatic term, which depends on temperature, is internally determined as part of the solution. Since the radiative relaxation time is short compared to the annual cycle, the departure of temperature from its radiatively determined value $T_r(y, z, t)$ must be small compared to the amplitude of the annual variation. Thus, an annually varying model without eddy forcing produces an annually varying temperature distribution close to radiative equilibrium (see Fig. 12.4).

We conclude that it is necessary to include the effects of eddy forcing to model the observed departure of the mean temperature from its radiatively determined state. For variations on the seasonal time scale the steady balance given in (12.4) is reasonably accurate. Thus, substituting from (12.5) into (12.4) we find that the departure from radiative equilibrium $\delta \bar{T} \equiv (\bar{T} - T_r)$ is given by

$$\frac{1}{f_0} \frac{\partial \bar{G}}{\partial y} = -\frac{1}{\rho_0} \frac{\partial}{\partial z} \left(\frac{\rho_0 R \alpha_r}{N^2 H} \delta \bar{T} \right) \approx \frac{R \alpha_r}{N^2 H^2} \delta \bar{T} \tag{12.7}$$

where to obtain the expression on the far right we have assumed that the exponential decrease of density with height dominates in the vertical derivative. Equation (12.7) implies that

$$\delta \bar{T} \sim \tau_r \left(\frac{N^2 H^2}{f_0 R} \right) \frac{\partial \bar{G}}{\partial y}$$

where τ_r designates the inverse of the radiative relaxation rate α_r. According to (12.7) the zonal force caused by the eddies drives temperature away from equilibrium. Radiation acts like a "spring" that relaxes the temperature toward an equilibrium state. Given the spatial and temporal distribution of the zonal force and the radiative relaxation rate, (12.7) can be solved for the temperature departure $\delta \bar{T}$. Equation (12.3) can then be used to determine the residual mean vertical motion, which should be a good approximation to the cross-isentropic mass transport.

A comparison of Figs. 12.2 and 12.4 shows that the largest departures from radiative equilibrium occur in the summer and winter mesophere and in the polar winter stratosphere. The zonal force in the mesophere is believed to be caused primarily by vertically propagating internal gravity waves. These transfer momentum from the troposphere into the mesosphere, where wave breaking produces a strong zonal force. The zonal force in the winter stratosphere is due primarily to stationary planetary Rossby waves. These, as will be discussed in Section 12.3, can propagate vertically provided that the mean zonal wind is westerly and less than a critical value that depends strongly on the wavelength of the eddies. Hence, in the extratropical stratosphere we expect a strong annual cycle in $\delta \bar{T}$, with large values (i.e., strong departure from radiative equilibrium) in winter and small in summer. This is indeed observed (see Figs. 12.2 and 12.4). Furthermore, since eddy forcing maintains the observed temperature above its radiative equilibrium in the extratropical stratosphere, there is radiative cooling and from (12.3) the residual vertical motion must be downward. By mass continuity it is then required that the residual vertical motion be upward in the tropics, implying that the temperature must be below radiative equilibrium in that region. Note that it is the dynamical driving by extratropical eddies, rather than local forcing, that is responsible for the upward residual motion and net radiative heating in the tropical stratosphere.

12.2.2 ZONAL MEAN TRANSPORT

Viewed from the TEM perspective, the overall meridional circulation in the winter stratosphere is qualitatively as pictured in Fig. 12.5. The residual circulation transfers mass and trace chemicals upward across the tropopause in the tropics and downward in the extratropics. This vertical circulation is closed in the lower stratosphere by a poleward meridional drift. That this schematic picture gives a qualitatively correct view can be ascertained by examination of the zonal mean mixing ratio distribution of any long-lived vertically stratified trace species. As an example, the distribution of N_2O is shown in Fig. 12.6. N_2O is produced at the surface and is uniformly mixed in the troposphere, but it decays with height in the stratosphere owing to photochemical dissociation. Thus, as shown in Fig. 12.6, the mixing ratio decreases upward in the stratosphere. Note, however, that surfaces of constant mixing ratio are displaced upward in the tropics and downward at higher latitudes, suggesting that the mean meridional mass transport is upward in the tropics and downward in the extratropics as suggested in Fig. 12.5.

In addition to the slow meridional drift by the residual meridional velocity shown in Fig. 12.5, tracers in the winter stratosphere are also subject to rapid quasi-isentropic transport and mixing by breaking planetary waves. Such eddy transport, which is highly variable in space and time, must be included for quantitatively accurate modeling of transport within the stratosphere.

In the mesosphere the residual circulation is dominated by a single circulation cell with upward motion in the summer polar region, meridional drift from the summer to the winter hemisphere, and subsidence in the winter polar region. This circulation, like the residual circulation in the stratosphere, is eddy driven. But in the mesosphere it appears that the

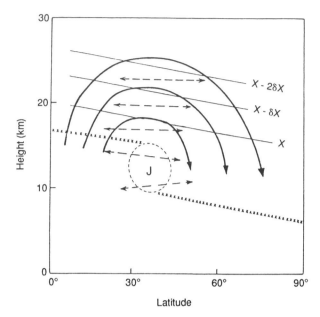

Fig. 12.5   Schematic cross section of transport in the stratosphere. Heavy lines show the mean
meridional mass circulation (approximately the residual circulation of the TEM
theory). Dashed lines indicate quasi-isentropic mixing by large-scale eddies. The
tropopause is indicated by crosses, and J indicates the location of the mean jet
stream core. Light lines labeled with mixing ratio values (X) show mean slope of
a long-lived vertically stratified tracer. (After Holton, 1986.)

dominant eddies are vertically propagating internal gravity waves, which
have shorter scales in space and time than the planetary waves that dominate
the eddy activity in the stratosphere.

12.3   Vertically Propagating Planetary Waves

In Section 12.1 we pointed out that the predominant eddy motions in the
stratosphere are vertically propagating quasi-stationary planetary waves and
that these waves are confined to the winter hemisphere. In order to understand the absence of synoptic-scale motions and the confinement of
planetary waves to the winter hemisphere, it is necessary to examine the
conditions under which planetary waves can propagate vertically.

To analyze planetary wave propagation in the stratosphere it is convenient
to write the equations in the log-pressure coordinate system introduced in
Section 8.4.1. For analysis of extratropical planetary wave motions in the
middle atmosphere we may refer the motions to the midlatitude β plane

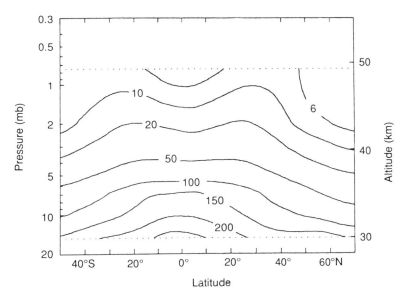

Fig. 12.6 Annually averaged cross section of N_2O (ppbv) from observations by the *Nimbus* 7 SAMS experiment. Note the strong vertical stratification owing to photochemical destruction in the stratosphere. The upward-bulging mixing ratio surfaces in the equatorial region are evidence of upward mass flow in the equatorial region.

and use the quasi-geostrophic potential vorticity equation (6.25), which in log-pressure coordinates can be written as follows:

$$\left(\frac{\partial}{\partial t} + \mathbf{V}_g \cdot \nabla\right) q = 0 \tag{12.8}$$

where

$$q \equiv \nabla^2 \psi + f + \frac{f_0^2}{\rho_0 N^2} \frac{\partial}{\partial z}\left(\rho_0 \frac{\partial \psi}{\partial z}\right)$$

[Compare with Eq. (6.26).] Here $\psi = \Phi/f_0$ is the geostrophic streamfunction and f_0 is a constant midlatitude reference value of the Coriolis parameter. We now assume that the motion consists of a small-amplitude disturbance superposed on a constant zonal mean flow. Thus, letting $\psi = -\bar{u}y + \psi'$, $q = \bar{q} + q'$, and linearizing (12.8) we find that the perturbation field must satisfy

$$\left(\frac{\partial}{\partial t} + \bar{u}\frac{\partial}{\partial x}\right) q' + \beta \frac{\partial \psi'}{\partial x} = 0 \tag{12.9}$$

where

$$q' \equiv \nabla^2 \psi' + \frac{f_0^2}{\rho_0 N^2} \frac{\partial}{\partial z} \left(\rho_0 \frac{\partial \psi'}{\partial z} \right)$$

and for this example $\partial \bar{q}/\partial y = \beta$.

Equation (12.9) has solutions in the form of harmonic waves with zonal and meridional wave numbers k and l and zonal phase speed c:

$$\psi'(x, y, z, t) = \Psi(z) e^{i(kx+ly-kct)+z/2H} \tag{12.10}$$

Here the factor $e^{z/2H}$ (which is proportional to $\rho_0^{-1/2}$) is included to simplify the equation for the vertical dependence. Substituting (12.10) into (12.9) we find that $\Psi(z)$ must satisfy

$$d^2\Psi/dz^2 + m^2\Psi = 0 \tag{12.11}$$

where

$$m^2 \equiv \frac{N^2}{f_0^2} \left[\frac{\beta}{(\bar{u}-c)} - (k^2+l^2) \right] - \frac{1}{4H^2} \tag{12.12}$$

Referring back to Section 7.4, we recall that $m^2 > 0$ is required for vertical propagation, and in that case m is the vertical wave number [i.e., solutions of (12.11) have the form $\Psi(z) = A e^{imz}$, where A is a constant amplitude and the sign of m is determined by requiring that the vertical component of group velocity be positive]. For stationary waves ($c = 0$) we see from (12.12) that vertically propagating modes can exist only for mean zonal flows that satisfy the condition

$$0 < \bar{u} < \beta[(k^2+l^2)+f_0^2/(4N^2H^2)]^{-1} \equiv U_c \tag{12.13}$$

where U_c is called the *Rossby critical velocity*. Thus, vertical propagation of stationary waves can occur only in the presence of westerly winds weaker than a critical value that depends on the horizontal scale of the waves. In the summer hemisphere the stratospheric mean zonal winds are easterly so that stationary planetary waves are all vertically trapped.

In the real atmosphere the mean zonal wind is not constant but depends on latitude and height. However, both observational and theoretical studies suggest that (12.13) still provides a qualitative guide for estimating vertical propagation of planetary waves even though the actual critical velocity may be larger than indicated by the β-plane theory.

12.4 Sudden Stratospheric Warmings

In the lower stratosphere the temperature is a minimum at the equator and has maxima at the summer pole and at about 45° latitude in the winter hemisphere (see Fig. 12.2). From thermal wind considerations the rapid decrease of temperature poleward of 45° in winter requires a zonal vortex with strong westerly shear with height.

In the Northern Hemisphere every other year or so this normal winter pattern of a cold polar stratosphere with a westerly vortex is interrupted in a spectacular manner in midwinter. Within the space of a few days the polar vortex becomes highly distorted and breaks down with an accompanying large-scale warming of the polar stratosphere, which can quickly reverse the meridional temperature gradient and (through thermal wind balance) create a circumpolar easterly current. Warmings of 40 K in a few days have occurred at the 50-mb level as shown in Fig. 12.7. Numerous observational studies of sudden warmings confirm that enhanced propagation of planetary waves from the troposphere, primarily zonal wave numbers 1 and 2, is essential for the development of the warmings. Since major warmings are observed only in the Northern Hemisphere, it is logical to conclude that the enhanced wave propagation into the stratosphere is due to topographically forced waves, which are much stronger in the Northern Hemisphere than in the Southern Hemisphere. Even in the Northern Hemisphere, however, it is apparently only in certain winters that conditions are right to produce sudden warmings.

It is generally accepted that sudden warmings are an example of transient mean-flow forcing owing to planetary wave driving. In Section 10.2.3 we showed that in order for planetary waves to decelerate the zonal mean circulation there must be a nonzero equatorward eddy potential vorticity flux (i.e., a net EP flux convergence). We further showed that for steady nondissipative waves, the divergence of the EP flux vanishes. For normal stationary planetary waves in the stratospheric polar night jet this constraint should be at least approximately satisfied, since radiative and frictional dissipation are rather small. Thus, the strong interaction between the waves and the mean flow that occurs during the course of a sudden warming must be associated with wave transience (i.e., change of amplitude with respect to time) and wave dissipation. Most of the dramatic mean flow deceleration that occurs during a sudden warming is caused by amplification of quasi-stationary planetary waves in the troposphere followed by propagation into the stratosphere.

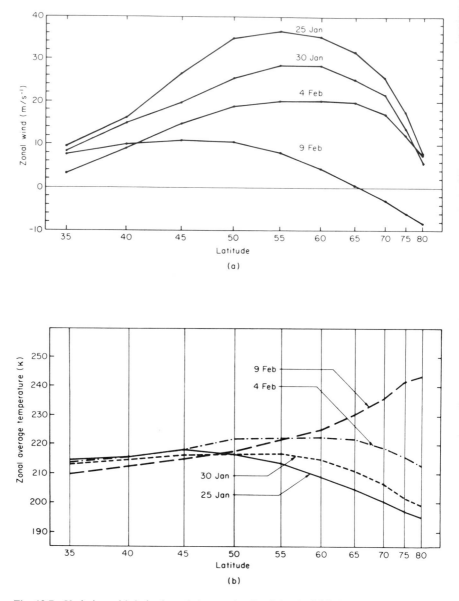

Fig. 12.7 Variation with latitude and time at the 50-mb level of (a) the zonal wind and (b) the zonal mean temperature during the sudden warming of 1957. (After Reed *et al.*, 1963. Reproduced with permission of the American Meteorological Society.)

This process can be understood by considering the interaction between the zonal mean and eddy potential vorticities. We assume that for the short time scales of the sudden warming diabatic processes and dissipation can be neglected. We also assume that the eddy motion is approximately governed by the linearized eddy potential vorticity equation and that the zonal mean is changed only by the eddy flux convergence. The eddies and mean flow are thus related by the quasi-linear system

$$\left(\frac{\partial}{\partial t} + \bar{u}\frac{\partial}{\partial x}\right) q' + v'\frac{\partial \bar{q}}{\partial y} = 0 \tag{12.14}$$

$$\partial \bar{q}/\partial t = -\partial(\overline{q'v'})/\partial y \tag{12.15}$$

[See (8.48) and (10.22).]

The role of transience in the interaction of eddies with the mean flow can be illustrated by multiplying through by q' in (12.14), averaging the result zonally, and solving for the eddy flux to obtain

$$\overline{q'v'} = -\frac{1}{2}\frac{\partial(\overline{q'^2})/\partial t}{\partial \bar{q}/\partial y} \tag{12.16}$$

Thus, for an amplifying wave ($\overline{q'^2}$ increasing in time) there will be an equatorward (negative) potential vorticity flux provided that the gradient of the basic state potential vorticity is positive. Since the potential vorticity flux must vanish at the pole this implies that the term on the right in the zonal mean potential vorticity equation (12.15) is negative between the pole and the latitude of maximum wave activity. Thus, the zonal mean potential vorticity at high latitudes must decrease in time, which implies that the westerly (cyclonic) mean flow must decrease in time.

The manner in which this sort of transient eddy mean-flow interaction can lead to a sudden warming can be illustrated by considering a situation in which a planetary wave is "turned on" at some initial time and begins to propagate upward. At a later time the wave will have reached the level z_0, so that the EP flux is constant with height below z_0 and goes to zero rapidly above that level. Thus, there must be strong EP flux convergence near z_0 and a rapid mean wind deceleration near that level (Fig. 12.8a). This deceleration is partly opposed by an induced residual meridional drift (Fig. 12.8b) in the region of EP flux convergence. The vertical circulation required by continuity will act to warm (cool) the polar (equatorial) stratosphere below the region of strong deceleration. As the mean flow decelerates the conditions for vertical propagation change, and the region of net EP flux convergence tends to spread downward in time. If the wave amplitude

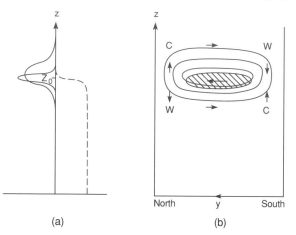

(a)   (b)

Fig. 12.8 Schematic of transient wave, mean-flow interactions occurring during a stratospheric warming. (a) Height profiles of EP flux (dashed), EP flux divergence (heavy line), and mean zonal flow acceleration (thin line); z_0 is the height reached by the leading edge of the wave packet at the time pictured. (b) Latitude–height cross section showing region where the EP flux is convergent (hatched), contours of induced zonal acceleration (thin lines), and induced residual circulation (arrows) Regions of warming (W) and cooling (C) are also shown. (After Andrews *et al.*, 1987.)

is sufficiently strong and the initial mean zonal wind distribution is favorable for propagation into the polar region, the interaction between the waves and mean flow can reverse the mean wind over much of the stratosphere poleward of 60°N.

In some cases the wave amplification may be large enough to produce a polar warming but insufficient to lead to reversal of the mean zonal wind in the polar region. Such "minor warmings" occur every winter and are generally followed by a quick return to the normal winter circulation. A "major warming" in which the mean zonal flow reverses at least as low as the 30-mb level in the polar region seems to occur only about once every couple of years. If the major warming occurs sufficiently late in the winter, the westerly polar vortex may not be restored at all before the normal seasonal circulation reversal.

12.5 Waves in the Equatorial Stratosphere

In Section 11.4 we discussed equatorially trapped waves in the context of shallow-water theory. Under some conditions, however, equatorial waves (both gravity and Rossby types) may propagate vertically, and the shallow-water model must be replaced by a continuously stratified atmosphere in order to examine the vertical structure. It turns out that vertically propagating equatorial waves share a number of physical properties with ordinary gravity modes. In Section 7.5 we discussed vertically propagating gravity waves in the presence of rotation for a simple situation in which the Coriolis parameter was assumed to be constant and the waves were assumed to be sinusoidal in both x and y. We found that such inertio-gravity waves can propagate vertically only when the wave frequency satisfies the inequality $f < \nu < N$. Thus, at middle latitudes waves with periods in the range of several days are generally vertically trapped (i.e., they are not able to propagate significantly into the stratosphere). As the equator is approached, however, the decreasing Coriolis frequency should allow vertical propagation to occur for lower-frequency waves. Thus, in the equatorial region there is the possibility for existence of long-period vertically propagating internal gravity waves.

As in Section 11.4 we consider linearized perturbations on an equatorial β plane. The linearized equations of motion, continuity equation, and first law of thermodynamics can then be expressed in log-pressure coordinates as follows:

$$\partial u'/\partial t - \beta y v' = -\partial \Phi'/\partial x \qquad (12.17)$$

$$\partial v'/\partial t + \beta y u' = -\partial \Phi'/\partial y \qquad (12.18)$$

$$\partial u'/\partial x + \partial v'/\partial y + \rho_0^{-1} \partial(\rho_0 w')/\partial z = 0 \qquad (12.19)$$

$$\partial^2 \Phi'/\partial t \, \partial z + w' N^2 = 0 \qquad (12.20)$$

We again assume that the perturbations are zonally propagating waves, but we now assume that they also propagate vertically with vertical wave number m. Owing to the basic state density stratification there will also be an amplitude growth in height proportional to $\rho_0^{-1/2}$. Thus, the x, y, z, and t dependences can be separated as

$$\begin{pmatrix} u' \\ v' \\ w' \\ \Phi' \end{pmatrix} = e^{z/2H} \begin{bmatrix} \hat{u}(y) \\ \hat{v}(y) \\ \hat{w}(y) \\ \hat{\Phi}(y) \end{bmatrix} \exp[i(kx + mz - \nu t)] \qquad (12.21)$$

Substituting from (12.21) into (12.17)–(12.20) yields a set of ordinary differential equations for the meridional structure:

$$-i\nu\hat{u} - \beta y\hat{v} = -ik\hat{\Phi} \qquad (12.22)$$

$$-iv\hat{v}+\beta y\hat{u}=-\partial\hat{\Phi}/\partial y \qquad (12.23)$$

$$(ik\hat{u}+\partial\hat{v}/\partial y)+i(m+i/2H)\hat{w}=0 \qquad (12.24)$$

$$v(m-i/2H)\hat{\Phi}+\hat{w}N^2=0 \qquad (12.25)$$

12.5.1   VERTICALLY PROPAGATING KELVIN WAVES

For Kelvin waves the above perturbation equations can be considerably simplified. Setting $\hat{v}=0$ and eliminating $\hat{w}$ between (12.24) and (12.25) we obtain

$$-iv\hat{u}=-ik\hat{\Phi} \qquad (12.26)$$

$$\beta y\hat{u}=-\partial\hat{\Phi}/\partial y \qquad (12.27)$$

$$-v(m^2+1/4H^2)\hat{\Phi}+\hat{u}kN^2=0 \qquad (12.28)$$

Equation (12.26) can be used to eliminate $\hat{\Phi}$ in (12.27) and (12.28). This yields two independent equations that the field of $\hat{u}$ must satisfy. The first of these determines the meridional distribution of $\hat{u}$ and is identical to (11.47). The second is simply the dispersion equation

$$c^2(m^2+1/4H^2)-N^2=0 \qquad (12.29)$$

where, as in Section 11.4, $c^2=(v^2/k^2)$.

If we assume that $m^2\gg 1/4H^2$, as is true for most observed stratospheric Kelvin waves, (12.29) reduces to the dispersion relationship for internal gravity waves (7.44) in the hydrostatic limit ($|k|\ll|m|$). For waves in the stratosphere that are forced by disturbances in the troposphere, the energy propagation (that is, the group velocity) must have an upward component. Therefore, according to the arguments of Section 7.4 the phase velocity must have a downward component. We showed in Section 11.4 that Kelvin waves must propagate eastward ($c>0$) if they are to be equatorially trapped. But eastward phase propagation requires $m<0$ for downward phase propagation. Thus, the vertically propagating Kelvin wave has phase lines that tilt eastward with height as shown in Fig. 12.9a.

12.5.2   VERTICALLY PROPAGATING ROSSBY-GRAVITY WAVES

For all other equatorial modes (12.22)–(12.25) can be combined in a manner exactly analogous to that described for the shallow-water equations in Section 11.4.1. The resulting meridional structure equation is identical to (11.38) if we again assume that $m^2\gg 1/4H^2$, and set

$$gh_e=N^2/m^2$$

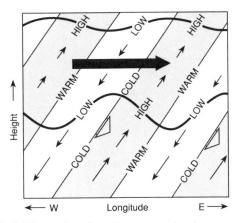

Fig. 12.9a Longitude–height section along the equator showing pressure, temperature, and wind perturbations for a thermally damped Kelvin wave. Heavy wavy lines indicate material lines; short blunt arrows show phase propagation. Areas of high pressure are shaded. Length of the small thin arrows is proportional to the wave amplitude, which decreases with height owing to damping. The large shaded arrow indicates the net mean flow acceleration owing to the wave stress divergence.

For the $n = 0$ mode (11.41) then implies that

$$|m| = N\nu^{-2}(\beta + \nu k) \tag{12.30}$$

When $\beta = 0$ we again recover the dispersion relationship for hydrostatic internal gravity waves. The role of the β effect in (12.30) is to break the symmetry between eastward ($\nu > 0$) and westward ($\nu < 0$) propagating waves. Eastward-propagating modes have shorter vertical wavelengths than westward-propagating modes. Vertically propagating $n = 0$ modes can exist only for $c = \nu/k > -\beta/k^2$. Since $k = s/a$, where s is the number of wavelengths around a latitude circle, this condition implies that for $\nu < 0$ solutions exist only for frequencies satisfying the inequality

$$|\nu| < 2\Omega/s \tag{12.31}$$

For frequencies that do not satisfy (12.31), the wave amplitude will not decay away from the equator, and it is not possible to satisfy boundary conditions at the pole.

After some algebraic manipulation the meridional structure of the horizontal velocity and geopotential perturbations for the $n = 0$ mode can be expressed as

$$
\begin{pmatrix} \hat{u} \\ \hat{v} \\ \hat{\Phi} \end{pmatrix} = \hat{v}
\begin{pmatrix} i|m|\nu N^{-1}y \\ 1 \\ i\nu y \end{pmatrix}
\exp\left(-\frac{\beta|m|y^2}{2N} \right) \tag{12.32}
$$

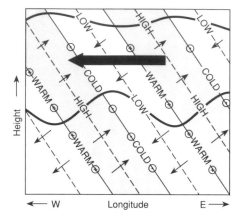

Fig. 12.9b Longitude–height section along a latitude circle north of the equator showing pressure, temperature, and wind perturbations for a thermally damped Rossby-gravity wave. Areas of high pressure are shaded. Small arrows indicate zonal and vertical wind perturbations with length proportional to the wave amplitude. Meridional wind perturbations are shown by arrows pointed into the page (northward) and out of the page (southward). The large shaded arrow indicates the net mean flow acceleration owing to the wave stress divergence.

The westward-propagating $n = 0$ mode is generally referred to as the Rossby–gravity mode.[2] For upward energy propagation this mode must have downward phase propagation ($m < 0$) just like an ordinary westward-propagating internal gravity wave. The resulting wave structure in the x, z plane at a latitude north of the equator is shown in Fig. 12.9b. Of particular interest is the fact that poleward (equatorward) moving air has positive (negative) temperature perturbations so that the eddy heat flux contribution to the vertical EP flux is positive.

12.5.3 Observed Equatorial Waves

Both Kelvin and Rossby–gravity modes have been identified in observational data from the equatorial stratosphere. The observed stratospheric Kelvin waves are primarily of zonal wave number $s = 1$ and have periods in the range of 12–20 d. An example of zonal wind oscillations caused by the passage of Kelvin waves at a station near the equator is shown in the form of a time–height section in Fig. 12.10. During the observational period shown in the figure the westerly phase of the quasi-biennial oscillation (see Section 12.6) is descending so that at each level there is a general increase in the mean zonal wind with increasing time. However, superposed on this secular trend is a large fluctuating component with a period between speed

---

[2] Some authors use this term to describe both eastward and westward $n = 0$ waves.

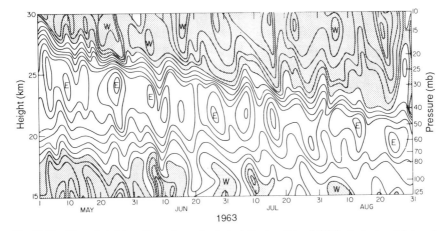

Fig. 12.10 Time–height section of zonal wind at Canton Island (3°S). Isotachs at intervals of 5 m s^{-1}. Westerlies are shaded. (Courtesy of J. M. Wallace and V. E. Kousky.)

maxima of about 12 days and a vertical wavelength (computed from the tilt of the oscillations with height) of about 10–12 km. Observations of the temperature field for the same period reveal that the temperature oscillation leads the zonal wind oscillation by 1/4 cycle (i.e., maximum temperature occurs prior to maximum westerlies), which is just the phase relationship required for upward-propagating Kelvin waves (see Fig. 12.9a). Additional observations from other stations indicate that these oscillations do propagate eastward at the theoretically predicted speed. Therefore, there can be little doubt that the observed oscillations are Kelvin waves.

The existence of the Rossby–gravity mode has been confirmed in observational data from the stratosphere in the equatorial Pacific. This mode is most easily identified in the meridional wind component, since v' is a maximum at the equator for the Rossby–gravity mode. The observed Rossby–gravity waves have $s = 4$ and a period range of 4–5 d. Both modes have significant amplitude only within about 20° latitude of the equator and have vertical wavelengths in the range of 6–12 km.

A more complete comparison of observed and theoretical properties of the Kelvin and Rossby–gravity modes is presented in Table 12.1. In comparing theory and observation it must be recalled that it is the frequency relative to the mean flow, not relative to the ground, that is dynamically relevant.

It appears that Kelvin and Rossby–gravity waves are excited by oscillations in the large-scale convective heating pattern in the equatorial troposphere. Although these waves do not contain much energy compared to typical tropospheric weather disturbances, they are the predominant disturbances of the equatorial stratosphere and through their vertical energy and

Table 12.1   *Characteristics of the Dominant Observed Planetary-Scale Waves in the Equatorial Lower Stratosphere*[a]

Theoretical description	Kelvin wave	Rossby–gravity wave
Discovered by	Wallace and Kousky (1968)	Yanai and Maruyama (1966)
Period (ground-based)$2\pi\omega^{-1}$	15 days	4–5 days
Zonal wave number s $= ka \cos \phi$	1–2	4
Vertical wavelength $2\pi m^{-1}$	6–10 km	4–8 km
Average phase speed relative to ground	$+25 \text{ m s}^{-1}$	-23 m s^{-1}
Observed when mean zonal flow is	Easterly (max. $\approx -25 \text{ m s}^{-1}$)	Westerly (max. $\approx +7 \text{ m s}^{-1}$)
Average phase speed relative to maximum zonal flow	$+50 \text{ m s}^{-1}$	-30 m s^{-1}
Approximate observed amplitudes		
u'	8 m s^{-1}	$2–3 \text{ m s}^{-1}$
v'	0	$2–3 \text{ m s}^{-1}$
T'	2–3 K	1 K
Approximate inferred amplitudes		
Φ'/g	30 m	4 m
w'	$1.5 \times 10^{-3} \text{ m s}^{-1}$	$1.5 \times 10^{-3} \text{ m s}^{-1}$
Approximate meridional scales $(2N/\beta|m|)^{1/2}$	1300–1700 km	1000–1500 km

[a] After Andrews *et al.* (1987).

momentum transport play a crucial role in the general circulation of the stratosphere. In addition to the stratospheric modes considered here, there are higher-speed Kelvin and Rossby–gravity modes, which are important in the upper stratosphere and mesosphere. There is also a broad spectrum of equatorial gravity waves, which appears to be important for the momentum balance of the equatorial middle atmosphere.

12.6   The Quasi-biennial Oscillation

The search for periodic oscillations in the atmosphere has a long history. Aside, however, from the externally forced diurnal and annual components and their harmonics, no compelling evidence exists for truly periodic atmospheric oscillations. The phenomenon that perhaps comes closest to exhibiting periodic behavior not associated with a periodic forcing function is the quasi-biennial oscillation (QBO) in the mean zonal winds of the equatorial stratosphere. This oscillation has the following observed features:

1.   Zonally symmetric easterly and westerly wind regimes alternate regularly with periods varying from about 24 to 30 months.

2. Successive regimes first appear above 30 km but propagate downward at a rate of 1 km month^{-1}.

3. The downward propagation occurs without loss of amplitude between 30 and 23 km, but there is rapid attenuation below 23 km.

4. The oscillation is symmetric about the equator with a maximum amplitude of about 20 m s^{-1} and an approximately Gaussian distribution in latitude with a half-width of about 12°.

This oscillation is best depicted by means of a time–height section of the zonal wind speed at the equator as shown in Fig. 12.11. It is apparent from the figure that the vertical shear of the wind is quite strong at the level where one regime is replacing the other. Because the QBO is zonally symmetric and causes only very small mean meridional and vertical motions, the QBO mean zonal wind and temperature fields satisfy the thermal wind balance equation. For the equatorial β- plane this has the form [compare (10.13)]

$$\beta y \, \partial \bar{u}/\partial z = -RH^{-1} \partial \bar{T}/\partial y$$

For equatorial symmetry $\partial \bar{T}/\partial y = 0$ at $y = 0$, and by L'Hôpital's rule thermal wind balance at the equator has the form

$$\partial \bar{u}/\partial z = -R(H\beta)^{-1} \partial^2 \bar{T}/\partial y^2 \tag{12.33}$$

Equation (12.33) may be used to estimate the magnitude of the QBO temperature perturbation at the equator. The observed magnitude of vertical shear of the mean zonal wind at the equator is ~ 5 m s^{-1} km^{-1}, and the meridional scale is ~ 1200 km, from which (12.33) shows that the temperature perturbation has an amplitude of ~ 3 K at the equator. Because the second derivative of temperature has the opposite sign to that of the temperature at the equator, the westerly and easterly shear zones have warm and cold equatorial temperature anomalies, respectively.

The main factors that a theoretical model of the QBO must explain are the approximate biennial periodicity, the downward propagation without loss of amplitude, and the occurrence of zonally symmetric westerlies at the equator. Because a zonal ring of air in westerly motion at the equator has an angular momentum per unit mass greater than that of the earth, no plausible zonally symmetric advection process could account for the westerly phase of the oscillation. Therefore, there must be a vertical transfer of momentum by eddies to produce the westerly accelerations in the downward-propagating shear zone of the QBO.

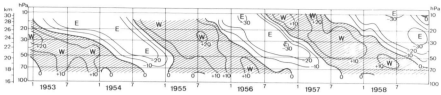

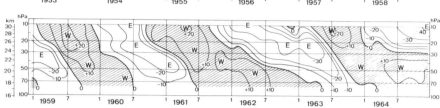

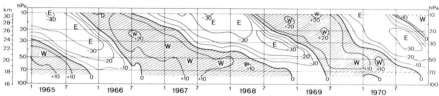

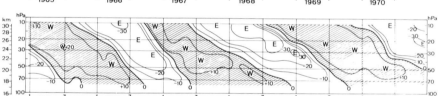

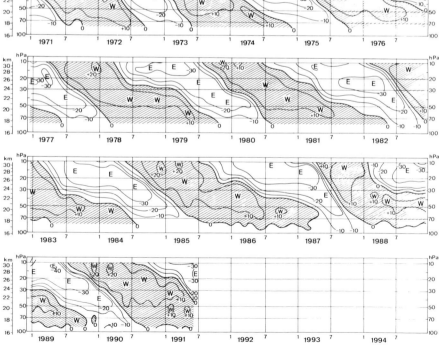

Observational and theoretical studies have confirmed that vertically propagating equatorial Kelvin and Rossby–gravity waves provide the zonal momentum sources necessary to drive the QBO. From Fig. 12.9a it is clear that Kelvin waves with upward energy propagation transfer westerly momentum upward (i.e., u' and w' are positively correlated so that $\overline{u'w'} > 0$). Thus, the Kelvin waves can provide the needed source of westerly momentum for the QBO.

Vertical momentum transfer by the Rossby–gravity mode requires special consideration. Examination of Fig. 12.9b shows that $\overline{u'w'} > 0$ also for the Rossby–gravity mode. The total effect of this wave on the mean flow cannot be ascertained from the vertical momentum flux alone, but rather the complete vertical EP flux must be considered. This mode has a strong poleward heat flux $(\overline{v'T'} > 0)$, which provides an upward-directed EP flux contribution. This dominates over the vertical momentum flux, and the net result is that the Rossby–gravity mode transfers easterly momentum upward and can provide the momentum source for the easterly phase of the QBO.

It was pointed out in Section 12.4 that quasi-geostrophic wave modes do not produce any net mean flow acceleration unless the waves are transient or they are mechanically or thermally damped. Similar considerations apply to the equatorial Kelvin and Rossby–gravity modes. Equatorial stratospheric waves are subject to thermal damping by infrared radiation and to both thermal and mechanical damping by small-scale turbulent motions. Such damping is strongly dependent on the Doppler-shifted frequency of the waves. As the Doppler-shifted frequency decreases, the vertical component of group velocity also decreases, and a longer time is available for the wave energy to be damped as it propagates through a given vertical distance. Thus, the westerly Kelvin waves tend to be damped preferentially in westerly shear zones, where their Doppler-shifted frequencies decrease with height. The momentum flux convergence associated with this damping provides a westerly acceleration of the mean flow and thus causes the westerly shear zone to descend. Similarly, the easterly Rossby–gravity waves are damped in easterly shear zones, thereby causing an easterly acceleration and descent of the easterly shear zone. We conclude that the QBO is indeed excited primarily by vertically propagating equatorial wave modes through wave transience and damping, which causes westerly accelerations in westerly shear zones and easterly accelerations in easterly shear zones.

Fig. 12.11 Time–height section of departure of monthly mean zonal winds (m s^{-1}) for each month from the long-term average for that month at equatorial stations. Note the alternating downward-propagating westerly (W) and easterly (E) regimes. (Updated from Naujokat, 1986, courtesy of B. Naujokat.)

This process of wave, mean-flow interaction can be elucidated by considering the heavy wave lines in Fig. 12.9. These lines indicate the vertical displacement of horizontal surfaces of fluid parcels (material surfaces) by the velocity field associated with the waves. (For sufficiently weak thermal damping they are approximately the same as isentropic surfaces.) The wavy lines show that the maximum upward displacement occurs one-quarter cycle after the maximum upward perturbation velocity. In the Kelvin wave case (Fig. 12.9a) positive pressure perturbations coincide with negative material surface slopes. Thus, the fluid below a wavy material line exerts an eastward-directed pressure force on the fluid above. Since the wave amplitude decreases with height, this force is larger for the lower of the two material lines in the figure. There will thus be a net westerly acceleration of the block of fluid contained between the two wavy material lines shown in Fig. 12.9a. For the Rossby–gravity wave, on the other hand, positive pressure perturbations coincide with positive slopes of the material lines. There is thus a westward-directed force exerted by the fluid below the lines on the fluid above. In this case the result is a net easterly acceleration of the fluid contained between the two wavy material lines shown in the figure. Thus, by considering the stresses acting across a material surface corrugated by the waves it is possible to deduce the mean flow acceleration caused by the waves without explicit reference to the EP fluxes of the waves.

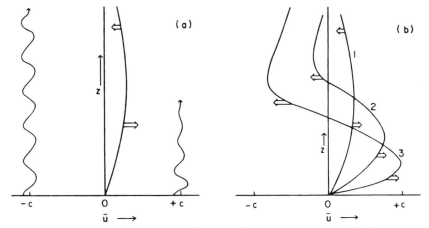

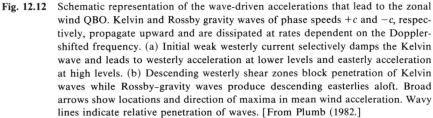

Fig. 12.12 Schematic representation of the wave-driven accelerations that lead to the zonal wind QBO. Kelvin and Rossby gravity waves of phase speeds $+c$ and $-c$, respectively, propagate upward and are dissipated at rates dependent on the Doppler-shifted frequency. (a) Initial weak westerly current selectively damps the Kelvin wave and leads to westerly acceleration at lower levels and easterly acceleration at high levels. (b) Descending westerly shear zones block penetration of Kelvin waves while Rossby–gravity waves produce descending easterlies aloft. Broad arrows show locations and direction of maxima in mean wind acceleration. Wavy lines indicate relative penetration of waves. [From Plumb (1982).]

How such a mechanism can cause a mean zonal flow oscillation when equal amounts of easterly and westerly momentum are transferred upward across the tropopause by the waves can be seen qualitatively by considering Fig. 12.12. If initially, as shown in panel (a), the mean zonal wind is weak and westerly, the Kelvin waves will be preferentially damped at lower altitude and produce westerlies, which move downward in time as the wave acceleration is concentrated at ever lower altitudes. The Rossby–gravity waves, on the other hand, penetrate to higher altitudes, where they produce easterlies, which also move downward in time. Eventually the westerlies are damped out as they approach the tropopause, and the stratosphere is dominated by easterlies so that the Kelvin waves can penetrate to high altitude and produce a new westerly phase. In this manner the mean zonal wind is forced to oscillate back and forth between westerlies and easterlies with a period that depends primarily on the vertical momentum transport and other properties of the waves, not on an oscillating external forcing.

12.7 The Ozone Layer

Above about 25 km altitude O_2 molecules are photodissociated by solar ultraviolet radiation of wavelengths less than 242 nm. The resulting oxygen atoms rapidly combine with O_2 to form ozone (O_3). Ozone itself strongly absorbs solar ultraviolet radiation at wavelengths below 300 nm. The resulting insolation warms the stratosphere until its temperature is high enough so that infrared cooling can balance the solar heating. Ultraviolet absorption by ozone thus causes the observed temperature inversion between the tropopause and the stratopause and thus is the fundamental reason for the existence of the stratosphere. In addition, the absorption of solar ultraviolet radiation by O_3 protects the biosphere from biologically harmful ultraviolet radiation. The existence of the ozone layer is thus crucial not only to the general circulation of the atmosphere but to the very existence of life itself.

It should be emphasized that, despite its fundamental importance, ozone is indeed a *trace constituent* in the atmosphere. Even at the peak of the ozone layer the ozone mixing ratio is only about 10 ppmv (parts per million by volume). If the entire atmospheric ozone content were brought to standard temperature and pressure (0°C, 101.325 kPa) the thickness of the ozone column would only be about 3 mm!

The concentration of ozone is highly variable in both space and time. This variability depends not only on the distribution of ozone sources and sinks but also importantly on transport of ozone by all scales of atmospheric motions. This dependence of ozone on atmospheric motions is illustrated in Fig. 12.13. Despite the fact that ozone is formed primarily in tropical latitudes above 25 km, the maximum column amounts are found near the North Pole in April and near 60° south in October. Minima occur in the

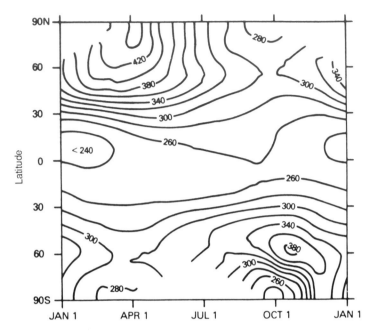

Fig. 12.13   Time–latitude section showing the seasonal variation of total ozone (Dobson units) based on total ozone mapping spectrometer (TOMS) data. Note the extratropical springtime maxima in both hemispheres and the minimum at 90°S centered on October 1. (After Bowman and Krueger, 1985.)

equatorial region at all seasons and in the Antarctic polar region in spring. The latter minimum, commonly called the *Antarctic ozone hole*, is due to human impact on the ozone layer owing to ozone destruction catalyzed by chlorine chemistry in association with polar stratospheric clouds.

The fact that the maximum ozone concentration occurs outside the source region indicates clearly that transport by atmospheric motions must play a fundamental role in determining the global ozone budget. The zonal mean ozone budget is most conveniently modeled within the TEM formulation introduced in Section 10.2.2. By analogy with the thermodynamic energy equation, the TEM version of the tracer continuity equation for an atmospheric constituent with mixing ratio χ_i may be expressed as

$$\left(\frac{\partial}{\partial t} + \bar{v}^* \frac{\partial}{\partial y} + \bar{w}^* \frac{\partial}{\partial z} \right) \bar{\chi}_i = \bar{S} + \bar{D} \tag{12.34}$$

where $\bar{S}$ represents the sum of the zonally averaged chemical sources and sinks and $\bar{D}$ represents the horizontal and vertical transport by eddies.

A complete understanding of the ozone budget requires a knowledge not only of the source distribution and transport by air motions but also of the sinks. In recent years great progress has been made in our understanding of the photochemistry of the ozone layer. It is now recognized that ozone is catalytically destroyed by a complex series of reactions involving (among other species) NO, ClO, and HO. The concentrations of these (and other) ozone-destroying chemicals are also influenced not only by chemistry but also by mass exchange between the stratosphere and troposphere and transport within the stratosphere.

The overall tropospheric–stratospheric mass exchange was discussed briefy in Section 12.2.2 and illustrated schematically in Fig. 12.5. The primary transport of mass from the troposphere into the stratosphere takes place in the equatorial region. It can be approximated by the residual vertical motion in the TEM formulation of the zonal mean equation of motion. According to the discussion in Section 10.2.2, this circulation arises in response to eddies that drive the stratosphere away from radiative equilibrium. Estimates of the strength of the cross-tropopause residual circulation indicate that nearly half of the mass above the 100-mb level is replaced in a year by flow across the tropical tropopause.

According to Fig. 12.5, the overall mean mass flow within the stratosphere is directed poleward and is downward outside the tropics. This pattern of mass flow accounts for the observed distribution of ozone shown in Fig. 12.13. The strong poleward and downward residual circulation during the winter transports ozone out of the chemically active region of the equatorial middle stratosphere into the high-latitude lower stratosphere, where the concentrations gradually build up during the winter to produce a springtime maximum.

The transport of ozone into the extratropical lower stratosphere by the residual circulation is balanced on average by transport into the troposphere. This cross-tropopause transport in middle latitudes is thought to be dominated by frontal scale processes associated with the upper-tropospheric jet stream. Studies of ozone, radioactive tracers, and potential vorticity indicate that considerable stratospheric air is mixed into the troposphere by intrusions of stratospheric air that occur in conjunction with upper-level frontogenesis. These intrusions, which occur in thin layers of ~100 km horizontal and ~1 km vertical scale, are quickly destroyed by irreversible vertical mixing in the troposphere, thus providing much of the sink to balance ozone production in the tropical stratosphere.

Problems

12.1. Suppose that temperature increases linearly with height in the layer between 20 and 50 km at a rate of 2 K km^{-1}. If the temperature is 200 K at 20 km, find the value of the scale height H for which the log-pressure height z coincides with actual height at 50 km. (Assume that z coincides with the actual height at 20 km and let g be a constant.)

12.2. Find the Rossby critical velocities for zonal wave numbers 1, 2, and 3 (i.e., for 1, 2, and 3 wavelengths around a latitude circle). Let the motion be referred to a β plane centered at 45°N, scale height $H = 7$ km, buoyancy frequency $N = 2 \times 10^{-2}\,\mathrm{s}^{-1}$, and infinite meridional scale ($l = 0$).

12.3. Find the geopotential and vertical velocity fluctuations for a Kelvin wave of zonal wave number 1, phase speed 40 m s^{-1}, and zonal velocity perturbation amplitude 5 m s^{-1}. Let $N^2 = 4 \times 10^{-4}\,\mathrm{s}^{-2}$.

12.4. For the situation of Problem 12.3 compute the vertical momentum flux $M \equiv \rho_0 \overline{u'w'}$. Show that M is constant with height.

12.5. Determine the form for the vertical velocity perturbation for the Rossby–gravity wave corresponding to the u', v', and Φ' perturbations given in (12.32).

12.6. For a Rossby–gravity wave of zonal wave number 4 and phase speed -20 m s^{-1} determine the latitude at which the vertical momentum flux $M \equiv \rho_0 \overline{u'w'}$ is a maximum.

12.7. Suppose that the mean zonal shear in the descending westerlies of the equatorial QBO can be represented analytically on the equatorial β plane in the form

$$\partial \bar{u}/\partial z = \Lambda \exp(-y^2/L^2)$$

where $L = 1200$ km. Determine the approximate meridional dependence of the corresponding temperature anomaly for $|y| \ll L$.

12.8. Estimate the TEM residual vertical velocity in the westerly shear zone of the equatorial QBO assuming that radiative cooling can be approximated by Newtonian cooling with a 20-day relaxation time, that the vertical shear is 20 m s^{-1} per 5 km, and that the meridional half-width is 12° latitude.

Suggested References

Brasseur and Solomon, *Aeronomy of the Middle Atmosphere*, has an excellent discussion of the chemistry of the stratosphere at an advanced level.

Andrews, Holton, and Leovy, *Middle Atmosphere Dynamics*, presents a graduate-level treatment of the dynamics of the stratosphere and mesosphere.

13 Numerical Modeling and Prediction

Dynamical meteorology provides the theoretical basis and methodology for modern weather forecasting. Stated simply, the objective of dynamical forecasting is to predict the future state of the atmospheric circulation from knowledge of its present state by use of numerical approximations to the dynamical equations. To fulfill this objective requires observations of the initial state of the field variables, a closed set of prediction equations relating the field variables, and a method of integrating the equations in time to obtain the future distribution of the field variables.

Numerical prediction is a highly specialized field, which is continually evolving. Operational forecast centers utilize complex prediction models that require the largest available supercomputers for their solution. It is difficult to provide more than a superficial introduction to such models in an introductory text. Fortunately, however, many aspects of numerical prediction can be illustrated using a simple model, such as the barotropic vorticity equation. In fact, this equation was the basis of the earliest operational numerical prediction models.

13.1   Historical Background

The first attempt to predict the weather numerically was due to the British scientist L. F. Richardson. His book, *Weather Prediction by Numerical Process*, published in 1922, is the classic treatise in this field. In his work Richardson showed how the differential equations governing atmospheric motions could be written approximately as a set of algebraic difference equations for values of the tendencies of various field variables at a finite number of points in space. Given the observed values of the field variables at these grid points, the tendencies could be calculated numerically by solving the algebraic difference equations. By extrapolating the computed tendencies ahead a small increment in time, an estimate of the fields at a short time in the future could be obtained. The new values of the field variables could then be used to recompute the tendencies, which could in turn be used to extrapolate further ahead in time, etc. Even for short-range forecasting over a small area of the earth, this procedure requires an enormous number of arithmetic operations. Richardson did not foresee the development of high-speed digital computers. He estimated that a work force of 64,000 people would be required just to keep up with the weather on a global basis!

Despite the tedious labor involved, Richardson worked out one example forecast for surface pressure tendencies at two grid points. Unfortunately, the results were very poor. Predicted pressure changes were an order of magnitude larger than those observed. At the time this failure was thought to be due primarily to the poor initial data available, especially the absence of upper-air soundings. However, it is now known that there were other even more serious problems with Richardson's scheme.

After Richardson's failure to obtain a reasonable forecast, numerical prediction was not again attempted for many years. Finally, after World War II interest in numerical prediction revived owing partly to the vast expansion of the meteorological observation network, which provided much improved initial data, but even more importantly to the development of digital computers, which made the enormous volume of arithmetic operations required in a numerical forecast feasible. At the same time it was realized that Richardson's scheme was not the simplest possible scheme for numerical prediction. His equations not only governed the slow-moving meteorologically important motions but also included high-speed sound and gravity waves as solutions. Such waves are in nature usually very weak in amplitude. However, for reasons that will be explained later, if Richardson had carried his numerical calculation beyond the initial time step, these oscillations would have amplified spuriously, thereby introducing so much "noise" in the solution that the meteorologically relevant disturbances would have been obscured.

The American meteorologist J. G. Charney showed in 1948 how the dynamical equations could be simplified by systematic introduction of the geostrophic and hydrostatic approximations so that the sound and gravity oscillations were filtered out. The equations that resulted from Charney's filtering approximations were essentially those of the quasi-geostrophic model. Thus, Charney's approach utilized the conservative properties of potential vorticity. A special case of this model, the *equivalent barotropic model*, was used in 1950 to make the first numerical forecast.

This model provided forecasts of the geopotential near 500 mb. Thus, it did not forecast "weather" in the usual sense. It could, however, be used by forecasters as an aid in predicting the local weather associated with large-scale circulations. Later multilevel versions of the quasi-geostrophic model provided explicit predictions of the surface pressure and temperature distributions, but the accuracy of such predictions was limited owing to the approximations inherent in the quasi-geostrophic model.

With the development of vastly more powerful computers and more sophisticated modeling techniques, numerical forecasting has now returned to models that are quite similar to Richardson's formulation and are potentially far more accurate than quasi-geostrophic models. Nevertheless, it is still worth considering the simplest filtered model, the barotropic vorticity equation, to illustrate some of the technical aspects of numerical prediction in a simple context.

13.2 Filtering Meteorological Noise

One difficulty in directly applying the unsimplified equations of motion to the prediction problem is that meteorologically important motions are easily lost in the noise introduced by large-amplitude sound and gravity waves, which may arise as a result of errors in the initial data, and then can spuriously amplify to dominate the forecast fields. As an example of how this problem might arise, consider the pressure and density fields. On the synoptic scale these are in hydrostatic balance to a very good approximation. As a consequence, vertical accelerations are extremely small. However, if the pressure and density fields were determined independently by observations, as would be the case if the complete dynamical equations were used, small errors in the observed fields would lead to large errors in the computed vertical acceleration simply because the vertical acceleration is a very small difference between two large forces (the vertical pressure gradient and buoyancy). Such spurious accelerations would appear in the computed solutions as high-speed sound waves of very large amplitude. In a similar fashion errors in the initial horizontal velocity and pressure fields would lead to spuriously large horizontal accelerations since the horizontal acceleration results from the small difference between the Coriolis and pressure gradient forces. Such spurious horizontal accelerations would excite both sound and gravity waves.

In order to overcome this problem, either the "observed" fields must be modified systematically to remove unrealistic force imbalances or the prediction equations must be modified to remove the physical mechanisms responsible for the occurrence of the unwanted oscillations, while still preserving the meteorologically important motions. In modern numerical forecasting gravity waves are usually controlled by suitably adjusting the initial data; sound waves, on the other hand, are generally filtered from the dynamical equations.

To help understand how sound and gravity wave noise can be filtered from the prediction equations it is useful to refer back to Section 7.3, in which the physical properties of such waves were discussed. If the pipe shown in Fig. 7.5 is tipped up in the vertical, it can be used to generate vertically traveling sound waves. If we now require that pressure be hydrostatic, the pressure at any point along the pipe is determined solely by the weight of the air above that point. Hence, the vertical pressure gradient cannot be influenced by adiabatic compression or expansion. Therefore, vertically propagating sound waves are not among the possible modes of oscillation in a hydrostatic system. Replacement of the vertical momentum equation by the hydrostatic approximation is thus sufficient to filter out ordinary sound waves. This approximation is used in virtually all models for forecasting medium- and large-scale motions.

A hydrostatically balanced atmosphere can still, however, support a special class of horizontally propagating acoustic waves. In this type of wave the vertical velocity is zero (neglecting orographic effects and departures of the basic state temperature from isothermal conditions). The pressure, horizontal velocity, and density, however, oscillate with the horizontal structure of the simple acoustic waves described in Section 7.3.1. These oscillations have maximum amplitude at the lower boundary and decay away from the boundary with the pressure and density fields remaining in hydrostatic balance everywhere. Since these oscillations, known as *Lamb waves*, have maximum pressure perturbation amplitude at the ground, they may be filtered simply by requiring that $\omega = Dp/Dt = 0$ at the lower boundary. This boundary condition is most easily applied by formulating the equations in isobaric coordinates. In that case the condition $\omega = 0$ at the lower boundary is a natural first approximation for geostrophically scaled motions over level terrain (see the discussion in Section 3.5.1).

The set of equations including the above approximations still has solutions in the form of internal gravity waves. Gravity waves may be important in accounting for mesoscale variability, and the zonal drag force owing to gravity waves must be included in the overall zonal-mean momentum balance. But gravity waves are not important for short-range forecasting of synoptic- and planetary-scale circulations. Owing, however, to their high frequency of oscillation they can create serious errors in numerical forecasts. One possible solution to this problem is to filter gravity waves from the forecast equations. From the discussion of Section 7.4 it should be apparent that gravity waves can be solutions only if the dynamical equations are second order in time. In physical terms this turns out to require that the local rate of change of the divergence of the horizontal velocity field must be implicitly included in the equations. One way to filter gravity waves is thus to divide the horizontal velocity into irrotational and nondivergent parts as discussed in Section 11.2 and to neglect the irrotational part of the motion in the velocity tendency term. The resulting prognostic equation is the barotropic vorticity equation (11.14). The streamfunction for the nondivergent flow can then be diagnostically related to the geopotential field by the nonlinear balance equation (11.15). For extratropical synoptic-scale motions (11.15) can be replaced by the geostrophic approximation on the midlatitude β plane, for which $\psi = \Phi/f_0$.

The barotropic vorticity equation can be used to compute the evolution of the flow at a single level in terms of the nondivergent flow at that level alone. No vertical coupling is involved. Thus, no explicit predictions for levels above or below the assumed nondivergent level are possible. The only predictions of surface weather possible with this model are those based on climatological relationships between the surface flow and flow at the nondivergent level. For extratropical flows it is possible to model the vertical structure explicitly and still filter gravity waves by utilizing the conservation of quasi-geostrophic potential vorticity (6.25), which is related to the geopotential field through the elliptic boundary value problem (6.26). In Sections 13.4 and 13.5 we will utilize the barotropic vorticity equation to illustrate the basic methodology of numerical weather prediction.

13.3 The Finite Difference Method

The equations of motion are an example of a general class of systems known as *initial value problems*. A system of differential equations is referred to as an initial value problem when the solution depends not only on boundary conditions but also on the values of the unknown fields or their derivatives at some initial time. Clearly, weather forecasting is a primary example of a *nonlinear* initial value problem. Owing to its nonlinearity, even the simplest forecast equation, the barotropic vorticity equation, is rather complicated to analyze. Fortunately, general aspects of the numerical solution of initial value problems can be illustrated using linearized proto-type equations that are much simpler than the barotropic vorticity equation.

13.3.1   FINITE DIFFERENCES

The equations of motions involve terms that are quadratic in the dependent variables (the advection terms). Such equations generally cannot be solved analytically. Rather they must be approximated by some suitable discretization and solved numerically. The simplest form of discretization is the *finite difference* method.

To introduce the concept of finite differencing we consider a field variable $\psi(x)$, which is a solution to some differential equation in the interval $0 \leq x \leq L$. If this interval is divided into J subintervals of length δx, then $\psi(x)$ can be approximated by a set of $J+1$ values as $\Psi_j = \psi(j\,\delta x)$, which are the values of the field at the $J+1$ grid points given by $x = j\,\delta x, j = 0, 1, 2, \ldots, J$, where $\delta x = L/J$. Provided that δx is sufficiently small compared to the scale on which ψ varies, the $J+1$ grid point values should provide good approximations to $\psi(x)$ and its derivatives.

The limits of accuracy of a finite difference representation of a continuous variable can be assessed by noting that the field can also be approximated by a finite Fourier series expansion:

$$\psi(x) = \frac{a_0}{2} + \sum_{m=1}^{J/2} \left[a_m \cos \frac{2\pi mx}{L} + b_m \sin \frac{2\pi mx}{L} \right] \qquad (13.1)$$

The available $J+1$ values of Ψ_j are sufficient to determine only $J+1$ coefficients in (13.1). That is, it is possible to determine a_0 plus a_m and b_m for wave numbers $m = 1, 2, 3, \ldots, J/2$. The shortest-wavelength component in (13.1) has wavelength $L/m = 2L/J = 2\delta x$. Thus, the shortest wave that can be resolved by finite differencing has a wavelength twice that of the grid increment. Accurate representation of derivatives is possible, however, only for wavelengths much greater than $2\delta x$.

We now consider how the grid point variable Ψ_j can be used to construct a finite difference approximation to a differential equation. That is, we wish to represent derivatives such as $d\psi/dx$ and $d^2\psi/dx^2$ in terms of the finite difference fields. We first consider the Taylor series expansions about the point x_0:

$$\psi(x_0 + \delta x) = \psi(x_0) + \psi'(x_0)\delta x + \psi''(x_0)\frac{\delta x^2}{2}$$

$$+ \psi'''(x_0)\frac{\delta x^3}{6} + O(\delta x^4) \qquad (13.2)$$

$$\psi(x_0 - \delta x) = \psi(x_0) - \psi'(x_0)\delta x + \psi''(x_0)\frac{\delta x^2}{2}$$

$$- \psi'''(x_0)\frac{\delta x^3}{6} + O(\delta x^4) \tag{13.3}$$

where the primes indicate differentiation with respect to x and $O(\delta x^4)$ means that terms with order of magnitude δx^4 or less are neglected.

Subtracting (13.3) from (13.2) and solving for $\psi'(x)$ gives a finite difference expression for the first derivative of the form

$$\psi'(x_0) = [\psi(x_0 + \delta x) - \psi(x_0 - \delta x)]/(2\delta x) + O(\delta x^2) \tag{13.4}$$

while adding the same two expressions and solving for $\psi''(x)$ gives a finite difference expression for the second derivative of the form

$$\psi''(x_0) = [\psi(x_0 + \delta x) - 2\psi(x_0) + \psi(x_0 - \delta x)]/(\delta x^2) + O(\delta x^2) \tag{13.5}$$

Since in (13.4) and (13.5) the difference approximations involve points at equal distances on either side of x_0, they are called *centered differences*. These approximations neglect terms of order δx^2. We thus say that the truncation error is order δx^2. Higher accuracy can be obtained by decreasing the grid interval, but at the cost of increasing the density of grid points. Alternatively, it is possible to obtain higher-order accuracy without decreasing the grid spacing by writing formulas analogous to (13.2)-(13.3) for the interval $2\delta x$ and using these together with (13.2)-(13.3) to eliminate error terms less than $O(\delta x^4)$. This approach, however, has the disadvantage of producing more complicated expressions and can be difficult to implement near boundary points.

13.3.2 CENTERED DIFFERENCES

As a prototype model we consider the linear one-dimensional advection equation

$$\partial q/\partial t + c\,\partial q/\partial x = 0 \tag{13.6}$$

with c a specified speed and $q(x, 0)$ a known initial condition. This equation can be approximated to second-order accuracy in x and t by the centered difference equation

$$[q(x, t+\delta t) - q(x, t-\delta t)]/(2\delta t) = -c[q(x+\delta x, t) - q(x-\delta x, t)]/(2\delta x)$$
$$(13.7)$$

The original differential equation (13.6) is thus replaced by a set of algebraic equations (13.7), which can be solved to determine solutions for a finite set of points that define a grid mesh in x and t (see Fig. 13.1). For notational convenience it is useful to identify points on the grid mesh by indices m, s. These are defined by letting $x = m\,\delta x$, $m = 0, 1, 2, 3, \ldots, M$, $t = s\,\delta t$, $s = 0, 1, 2, 3, \ldots, S$, and writing $\hat{q}_{m,s} \equiv q(m\,\delta x, s\,\delta t)$. The difference equation (13.7) can then be expressed as

$$\hat{q}_{m,s+1} - \hat{q}_{m,s-1} \equiv -\sigma(\hat{q}_{m+1,s} - \hat{q}_{m-1,s}) \qquad (13.8)$$

where $\sigma \equiv c\,\delta t/\delta x$. This form of time differencing is referred to as the *leapfrog* method since the tendency at time step s is given by the difference in values computed for time steps $s+1$ and $s-1$ (i.e., by leaping across point s).

Leapfrog differencing cannot be used for the initial time $t = 0$ ($s = 0$) since $q_{m,-1}$ is not known. For the first time step an alternative method is

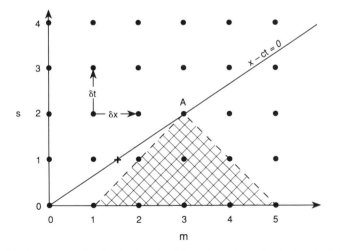

Fig. 13.1 Grid in x-t space showing the domain of dependence (cross-hatched area) of the explicit finite-difference solution at $m = 3$ and $s = 2$ for the one-dimensional linear advection equation. Solid circles show grid points. The sloping line is a characteristic curve along which $q(x, t) = q(x, 0)$, and the cross shows an interpolated point for the semi-Lagrangian differencing scheme.

required such as the forward difference approximation

$$\hat{q}_{m,1} - \hat{q}_{m,0} \equiv -(\sigma/2)(\hat{q}_{m+1,0} - \hat{q}_{m-1,0}) \tag{13.9}$$

13.3.3   COMPUTATIONAL STABILITY

Experience shows that solutions to finite difference approximations such as (13.8) will not always resemble solutions to the original differential equations even when the finite difference increments in space and time are very small. It turns out that the character of the solutions depends critically on the *computational stability* of the difference equations. If the difference equations are not stable, numerical solutions will exhibit exponential growth even when, as in the case of the linear advection equation, the original differential equation system has solutions whose amplitudes remain constant in time. In the example (13.8) stability considerations place stringent limitations on the value of the parameter σ, as we show below.

If the initial condition is specified as

$$q(x, 0) = \text{Re}[\exp(ikx)] = \cos(kx)$$

The analytic solution of (13.6) that satisfies this initial condition is

$$q(x, t) = \text{Re}\{\exp[ik(x - ct)]\} = \cos(kx - ct) \tag{13.10}$$

We now compare (13.10) with the solution of the finite difference system (13.8) and (13.9). In finite difference form the initial condition is

$$\hat{q}_{m,0} = \exp(ikm\,\delta x) = \exp(ipm) \tag{13.11}$$

where $p \equiv k\,\delta x$. Noting that the analytic solution (13.10) is separable in x and t, we consider solutions of (13.8)–(13.9) of the form

$$\hat{q}_{m,s} = B^s \exp(ipm) \tag{13.12}$$

where B is a complex constant. Substituting into (13.8) and dividing through by the common factor B^{s-1}, we obtain a quadratic equation in B:

$$B^2 + (2i\sin\theta_p)B - 1 = 0 \tag{13.13}$$

where $\sin\theta_p \equiv \sigma\sin p$. Equation (13.13) has two roots, which may be expressed in the form

$$B_1 = \exp(-i\theta_p), \qquad B_2 = -\exp(+i\theta_p)$$

The general solution of the finite difference equation is thus

$$\hat{q}_{m,s} = [CB_1^s + DB_2^s]\exp(ipm) = Ce^{i(pm-\theta_p s)} + D(-1)^s e^{i(pm+\theta_p s)} \quad (13.14)$$

where C and D are constants to be determined by the initial conditions (13.11) and the first time step (13.9). The former gives $C + D = 1$, while the latter gives

$$Ce^{-i\theta_p} - De^{+i\theta_p} = 1 - i\sin\theta_p \quad (13.15)$$

Thus,

$$C = \frac{1 + \cos\theta_p}{2\cos\theta_p}, \quad D = -\frac{1 - \cos\theta_p}{2\cos\theta_p} \quad (13.16)$$

The definition of θ_p shows that for $\sigma > 1$, θ_p must be imaginary, at least for the shortest waves. From inspection it is clear that the solution will remain finite for $s \to \infty$ provided that θ_p is real. If θ_p is imaginary then one term in (13.14) will grow exponentially and the solution will be unbounded for $s \to \infty$. It is this sort of behavior that is referred to as *computational instability*. Now since

$$\theta_p = \sin^{-1}(\sigma \sin p) \quad (13.17)$$

θ_p will be real only for $|\sigma \sin p| \leq 1$, which can be valid for all waves (i.e., for all p) only if $\sigma \leq 1$. Thus, for the difference equation (13.8), stability requires that

$$\sigma = c \, \delta t / \delta x \leq 1 \quad (13.18)$$

which is referred to as the Courant–Friedrichs–Levy (CFL) stability criterion.

The CFL criterion for this example states that for a given space increment δx the time step δt must be chosen so that the dependent field will be advected a distance less than one grid length per time step. Other methods of finite differencing the one-dimensional advection equation may lead to more stringent limits on σ than given in (13.18). Similar considerations apply in two dimensions; there, if the grid has uniform spacing $\delta x = \delta y$, the CFL condition is $c \, \delta t / \delta x \leq 1/\sqrt{2}$, where c is the magnitude of the maximum velocity in the domain.

The restriction on σ required for stability of the advection equation can be understood physically by considering the character of the solution in the x, t plane as shown in Fig. 13.1. In the situation shown in the figure, $\sigma = 1.5$. Examination of the centered difference system (13.14) shows that the numerical solution at point A in the figure depends only on grid points within the shaded region of the figure. But, since A lies on the *characteristic* line $x - ct = 0$, the true solution at point A depends only on the initial condition at $x = 0$ (i.e., a parcel which is initially at the origin will be advected to the point $3\delta x$ in time $2\delta t$). The point $x = 0$ is outside the *domain of influence* of the numerical solution. Hence, the numerical solution cannot possibly faithfully reproduce the solution to the original differential equation since, as shown in Fig. 13.1, the value at point A in the numerical solution has no dependence on the conditions at $x = 0$. Only when the CFL criterion is satisfied will the domain of influence of the numerical solution include the characteristic lines of the analytic solution.

The existence of computational instability is one of the prime motivations for using filtered equations. In the quasi-geostrophic system no gravity or sound waves occur. Thus, the speed c in (13.18) is just the maximum wind speed. Typically, $c < 100 \text{ m s}^{-1}$ so that for a grid interval of 200 km a time increment of over 30 min is permissible. On the other hand, if we were to use the complete nonhydrostatic equations, the solution would have characteristics corresponding to acoustic modes, and to assure that the domain of influence included such characteristics c would need to be set equal to the speed of sound, which is the fastest wave described in that set of equations. In that case $c \approx 300 \text{ m s}^{-1}$ and for a 200-km grid interval a time step of only a few minutes would be permitted.

13.3.4 IMPLICIT TIME DIFFERENCING

The centered difference scheme for the advection equation (13.8) is an example of an *explicit* time differencing scheme. In an explicit difference scheme the value of the predicted field at a given grid point for time step $s + 1$ depends only on the known values of the field at time steps s and $s - 1$. The difference equation can then be solved simply by marching through the grid and obtaining the solution for each point in turn. The explicit leapfrog scheme is thus simple to solve. However, as we saw in the previous section, it has disadvantages in that it introduces a spurious mode [proportional to D in (13.14)] and suffers computational instability if the time step violates the CFL criterion.

These two difficulties can be avoided by utilizing an alternative finite differencing scheme called the *trapezoidal implicit scheme*. For the linear advection equation (13.6) this scheme can be written in the form

$$\frac{\hat{q}_{m,s+1} - \hat{q}_{m,s}}{\delta t} = -\frac{c}{2}\left[\frac{\hat{q}_{m+1,s+1} - \hat{q}_{m-1,s+1}}{2\delta x} + \frac{\hat{q}_{m+1,s} - \hat{q}_{m-1,s}}{2\delta x}\right] \quad (13.19)$$

Substituting the trial solution (13.12) into (13.19) yields

$$B^{s+1} = \left[\frac{1 - i(\sigma/2)\sin p}{1 + i(\sigma/2)\sin p}\right]B^s \quad (13.20)$$

where, as before, $\sigma = c\,\delta t/\delta x$ and $p = k\,\delta x$. Defining

$$\tan\theta_p \equiv (\sigma/2)\sin p \quad (13.21)$$

and eliminating the common term B^s in (13.20), it can be shown that

$$B = \frac{1 - i\tan\theta_p}{1 + i\tan\theta_p} = \exp(-2i\theta_p) \quad (13.22)$$

so that the solution may be expressed simply as

$$\hat{q}_{m,s} = A\exp[ik(m\,\delta x - 2\theta_p s/k)] \quad (13.23)$$

Equation (13.19) involves only two time levels. Hence, unlike (13.14), the solution yields only a single mode, which has phase speed $c' = 2\theta_p/(k\,\delta t)$. According to (13.21) θ_p remains real for all values of δt. [This should be contrasted to the situation for the explicit scheme given by (13.17).] Thus, the implicit scheme is absolutely stable. The truncation errors, however, can become large if θ_p is not kept sufficiently small (see Problem 13.9). A disadvantage of the implicit scheme is that the integration cannot proceed by marching through the grid, as in the explicit case. In (13.19) there are terms involving the $s+1$ time level on both sides of the equal sign, and these involve the values at a total of three grid points. Thus, the system (13.19) must be solved simultaneously for all points on the grid. If the grid is large this may involve inverting a very large matrix, so it is computationally intensive. Furthermore, it is usually not feasible to utilize the implicit approach for nonlinear terms, such as the advection terms in the momentum equations. Semi-implicit schemes in which the linear terms are treated implicitly and the nonlinear terms explicitly have, however, become popular in modern forecasting models. These will be discussed briefly in Section 13.6.

13.3.5   THE SEMI-LAGRANGIAN INTEGRATION METHOD

The differencing schemes discussed above are Eulerian schemes in which the time integration is carried out by computing the tendencies of the predicted fields at a set of grid points fixed in space. Although it would be possible in theory to carry out predictions in a Lagrangian framework by following a set of marked fluid parcels, in practice this is not a viable alternative since shear and stretching deformations tend to concentrate marked parcels in a few regions so that it is difficult to maintain uniform resolution over the forecast region. It is possible to take advantage of the conservative properties of Lagrangian schemes, while maintaining uniform resolution, by employing a semi-Lagrangian technique. This approach permits relatively long time steps while retaining numerical stability and high accuracy. The semi-Lagrangian method can be illustrated in a very simple fashion with the one-dimensional advection equation (13.6). According to this equation the field q is conserved following the zonal flow at speed c. Thus, for any grid point $x = m\,\delta x$,

$$q(x, t + \delta t) = q(x - c\,\delta t, t) \tag{13.24}$$

Now, in general the position $x - c\,\delta t$ does not lie on a gridpoint (see Fig. 13.1), so evaluation of the right-hand side of (13.24) requires interpolating from the grid point values at time t. If linear interpolation is used

$$q(x - c\,\delta t, t) = \sigma q(x - \delta x, t) + (1 - \sigma)q(x, t)$$

In an actual prediction model the velocity field is predicted rather than known as in this simple example. Thus, for a two-dimensional field

$$q(x, y, t + \delta t) = q(x - u\delta t, y - v\,\delta t, t) \tag{13.25}$$

where the velocity components at time t can be used to estimate the fields at $t + \delta t$; once these are obtained they can be used to provide more accurate approximations to the velocities on the right in (13.25). The right side in (13.25) is again estimated by interpolation, which now must be carried out in two dimensions.

As shown in Fig. 13.1, the semi-Lagrangian scheme guarantees that the domain of influence in the numerical solution corresponds to that of the physical problem. Thus, the scheme is computationally stable for time steps much longer than possible with an explicit Eulerian scheme. The semi-Lagrangian scheme also preserves the values of conservative properties quite accurately and is particularly useful for accurately advecting trace constituents such as water vapor.

13.3.6 TRUNCATION ERROR

To be useful it is necessary not only that a numerical solution be stable but also that it provide an accurate approximation to the true solution. The difference between the numerical solution to a finite difference equation and the solution to the corresponding differential equation is called the *discretization error.* If this error approaches zero with δt and δx the solution is called *convergent.* The difference between a differential equation and the finite difference analog to it is referred to as *truncation error* since it arises from truncating the Taylor series approximation to the derivatives. If this error approaches zero as δt and δx go to zero the scheme is called *consistent.* According to the Lax equivalence theorem,[1] if the finite difference formulation satisfies the consistency condition, then stability is the necessary and sufficient condition for convergence. Thus, if a finite difference approximation is consistent and stable, one can be certain that the discretization error will decrease as the difference intervals are decreased, even if it is not possible to determine the error exactly.

Because numerical solutions are as a rule sought only when analytic solutions are unavailable, it is usually not possible to determine the accuracy of a solution directly. For the linear advection equation with constant advection speed considered in the above subsection it is possible, however, to compare the solutions of the finite difference equation (13.8) and the original differential equation (13.6). We can then use this example to investigate the accuracy of the difference method introduced above.

From the above discussion we already can conclude that the magnitude of the truncation error in the present case will be of order δx^2 and δt^2. It is possible to obtain more precise information on accuracy from examination of the solution (13.14). Note that for $\theta_p \to 0$, $C \to 1$ and $D \to 0$. The part of the solution proportional to C is the *physical mode.* The part proportional to D is called the *computational mode* since it has no counterpart in the analytic solution to the original differential equation. It arises because centered time differencing has turned a differential equation that is first order in time into a second-order finite difference equation. The accuracy of the finite difference solution depends not only on the smallness of D and the closeness of C to unity but also on the match between the phase speed of the physical mode and the phase speed in the analytic solution. The phase of the physical mode is given in (13.14) by

$$pm - \theta_p s = (p/\delta x)(m\,\delta x - \theta_p s\,\delta x/p) = k(x - c't) \qquad (13.24)$$

[1] See Richtmeyer and Morton, 1967.

Table 13.1   *Phase and Amplitude Accuracy of Centered Difference Solution of the Advection Equation as a Function of Resolution*

$L/\delta x$	p	θ_p	c'/c	$|D|/|C|$
2	π	π	—	∞
4	$\pi/2$	0.848	0.720	0.204
8	$\pi/4$	0.559	0.949	0.082
16	$\pi/8$	0.291	0.988	0.021
32	$\pi/16$	0.147	0.997	0.005

where $c' = \theta_p \, \delta x/(p \, \delta t)$ is the phase speed of the physical mode. Its ratio to the true phase speed is

$$c'/c = \theta_p \, \delta x/(pc \, \delta t) = \sin^{-1}(\sigma \sin p)/(\sigma p) \tag{13.25}$$

so that $c'/c \to 1$ as $\sigma p \to 0$. The dependence of c'/c and $|D|/|C|$ on wavelength is shown in Table 13.1 for the particular case where $\sigma = 0.75$.

It is clear from Table 13.1 that phase speed and amplitude errors both increase as the wavelength decreases. Short waves move slower than long waves in the finite difference solution, even though in the original equation all waves move at speed c. This dependence of phase speed on wavelength in the difference solution is called *numerical dispersion*. It is a serious problem for numerical modeling of any advected field that has sharp gradients (and hence large-amplitude short-wave components).

Short waves also suffer from having significant amplitude in the computational mode. This mode, which has no counterpart in the solution to the original differential equation, propagates opposite to the direction of the physical mode and changes sign from one time step to the next. This behavior makes it easy to recognize when the computational mode has significant amplitude.

13.4   The Barotropic Vorticity Equation in Finite Differences

The simplest example of a dynamical forecast model is the barotropic vorticity equation (11.14), which for a Cartesian β plane can be written in the form

$$\frac{\partial \zeta}{\partial t} = -F(x, y, t) \tag{13.26}$$

where

$$F(x, y, t) = \mathbf{V}_\psi \cdot \mathbf{\nabla}(\zeta + f) = \frac{\partial}{\partial x}(u_\psi \zeta) + \frac{\partial}{\partial y}(v_\psi \zeta) + \beta v_\psi \qquad (13.27)$$

and $u_\psi = -\partial\psi/\partial y$, $v_\psi = \partial\psi/\partial x$, $\zeta = \nabla^2\psi$. We have here used the fact that the horizontal velocity is nondivergent ($\partial u_\psi/\partial x + \partial v_\psi/\partial y = 0$) to write the advection term in flux form. The advection of absolute vorticity $F(x, y, t)$ may be calculated provided that we know the field of $\psi(x, y, t)$. Equation (13.26) can then be integrated forward in time to yield a prediction for ζ. It is then necessary to solve the Poisson equation $\zeta = \nabla^2\psi$ to predict the streamfunction.

A straightforward solution method is the leapfrog scheme discussed in Section 13.3.2. This requires writing (13.27) in finite difference form. Suppose that the horizontal x, y space is divided into a grid of $(M+1) \times (N+1)$ points separated by distance increments δx and δy. Then we can write the coordinate position of a given grid point as $x = m\delta x$, $y = n\delta y$, where $m = 0, 1, 2, \ldots, M$ and $n = 0, 1, 2, \ldots, N$. Thus any point on the grid is uniquely identified by the indices (m, n). A portion of such a grid space is shown in Fig. 13.2.

Centered difference formulas of the type (13.4) can then be used to approximate derivatives in the expression $F(x, y, t)$. For example, if we assume that $\delta x = \delta y \equiv d$.

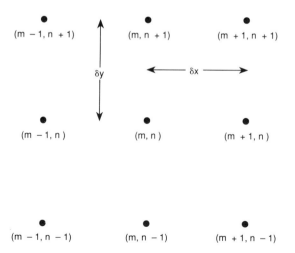

Fig. 13.2 A portion of a two-dimensional (x, y) grid mesh for solution of the barotropic vorticity equation.

$$u_\psi \approx u_{m,n} = -(\psi_{m,n+1} - \psi_{m,n-1})/2d$$

$$v_\psi \approx v_{m,n} = (\psi_{m+1,n} - \psi_{m-1,n})/2d$$

$$(13.28)$$

Similarly, with the aid of (13.5) we find that the horizontal Laplacian can be approximated as

$$\nabla^2 \psi \approx (\psi_{m+1,n} + \psi_{m-1,n} + \psi_{m,n+1} + \psi_{m,n-1} - 4\psi_{m,n})/d^2 = \zeta_{m,n} \quad (13.29)$$

The finite difference form of the Laplacian is proportional to the difference between the value of the function at the central point and the average value at the four surrounding grid points. If there are $(M-1) \times (N-1)$ interior grid points, then (13.29) yields a set of $(M-1) \times (N-1)$ simultaneous equations, which together with suitable boundary conditions determine $\psi_{m,n}$ for a given array $\zeta_{m,n}$. This set can be solved by standard methods of matrix inversion.

Before expressing the advection term $F(x, y, t)$ in finite difference form it is worth noting that if ψ is taken to be constant on the northern and southern boundaries of the β-plane channel, it is easily shown by integration over the area of the channel that the average value of F is zero. This implies that the mean vorticity is conserved for the channel. It is also possible, with a little more algebra, to show that the mean kinetic energy and the mean square vorticity (called the *enstrophy*) are conserved.

For accuracy of long-term integrations it is desirable that any finite difference approximation to F satisfy the same conservation constraints as the original differential form; otherwise the finite difference solution would not be conservative. The mean vorticity, for example, might then drift systematically in time purely owing to the nature of the finite difference solution. Finite difference schemes that simultaneously conserve vorticity, kinetic energy, and enstrophy have been designed. They are, however, rather complex. For our purposes it is sufficient to note that by writing the advection in the flux form (13.27) and using centered space differences we can conserve both mean vorticity and mean kinetic energy:

$$F_{m,n} = \frac{1}{2d} [(u_{m+1,n}\zeta_{m+1,n} - u_{m-1,n}\zeta_{m-1,n}) + (v_{m,n+1}\zeta_{m,n+1} - v_{m,n-1}\zeta_{m,n-1})]$$

$$+ \beta v_{m,n} \quad (13.30)$$

It is readily verified that if ψ is constant on the boundaries there is a cancellation of terms when (13.30) is summed over the domain. Thus,

$$\sum_{m=1}^{M-1} \sum_{n=1}^{N-1} F_{m,n} = 0 \quad (13.31)$$

Therefore, mean vorticity is conserved (except for errors introduced by time differencing) when (13.30) is used as the finite difference form of the advection term. This form also conserves mean kinetic energy (see Problem 13.2). Enstrophy is not conserved in this difference formulation, and it is conventional to add a small diffusion term in order to control any numerically generated increase in enstrophy.

The procedure for preparing a numerical forecast with the barotropic vorticity equation can now be summarized as follows:

1. The observed geopotential field at the initial time is used to compute the initial streamfunction $\psi_{m,n}(t = 0)$ at all grid points.
2. $F_{m,n}$ is evaluated at all grid points.
3. $\zeta_{m,n}(t + \delta t)$ is determined using centered differencing except at the first time step, when a forward difference must be used.
4. The simultaneous set (13.29) is solved for $\psi_{m,n}(t + \delta t)$.
5. The predicted array of $\psi_{m,n}$ is used as data and steps 2–4 are repeated until the desired forecast period is reached. For example, a 24-h forecast with 30-min time increments would require 48 cycles.

13.5   The Spectral Method

In the finite difference method discussed above, the dependent variables are specified on a set of grid points in space and time, and derivatives are approximated using finite differences. An alternative approach, referred to as the spectral method, involves representing the spatial variations of the dependent variables in terms of finite series of orthogonal functions called *basis functions*. For the Cartesian geometry of a midlatitude β plane channel the appropriate set of basis functions is a double Fourier series in x and y. For the spherical earth, on the other hand, the appropriate basis functions are the spherical harmonics.

A finite difference approximation is *local* in the sense that the finite difference variable $\Psi_{m,n}$ represents the value of $\psi(x, y)$ at a particular point in space, and the finite difference equations determine the evolution of the $\Psi_{m,n}$ for all grid points. The spectral approach, on the other hand, is based on *global* functions, i.e., the individual components of the appropriate series of basis functions. In the case of Cartesian geometry, for example, these components determine the amplitudes and phases of the sinusoidal waves that when summed determine the spatial distribution of the dependent variable. The solution proceeds by determining the evolution of a finite number of Fourier coefficients. Because the distribution in wave number space of the Fourier coefficients for a given function is referred to as its *spectrum*, it is appropriate to call this approach the spectral method.

At low resolution the spectral method is generally more accurate than the grid point method, partly because for linear advection the numerical dispersion discussed in Section 13.3.4 can be severe in a grid point model but does not occur in a properly formulated spectral model. For the range of resolutions commonly used in forecast models the two approaches are comparable in accuracy, and each has its advocates.

13.5.1 THE BAROTROPIC VORTICITY EQUATION IN SPHERICAL COORDINATES

The spectral method is particularly advantageous for solution of the vorticity equation. When the proper set of basis functions is chosen it is trivial to solve the Poisson equation for the streamfunction. This property of the spectral method not only saves computer time but also eliminates the truncation error associated with finite differencing the Laplacian operator.

In practice, the spectral method is most frequently applied to global models. This requires use of spherical harmonics, which are more complicated than Fourier series. In order to keep the discussion as simple as possible, it is again useful to consider the barotropic vorticity equation as a prototype forecast model in order to illustrate the spectral method on the sphere.

The barotropic vorticity equation in spherical coordinates may be expressed as

$$\frac{D}{Dt}(\zeta + 2\Omega \sin \phi) = 0 \tag{13.32}$$

where as usual $\zeta = \nabla^2 \psi$, with ψ a streamfunction, and

$$\frac{D}{Dt} \equiv \frac{\partial}{\partial t} + \frac{u}{a \cos \phi} \frac{\partial}{\partial \lambda} + \frac{v}{a} \frac{\partial}{\partial \phi} \tag{13.33}$$

It turns out to be convenient to use $\mu \equiv \sin \phi$ as the latitudinal coordinate, in which case the continuity equation can be written

$$\frac{1}{a} \frac{\partial}{\partial \lambda} \left(\frac{u}{\cos \phi} \right) + \frac{1}{a} \frac{\partial}{\partial \mu} (v \cos \phi) = 0 \tag{13.34}$$

so that the streamfunction is related to the zonal and meridional velocities according to

$$\frac{u}{\cos \phi} = -\frac{1}{a} \frac{\partial \psi}{\partial \mu}, \qquad v \cos \phi = \frac{1}{a} \frac{\partial \psi}{\partial \lambda} \tag{13.35}$$

The vorticity equation can then be expressed as

$$\frac{\partial \nabla^2 \psi}{\partial t} = \frac{1}{a^2}\left[\frac{\partial \psi}{\partial \mu}\frac{\partial \nabla^2 \psi}{\partial \lambda} - \frac{\partial \psi}{\partial \lambda}\frac{\partial \nabla^2 \psi}{\partial \mu}\right] - \frac{2\Omega}{a^2}\frac{\partial \psi}{\partial \lambda} \tag{13.36}$$

where

$$\nabla^2 \psi = \frac{1}{a^2}\left\{\frac{\partial}{\partial \mu}\left[(1-\mu^2)\frac{\partial \psi}{\partial \mu}\right] + \frac{1}{1-\mu^2}\frac{\partial^2 \psi}{\partial \lambda^2}\right\} \tag{13.37}$$

The appropriate orthogonal basis functions are the *spherical harmonics*, which are defined as

$$Y_\gamma(\mu, \lambda) \equiv P_\gamma(\mu)e^{im\lambda} \tag{13.38}$$

where $\gamma \equiv (n, m)$ is a vector containing the integer indices for the spherical harmonics. These are given by $m = 0, \pm 1, \pm 2, \pm 3, \ldots, n = 1, 2, 3, \ldots$, where it is required that $|m| \le n$. Here, P_γ designates an associated Legendre function of the first kind of degree n. From (13.38) it is clear that m designates the zonal wave number. It can be shown[2] that $n - |m|$ designates the number of nodes of P_γ in the interval $-1 < \mu < 1$ (i.e., between the poles) and thus measures the meridional scale of the spherical harmonic. The structures of a few spherical harmonics are shown in Fig. 13.3.

An important property of the spherical harmonics is that they satisfy the relationship

$$\nabla^2 Y_\gamma = -\frac{n(n+1)}{a^2}Y_\gamma \tag{13.39}$$

so that the Laplacian of a spherical harmonic is proportional to the function itself, which implies that the vorticity associated with a particular spherical harmonic component is simply proportional to the streamfunction for the same component.

In the spectral method on the sphere the streamfunction is expanded in a finite series of spherical harmonics by letting

$$\psi(\lambda, \mu, t) = \sum_\gamma \psi_\gamma(t) Y_\gamma(\mu, \lambda) \tag{13.40}$$

[2] See Washington and Parkinson (1986) for a discussion of the properties of the Legendre function.

Antisymmetric   Symmetric

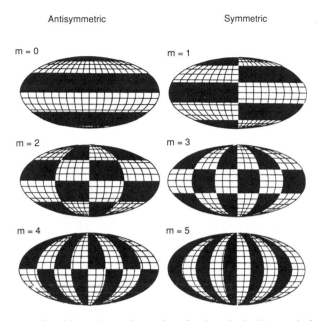

m = 0   m = 1

m = 2   m = 3

m = 4   m = 5

Fig. 13.3 Patterns of positive and negative regions for the spherical harmonic functions with $n = 5$ and $m = 0, 1, 2, 3, 4, 5$. (After Washington and Parkinson, 1986, adapted from Baer, 1972.)

where ψ_γ is the complex amplitude for the Y_γ spherical harmonic and the summation is over both n and m. The individual spherical harmonic coefficients ψ_γ are related to the streamfunction $\psi(\lambda, \mu)$ through the inverse transform

$$\psi_\gamma(t) = \frac{1}{4\pi} \int_S Y_\gamma^* \psi(\lambda, \mu, t) \, dS \tag{13.41}$$

where $dS = d\mu \, d\lambda$, and Y_γ^* designates the complex conjugate of Y_γ.

13.5.2 ROSSBY-HAURWITZ WAVES

Before considering numerical solution of the barotropic vorticity equation it is worth noting that an exact analytic solution of the nonlinear equation can be obtained in the special case where the streamfunction is equal to a single spherical harmonic mode. Thus, we let

$$\psi(\lambda, \mu, t) = \psi_\gamma(t) e^{im\lambda} P_\gamma(\mu) \tag{13.42}$$

Substituting from (13.42) into (13.36) and applying (13.39) we find that the nonlinear advection term is identically zero so that the amplitude coefficient satisfies the ordinary linear equation

$$-n(n+1)\, d\psi_\gamma/dt = -2\Omega i m \psi_\gamma \tag{13.43}$$

which has the solution $\psi_\gamma(t) = \psi_\gamma(0)\,\exp(i\nu_\gamma t)$, where

$$\nu_\gamma = 2\Omega m/[n(n+1)] \tag{13.44}$$

is the dispersion relationship for *Rossby–Haurwitz waves*, which is the name given to planetary waves on a sphere. [This should be compared to (7.91), which is the equivalent expression for a midlatitude β plane.] Since the horizontal scale of a spherical harmonic mode is proportional to n^{-1}, (13.44) shows that a single mode propagates westward on a sphere at a speed that is approximately proportional to the square of the horizontal scale. This solution also suggests why for some problems the spectral method is superior to the finite difference method at coarse resolution. A model containing even a single Fourier component can represent a realistic meteorological field (the Rossby wave), while many grid points are required for an equivalent representation in finite differences.

13.5.3   THE SPECTRAL TRANSFORM METHOD

When many spherical harmonic modes are present, solution of (13.36) by a purely spectral method requires evaluation of the nonlinear interactions among various modes owing to the advection term. It turns out that the number of interaction terms increases as the square of the number of modes retained in the series (13.40), so that this approach becomes computationally inefficient for models with the sort of spatial resolution required for prediction of synoptic-scale weather disturbances. The spectral transform method overcomes this problem by transforming between spherical harmonic wave number space and a latitude–longitude grid at every time step and carrying out the multiplications in the advection term in grid space so that it is never necessary to compute products of spectral functions.

To illustrate this method, it is useful to rewrite the barotropic vorticity equation in the form

$$\frac{\partial \nabla^2 \psi}{\partial t} = -\frac{1}{a^2}\left[2\Omega\frac{\partial\psi}{\partial\lambda} + A(\lambda,\mu)\right] \tag{13.45}$$

where

$$A(\lambda,\mu) \equiv \left[-\frac{\partial\psi}{\partial\mu}\frac{\partial\nabla^2\psi}{\partial\lambda} + \frac{\partial\psi}{\partial\lambda}\frac{\partial\nabla^2\psi}{\partial\mu}\right] \tag{13.46}$$

Substituting from (13.40) into (13.45) then yields for the spherical harmonic coefficients

$$d\psi_\gamma/dt = i\nu_\gamma\psi_\gamma + A_\gamma[n(n+1)]^{-1} \tag{13.47}$$

where A_γ is the γ component of the transform of $A(\lambda, \mu)$;

$$A_\gamma = \frac{1}{4\pi}\int_0^{2\pi}\int_{-1}^{+1} A(\lambda, \mu)Y_\gamma^* \, d\lambda \, d\mu \tag{13.48}$$

The transform method utilizes the fact that if the sum $\gamma = (n, m)$ is taken over a finite number of modes, the integral in the transform (13.48) can be evaluated exactly by numerical quadrature [i.e., by summing appropriately weighted values of $A(\lambda, \mu)$ evaluated at the grid points (λ_j, μ_k) of a latitude-longitude grid mesh]. Computation of the distribution of $A(\lambda, \mu)$ for all grid-points can be carried out without need to introduce finite differences for derivatives by noting that we can express the advection term in the form

$$A(\lambda_j, \mu_k) = (1 - \mu^2)^{-1}(F_1F_2 + F_3F_4] \tag{13.49}$$

where

$$F_1 = -(1 - \mu^2)\, \partial\psi/\partial\mu, \qquad F_2 = \partial\nabla^2\psi/\partial\lambda$$

$$F_3 = (1 - \mu^2)\, \partial\nabla^2\psi/\partial\mu, \qquad F_4 = \partial\psi/\partial\lambda$$

The quantities $F_1 - F_4$ can be computed exactly for each gridpoint using the spectral coefficients ψ_γ and the known differential properties of the spherical harmonics. For example,

$$F_4 = \partial\psi/\partial\lambda = \sum_\gamma im\psi_\gamma Y_\gamma(\lambda_j, \mu_k)$$

Once these terms have been computed for all grid points, $A(\lambda, \mu)$ can be computed at the grid points by forming the products F_1F_2 and F_3F_4; no finite difference approximations to the derivatives are required in this procedure. The transform (13.48) is then evaluated by numerical quadrature to compute the spherical harmonic components A_γ. Finally, (13.47) can be stepped ahead by a time increment δt in order to obtain new estimates of the spherical harmonic components of the streamfunction. The whole process is then repeated until the desired forecast period is reached. The steps in forecasting with the barotropic vorticity equation using the spectral transform method are summarized in schematic form in Fig. 13.4.

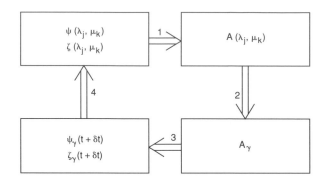

Fig. 13.4 Steps in the prediction cycle for the spectral transform method of solution of the barotropic vorticity equation.

13.6 Primitive Equation Models

Modern numerical forecast models are based on a formulation of the dynamical equations, referred to as the *primitive equations,* which is essentially the formulation proposed by Richardson. The primitive equations differ from the complete momentum equations (2.19)–(2.21) in that the vertical momentum equation is replaced by the hydrostatic approximation, and the small terms in the horizontal momentum equations given in columns C and D in Table 2.1 are neglected. In most models some version of the sigma-coordinate system introduced in Section 10.3.1 is used, and the vertical dependence is represented by dividing the atmosphere into a number of levels and utilizing finite difference expressions for vertical derivatives. Both finite differencing and the spectral method have been used for horizontal discretization in operational primitive equation forecast models. Excellent examples of such models are the grid point and spectral models developed at the European Centre for Medium-Range Weather Forecasts (ECMWF).

13.6.1 The ECMWF Grid Point Model

The earliest version of the ECMWF operational model was a global grid point model with second-order horizontal differencing on a uniform grid with intervals of 1.875° in both latitude and longitude. For computational efficiency and noise control the grid was staggered in space, as shown in Fig. 13.5, so that not all variables were carried at the same points in λ and ϕ. Variables were also staggered vertically, as shown in Fig. 13.6, and there were in total 15 unequally spaced σ layers, arranged to provide the finest resolution near the ground, the coarsest resolution in the stratosphere, and an uppermost layer near the 25-mb level.

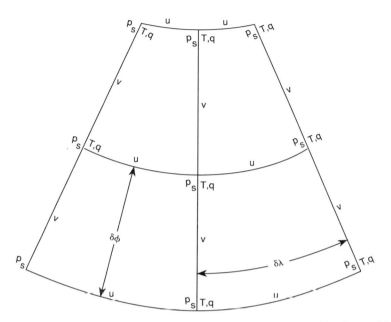

Fig. 13.5 Horizontal staggered grid arrangement in the ECMWF grid point model.

The difference equations on the staggered grid can be expressed most conveniently by defining a special operator notation as follows:

$$\bar{u}^\lambda = [u(\lambda + \delta\lambda/2) + u(\lambda - \delta\lambda/2)]/2$$

$$\delta_\lambda u = [u(\lambda + \delta\lambda/2) - u(\lambda - \delta\lambda/2)]/\delta\lambda$$

(13.50)

These represent, respectively, the averaging and differencing of adjacent grid point values of the field u on a latitude circle. Similar expressions can be defined for the meridional and vertical directions. The continuity equation, for example, can then be expressed in the form

$$\frac{\partial p_s}{\partial t} + \frac{1}{a \cos \phi} [\delta_\lambda U + \delta_\phi (V \cos \phi)] + \delta_\sigma (p_s \dot{\sigma}) = 0$$

(13.51)

where $U = \bar{p}_s^\lambda u$ and $V = \bar{p}_s^\phi v$ are horizontal mass fluxes (see Section 10.3.2). Owing to the arrangement of variables on the staggered grid of Fig. 13.5, the horizontal mass flux divergence evaluated by taking differences over the increments $\delta\lambda$, $\delta\phi$ in (13.51) is centered at the point where p_s is defined. This provides an improvement on the accuracy possible with a nonstaggered grid, in which these differences would need to be evaluated over grid distances of $2\delta\lambda$, $2\delta\phi$.

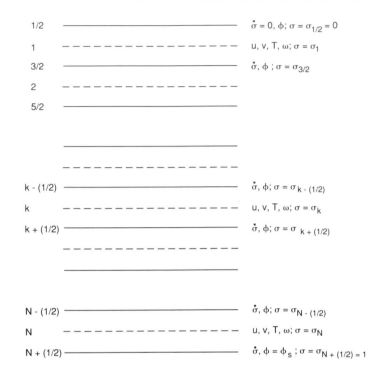

Fig. 13.6   Vertical disposition of variables in the ECMWF grid point model. (After Simmons *et al.*, 1989.)

Grid staggering provides a similar gain in accuracy in computation of the pressure gradient terms in the momentum equations. Thus,

$$\frac{\partial u}{\partial t} = \frac{-1}{a \cos \phi} [\delta_\lambda \bar{\Phi}^\sigma + R \bar{T}^\lambda \delta_\lambda (\ln p_s) + \cdots] \tag{13.52}$$

$$\frac{\partial v}{\partial t} = -\frac{1}{a} [\delta_\phi \bar{\Phi}^\sigma + R \bar{T}^\phi \delta_\phi (\ln p_s) + \cdots] \tag{13.53}$$

In this case, however, the vertical average of Φ is required since, as indicated in Fig. 13.6, Φ is staggered vertically with respect to u, v, T. The complete finite difference equations for this model, including the required special treatment of polar regions, are quite complex and will not be given here. They are discussed in Simmons *et al.* (1989).

The vertical finite differences in this model are defined so that the model can be run in either the σ coordinates discussed in Section 10.3.1 or in a *hybrid coordinate* system in which the coordinates vary from pure sigma coordinates in the troposphere to isobaric coordinates in the stratosphere. The bounding surfaces for the layers in such a system can be defined in the form

$$p_{k+1/2} = A_{k+1/2} + B_{k+1/2} p_s \qquad (13.54)$$

for $k = 0, 1, 2, \ldots, N$. Here, $A_{k+1/2}$ and $B_{k+1/2}$ are constants which satisfy the requirements that $p_{N+1/2} = p_s$ and $p_{1/2} = 0$. Thus, $A_{1/2} = B_{1/2} = 0$, $A_{N+1/2} = 0$, and $B_{N+1/2} = 1$.

In conventional σ coordinates (as used in the operational ECMWF grid point model) $A_{k+1/2} = 0$ and the values of $B_{k+1/2}$ are specified to give the desired vertical spacing of the model layers. More recent formulations of the ECMWF spectral model have employed hybrid coordinates that provide approximately level coordinate surfaces in the stratosphere. The two systems are compared in Fig. 13.7.

13.6.2 SPECTRAL MODELS

At the time of writing, ECMWF and most other operational forecast centers utilize spectral models as their primary hemispheric or global forecast models. Grid point models are generally employed for fine-scale limited-area models and in the future are expected to become more common for very high resolution global models, because the number of operations increases linearly with the number of grid points but increases as $n \log n$ in a spectral model (where n is the number of modes). Operational spectral models employ a primitive equation version of the spectral transform method described in Section 13.5. In this method the values of all meteorological fields are available in both spectral and grid point domains at each time step. Vorticity and divergence are employed as predictive variables rather than u and v. Physical computations involving such processes as radiative heating and cooling, condensation and precipitation, and convective overturning are calculated in physical space on the grid mesh while differential dynamical quantities such as pressure gradients and velocity gradients are evaluated exactly in spectral space. This combination preserves the simplicity of the grid point representation for physical processes that are local in nature, while retaining the superior accuracy of the spectral method for dynamical calculations.

In one version of the ECMWF model, used operationally from 1985 to 1991, the series of spherical harmonics is truncated at $M = N = 106$, where (N, M) are the maximum retained values of (n, m). This truncation is referred to as a *triangular truncation* (T106) because on a plot of m versus n the retained modes occupy a triangular area. Another popular spectral

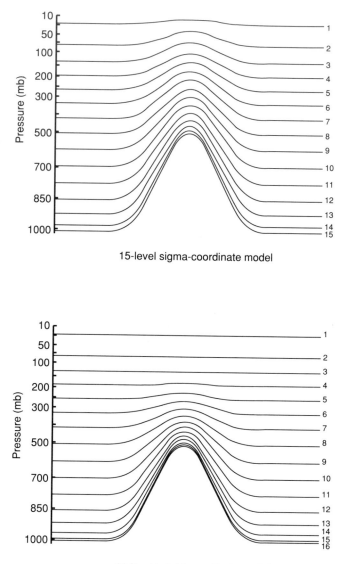

Fig. 13.7   Comparison of the ECMWF 15-level σ-coordinate model and the 16-level hybrid-coordinate model. (After Simmons *et al.*, 1989).

truncation is *rhomboidal truncation*, which has $N = |m| + M$. Both of these truncations are shown schematically in Fig. 13.8. In triangular truncation the horizontal resolution in the zonal and meridional directions is nearly the same. In rhomboidal truncation, on the other hand, the latitudinal resolution is the same for every zonal wave number. Rhomboidal truncation has some advantages for low-resolution models, but at high resolution the triangular truncation appears to be superior. A T106 spectral model has a smallest resolved half-wavelength of about 200 km and uses a grid with latitude–longitude resolution of about 1.21°.

Owing to the high resolution of the ECMWF T106 model, a time integration scheme based on the centered difference leapfrog method or other explicit scheme would require very small time steps in order to satisfy the CFL condition (13.18) for the fastest-propagating gravity wave modes. To avoid the computational overhead of such a model the time integration is generally carried out using a semi-implicit approach. In this method the linear terms associated with gravity wave propagation are handled implicitly; the nonlinear advection terms are handled explicitly. Equation (13.18) then need only be satisfied for the fastest zonal wind speed, which is much smaller than the phase speed of the Lamb mode external gravity wave.

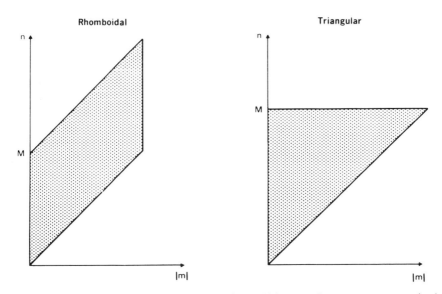

Fig. 13.8 Regions of wave number space (n, m) in which spectral components are retained for the rhomboidal truncation (left) and triangular truncation (right). (After Simmons and Bengtsson, 1984.)

13.6.3   PHYSICAL PARAMETERIZATIONS

The physical processes included in modern operational forecast models are generally the same as those included in general circulation models and were shown schematically in Fig. 10.21. Inclusion of such processes as boundary layer fluxes, vertical mixing by dry and moist convection, formation of clouds and precipitation, and the interaction of cloud and radiation fields requires that the relevant subgrid scale processes be represented in terms of model predicted fields. The approximation of unresolved processes in terms of resolved variables is referred to as *parameterization*; it is probably the most difficult and controversial area of weather and climate modeling.

Perhaps the most important physical process that must be parameterized is convection. Vertical heat transfer by convection is essential to maintain the observed tropospheric lapse rate and moisture distribution. The simplest way to mimic this effect of unresolved convective motions is through *convective adjustment*. In a simple version of this method the relative humidity and lapse rate in each grid column are examined at the end of each time step; if the lapse rate is superadiabatic the temperature profile is adjusted to dry static neutrality in a manner that conserves energy; if the column is conditionally unstable and the humidity exceeds a specified value, the column is adjusted to moist static neutrality. More sophisticated schemes utilize the fact that moist convection is dependent on low-level moisture convergence and include a moisture budget as part of the parameterization (see Section 11.3).

Various methods are used to relate clouds to the resolved humidity, temperature, and wind fields. None is completely satisfactory. In many models the model-predicted cloud fields are used only in the portion of the model concerned with precipitation. Cloud–radiation interactions are parameterized using specified climatologies for the clouds.

13.7   Data Assimilation

The capability of a numerical forecasting model to produce useful forecasts depends not only on the resolution of the model and the accuracy with which dynamical and physical processes are represented; It is also critically dependent on the initial conditions employed for integrating the model. As Richardson's early attempt to forecast weather numerically showed, observations cannot be used directly to initialize a numerical forecast. It turns out that the observational data must be modified in a dynamically consistent fashion in order to obtain a suitable data set for model initialization. This process is usually referred to as *data assimilation*. Traditionally, data assimilation is divided into two processes: *objective* analysis of the observations and *data initialization.* In the objective analysis step all data acquired for a given time (generally 00Z or 12Z) from the irregularly spaced observational network of surface and upper air stations are checked for accuracy and converted to analyses of the meteorological fields on a regular latitude–longitude grid at standard pressure levels by using a suitable interpolation scheme. Such objectively analyzed data still contain noise that is likely to be interpreted as spuriously large gravity waves when the data are used as initial data in a numerical model. In the initialization step the objectively analyzed data are modified in order to minimize the gravity wave noise and hence reduce the magnitude of initial velocity and pressure tendencies.

13.7.1 THE INITIALIZATION PROBLEM

The importance of the initialization step in preparing the initial data for forecasting with primitive equation models can be illustrated by considering the relative magnitudes of the various terms in the momentum equation in pressure coordinates:

$$\frac{\partial \mathbf{V}}{\partial t} + (\mathbf{V} \cdot \boldsymbol{\nabla})\mathbf{V} + \omega \frac{\partial \mathbf{V}}{\partial p} = -f\mathbf{k} \times \mathbf{V} - \boldsymbol{\nabla}\Phi \qquad (13.55)$$

For synoptic-scale motions in extratropical latitudes the wind and pressure fields are in approximate geostrophic balance. Thus, the acceleration following the motion is measured by the small difference between the two nearly equal terms $f\mathbf{k} \times \mathbf{V}$ and $-\boldsymbol{\nabla}\Phi$. Although the Φ field can be determined observationally with quite good accuracy, observed winds are often 10–20% in error. Such observational errors inevitably affect the objectively analyzed fields, even though a suitable analysis procedure may be able to reduce the magnitudes of the errors. Nevertheless, using the objectively analyzed winds in the initial data can easily lead to an estimate of the Coriolis force that is 10% in error at the initial time. Since the acceleration is normally only about 10% of the Coriolis force in magnitude, an acceleration computed using the observed wind and geopotential fields will generally be 100% in error. Such spurious accelerations not only lead to poor estimates of the initial pressure and velocity tendencies but also tend to produce large-amplitude gravity wave oscillations as the flow attempts to adjust from the initial unbalanced state back toward a state of quasi-geostrophic balance. These strong gravity waves are not present on the synoptic scale in nature, and their presence in the solution of the model equations quickly spoils any chance of a reasonable forecast.

One possible approach to avoid this problem might be to neglect the observed wind data and derive a wind field from the observed Φ field as part of the objective analysis process. The simplest scheme of this type would be to assume that the initial wind field was in geostrophic balance. However, the error in the computed initial local acceleration would still be ~100%. This can be seen by considering the example of zonally symmetric balanced flow about a circularly symmetric pressure system in a model with constant f. In that case $\partial V/\partial t = 0$ so that if we neglect the effects of vertical advection the flow is in gradient wind balance (see Section 3.2.4). In vectorial form this balance is simply

$$(\mathbf{V} \cdot \nabla)\mathbf{V} + f\,\mathbf{k} \times \mathbf{V} = -\nabla(\Phi) \tag{13.56}$$

However, if $\mathbf{V}$ were replaced by $\mathbf{V_g}$ in (13.56) the Coriolis and pressure gradient forces would identically balance. Thus, the inertial force would be unbalanced and

$$(\partial \mathbf{V}/\partial t)_{t=0} \approx -(\mathbf{V_g} \cdot \nabla)\mathbf{V_g} \tag{13.57}$$

Therefore, by assuming that initially the wind is in geostrophic balance, rather than gradient balance, we compute a local acceleration that is completely erroneous!

It should now be clear that in order to avoid large errors in the initial acceleration, the initial wind field in the above example should be determined by the gradient wind balance, not by the geostrophic approximation or direct observations. The gradient wind formula (3.15) is not in itself a suitable balance condition to use, because the radius of curvature must be computed for parcel trajectories. An appropriate balance condition (which is equivalent to the gradient wind in the special case of stationary circularly symmetric flow) can be obtained by taking the divergence of the horizontal momentum equation, and assuming that at the initial time

$$(\partial \nabla \cdot \mathbf{V}/\partial t)_{t=0} = 0 \quad \text{and} \quad (\nabla \cdot \mathbf{V})_{t=0} = 0$$

Then the initial wind field is nondivergent and is related to the geopotential field by the balance equation (11.15).

The balance equation, however, is quadratic in the streamfunction ψ and hence does not provide a unique horizontal velocity field for a given Φ field. Furthermore, realistic baroclinic motions in the atmosphere are not exactly nondivergent, but rather have slowly evolving divergent wind components associated with the secondary circulation discussed in Chapter 6, which are required to maintain the delicate balance between the mass and velocity fields that is maintained as the flow evolves.

Nevertheless, the traditional approach to data assimilation works well as long as the prediction model is limited to a region of adequate data coverage (such as the North American continent) and all observations are made at the standard 00Z and 12Z times so that they can be directly incorporated into synoptic analyses for those times. For global analysis it is less successful since in many oceanic regions very little conventional data is available, and it is necessary to rely on asynoptic data (i.e., observations at nonstandard times, such as those from aircraft and satellites). Such data have proved invaluable, especially over the oceans and in the Southern Hemisphere. But they are not easily incorporated into traditional objective analysis schemes.

The deficiencies of the traditional approach have led the major forecast centers to adopt assimilation procedures in which the separate processes of objective analysis and data initialization are combined into a continuous cycle of data assimilation, which is often referred to as *four-dimensional assimilation*. An example of a modern data assimilation procedure, as used in the 1980s at ECMWF, is shown schematically in Fig. 13.9. In the ECMWF scheme an assimilation cycle is begun by using a 6-h forecast from the previous initial state as the first-guess "observed" state for the 18Z analysis time. All available data within a window of ±3 h of that time are then used to update the analysis defined by the first guess. The analyzed data are then subjected to an initialization process and used in a 6-h forecast, which is then used as the first guess for the 24Z analysis time. This analysis-initialization cycle is then repeated. Any of the initialized states can be used

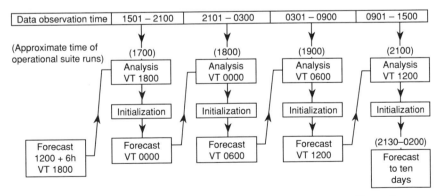

Fig. 13.9 The ECMWF data assimilation cycle, valid in February 1984. Times shown are GMT. (After Bengtsson, 1985.)

to start a long forecast run. In the example shown in Fig. 13.9 forecasts are carried out for a 10-day period once each day starting from the initialized state for 12Z.

13.7.2 NONLINEAR NORMAL MODE INITIALIZATION

Owing to the important role of the 6-h forecast fields in the four-dimensional assimilation process outlined above, it is crucial that these forecasts be free of spurious gravity wave noise introduced by initial imbalances. Thus, it is important to employ an initialization procedure that efficiently eliminates high-frequency gravity modes, while retaining the relationships among the dynamical and thermodynamic fields that describe the slowly evolving large-scale flow. The method that is most successful in meeting these objectives is called *normal mode initialization.*

The normal modes of a dynamical system are the free modes of oscillation of the system. If the discretized form of the primitive equations used in a prediction model are linearized about a state of rest, the normal modes can be calculated. For each grid point and pressure level there are three normal modes—an eastward- and westward-propagating gravity mode and a westward-propagating Rossby mode. The analyzed fields can be represented by a combination of these normal mode solutions in a manner analogous to the expansion of a field in terms of a finite series of Fourier components or spherical harmonics. Such a representation is referred to as a *projection* onto the normal modes.

This method can be illustrated in a simple fashion using the shallow-water equations. We first must rewrite (7.69)–(7.71) to include the nonlinear terms neglected in Section 7.6:

$$\frac{\partial u}{\partial t} - fv + \frac{\partial \Phi}{\partial x} = -\varepsilon \left[u\frac{\partial u}{\partial x} + v\frac{\partial u}{\partial y} \right] \tag{13.58}$$

$$\frac{\partial v}{\partial t} + fu + \frac{\partial \Phi}{\partial y} = -\varepsilon \left[u\frac{\partial v}{\partial x} + v\frac{\partial v}{\partial y} \right] \tag{13.59}$$

$$\frac{\partial \Phi}{\partial t} + c^2\left(\frac{\partial u}{\partial x} + \frac{\partial v}{\partial y}\right) = -\varepsilon \left[\frac{\partial(\Phi u)}{\partial x} + \frac{\partial(\Phi v)}{\partial y} \right] \tag{13.60}$$

where $c^2 = gH$ with H the mean depth, $\Phi = gh$ with h the local deviation of depth from the mean, and ε an indicator function with value 0 or 1.

If the nonlinear terms on the right in (13.58)–(13.60) are omitted by letting $\varepsilon = 0$, normal mode solutions can be obtained of the form

$$\begin{pmatrix} u \\ v \\ \Phi \end{pmatrix} = \begin{pmatrix} \hat{u} \\ \hat{v} \\ \hat{\Phi} \end{pmatrix} \exp[i(kx + ly - \nu t)] \tag{13.61}$$

where for this simple f-plane model it can be shown (Problem 13.10) that the Rossby normal mode has $\nu_R = 0$ and the eastward and westward gravity modes satisfy $\nu_{G\pm} = \pm[f^2 + c^2(k^2 + l^2)]^{1/2}$. Substituting (13.61) into (13.58)–(13.60) and neglecting the nonlinear terms then yields the relationships of velocity to geopotential for the three normal mode solutions:

$$\hat{u}_R = -ilf^{-1}\hat{\Phi}_R, \qquad \hat{v}_R = +ikf^{-1}\hat{\Phi}_R \tag{13.62}$$

$$\hat{u}_{G\pm} = (\nu_{G\pm}^2 - f^2)^{-1}(k\nu_{G\pm} + ilf]\hat{\Phi}_{G\pm}$$

$$\hat{v}_{G\pm} - (\nu_{G\pm}^2 - f^2)^{-1}[l\nu_{G\pm} - ikf]\hat{\Phi}_{G\pm} \tag{13.63}$$

Here the subscripts R and G± indicate the Rossby (geostrophic) mode and the eastward and westward gravity modes, respectively.

The normal mode solutions thus can be expressed in terms of the three independent amplitude coefficients, $\hat{\Phi}_R$, $\hat{\Phi}_{G+}$, and $\hat{\Phi}_{G-}$, respectively.

Suppose that the observed fields at $t = 0$ are given by

$$\begin{pmatrix} u \\ v \\ \Phi \end{pmatrix} = \begin{pmatrix} \hat{u}_0 \\ \hat{v}_0 \\ \hat{\Phi}_0 \end{pmatrix} \exp(i(kx + ly)] \tag{13.64}$$

Thus, the observations involve three independent amplitude coefficients. These can be represented as sums over the normal modes:

$$\hat{u}_R + \hat{u}_{G+} + \hat{u}_{G-} = \hat{u}_0$$

$$\hat{v}_R + \hat{v}_{G+} + \hat{v}_{G-} = \hat{v}_0 \tag{13.65}$$

$$\hat{\Phi}_R + \hat{\Phi}_{G+} + \hat{\Phi}_{G-} = \hat{\Phi}_0$$

Upon substituting from (13.62) and (13.63) into (13.65) we obtain a set of three inhomogeneous linear equations, which can be solved for the relative geopotential amplitudes of the three normal modes when projected onto the observed fields. The result is

$$\hat{\Phi}_R = \frac{c^2}{\nu_{G+}^2}\left[if(l\hat{u}_0 - k\hat{v}_0) + \frac{f^2}{c^2}\hat{\Phi}_0 \right] \tag{13.66}$$

$$\hat{\Phi}_{G+} = \frac{c^2}{2\nu_{G+}^2}[(k\nu_{G+} - ifl)\hat{u}_0 + (l\nu_{G+} + ifk)\hat{v}_0 + (k^2 + l^2)\hat{\Phi}_0] \quad (13.67)$$

$$\hat{\Phi}_{G-} = \frac{c^2}{2\nu_{G-}^2}[(k\nu_{G-} - ifl)\hat{u}_0 + (l\nu_{G-} + ifk)\hat{v}_0 + (k^2 + l^2)\hat{\Phi}_0] \quad (13.68)$$

It is readily verified that if $\hat{u}_0$ and $\hat{v}_0$ are in exact geostrophic balance with $\hat{\Phi}_0$ then $\hat{\Phi}_R = \hat{\Phi}_0$ and $\hat{\Phi}_{G+} = \hat{\Phi}_{G-} = 0$. When the observed velocities are not exactly in geostrophic balance (13.66)–(13.68) give the relative weightings for the projections onto the Rossby mode and the two gravity modes. For the linear system it is possible to set the gravity modes to zero and to initialize with the projection of the observed fields onto the Rossby mode. Note from (13.66) that the relative weighting of the velocity observations (which appear in the form of relative vorticity) increases as the scale of the disturbance decreases.

As explained in the previous section, merely setting the gravity modes to zero does not work for the nonlinear system. In that case when the normal mode projection defined by (13.62) and (13.66) is used to determine the amplitude coefficients in (13.61) and the result is substituted into (13.58)–(13.60) the unbalanced advective terms will lead to large initial accelerations. This problem can be solved by including the gravity modes, but with their amplitudes adjusted so that their initial tendencies vanish. Since the equations are nonlinear it is necessary to carry out the solution iteratively, using the linear solution as a first guess.

In practice, the nonlinear normal modes method works very well when applied to primitive equation forecast models. The method can be modified to incorporate not only nonlinear advection but also the effects of diabatic heating in the tropics in order to preserve the Hadley circulation in the initial data. Although further improvements will no doubt be made in data assimilation, it appears that nonlinear normal mode initialization is a very satisfactory solution to the long-standing problem of model initialization.

13.8   Predictability

For short-range forecasts (1 or 2 days) of the midlatitude 500-mb flow, it is possible to neglect diabatic heating and frictional dissipation. It is important, however, to have good initial data in the region of interest because on this short time scale forecasting ability depends primarily on the proper advection of the initial vorticity field. As the length of the forecast period increases, the effects of propagation of influences from other regions and changes due to various sources and sinks of momentum and energy become increasingly important. Therefore, the flow at a point in the middle-latitude troposphere will depend on initial conditions for an increasing domain and on accurate representation of various physical processes as the period of the forecast is extended. In fact, according to the estimate of Smagorinsky shown in Fig. 13.10, for periods greater than 1 week it is necessary to know the initial state of the entire global atmosphere from the stratosphere to the surface, as well as the state of the upper layers of the oceans.

However, even if an ideal data network were available to specify the initial state on a global scale, there still would be a time limit beyond which a useful forecast would not be possible. The atmosphere is a continuum with a continuous spectrum of scales of motion. No matter how fine the grid resolution is made, there will always be motions whose scales are too small to be properly observed and represented in the model. Thus, there is an unavoidable level of error in the determination of the initial state. The nonlinearity and instability of atmospheric flow will inevitably cause the small inherent errors in the initial data to grow and gradually affect the

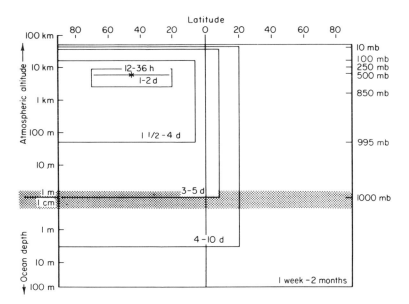

Fig. 13.10 Schematic diagram of the domain of initial dependence for a prediction point in midtroposphere at midlatitudes (denoted by an asterisk) as a function of forecast time span. The atmospheric and ocean elevations are given on a logarithmic scale, increasing upward and downward, respectively. The stripped area is the interface zone. (After Smagorinsky, 1967. Reproduced with permission of the American Meteorological Society.)

larger scales of motion so that the forecast flow field will eventually evolve differently from the actual flow.

A very simple example of this process of error growth was presented by Lorenz (1984). Lorenz illustrated the general problem of predictability by considering the first-order quadratic difference equation

$$Y_{s+1} = aY_s - Y_s^2 \tag{13.69}$$

which may be solved for Y_{s+1} by iteration if the constant a and the initial condition Y_0 are specified. The solid line in Fig. 13.11 shows a portion of the solution sequence with $a = 3.75$ and $Y_0 = 1.5$. This line represents a control run that may be regarded as defining the "observations." The dashed line is the solution for a case in which the initial value Y_0 is perturbed by 0.001, while the dash–dot line shows a solution in which the original initial data are used but the coefficient a is perturbed by 0.001.

For the first several steps the two "predictions" are very close to the observed values. But after about 15 steps they begin to diverge. Thus, errors in either the initial conditions or the governing equation can produce comparable results—the departure of a predicted solution from the observed sequence increases rapidly. For the case shown here only about 20 steps are required for the three sequences to lose all resemblance to each other. Such behavior is characteristic of a wide variety of systems governed by deterministic equations, including atmospheric flows.

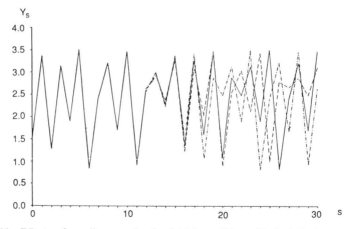

Fig. 13.11 Effects of small errors in the initial conditions (dashed line) and imperfect knowledge of the governing equation (dash–dotted line) on the prediction of a time series generated by the difference equation (13.69). (Courtesy of E. Källén.)

Estimates of how this sort of error growth limits the inherent predictability of the atmosphere can be made using primitive equation forecast models. In these predictability experiments a "control" run is made using initial data corresponding to the observed flow at a given time. The initial data are then perturbed by introducing small random errors and the model is run again. The growth of inherent error can then be estimated by comparing the second "forecast" run with the control. Results from a number of such studies indicate that the doubling time for the root-mean-square geopotential height error is about 2–3 days for small errors and somewhat greater for large errors. Thus, the theoretical limit for predictability on the synoptic scale is probably about 2 weeks.

Actual predictive skill with present models is, however, less than the theoretical limit imposed by inherent error growth. An indication of the skill of the present generation of global forecast models is given in Fig. 13.12. In this figure forecast skill is plotted in terms of the *anomaly correlation* of the geopotential height field, defined as the correlation between observed and predicted deviations from climatology at one or more levels. Subjective evaluations suggest that useful forecasts are obtained when the anomaly correlation is greater than 0.6. Thus, mean predictive skill out to about 5–6 days is implied by the 1983 ECMWF results of Fig. 13.12, although there

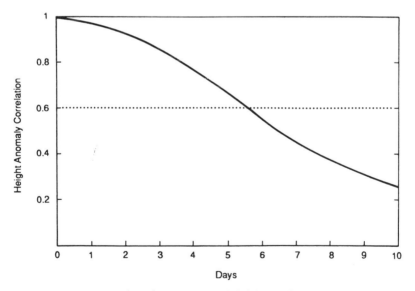

Fig. 13.12 Anomaly correlation of 1000- to 200-mb height as a function of forecast range for the ECMWF operational model for the area from 20°N to 82.5°N averaged over all operational forecasts carried out in 1983. (After Simmons, 1986.)

can be quite wide variations in skill from one situation to another. These may reflect variations in the degree of predictability from one atmospheric flow regime to another.

A number of factors are probably responsible for the failure of current models to reach the theoretical forecast skill limits. These include observational and analysis errors in the initial data, inadequate model resolution, and unsatisfactory representation of physical processes involving radiation, clouds, precipitation, and boundary layer fluxes. Forecast error cannot be attributed to any single cause but apparently results from the combined effects of a number of factors, both physical and computational.

Problems

13.1. Show that for the barotropic vorticity equation on the Cartesian β plane (13.26) enstrophy and kinetic energy are conserved when averaged over the whole domain, i.e., that the following integral constraints are satisfied:

$$\frac{d}{dt} \int \int \frac{\zeta^2}{2} \, dx \, dy = 0, \qquad \frac{d}{dt} \int \int \frac{\nabla \psi \cdot \nabla \psi}{2} \, dx \, dy = 0$$

Hint: to prove energy conservation multiply (13.26) through by $-\psi$ and use the chain rule of differentiation.

13.2. Verify the expression (13.31) in the text.

13.3. The Euler backward method of finite differencing the advection equation is a two-step method consisting of a forward prediction step followed by a backward corrector step. A complete cycle is thus defined by

$$\hat{q}_m^* - \hat{q}_{m,s} = -\frac{\sigma}{2} (\hat{q}_{m+1,s} - \hat{q}_{m-1,s})$$

$$\hat{q}_{m,s+1} - \hat{q}_{m,s} = -\frac{\sigma}{2} (\hat{q}_{m+1}^* - \hat{q}_{m-1}^*)$$

where $\hat{q}_m^*$ is the first guess for time step $s+1$. Use the method of Section 13.3.3 to determine the necessary condition for stability of this method.

13.4. Carry out truncation error analyses analogous to that of Table 13.1 for the cases $\sigma = 0.95$ and $\sigma = 0.25$.

13.5. Suppose that the streamfunction ψ is given by a single sinusoidal wave $\psi(x) = A \sin kx$. Find an expression for the error of the finite difference approximation

$$\frac{\partial^2 \psi}{\partial x^2} \approx \frac{\psi_{m+1} - 2\psi_m + \psi_{m-1}}{\delta x^2}$$

for $k \,\delta x = \pi/8$, $\pi/4$, $\pi/2$, and π. Here $x = m \,\delta x$ with $m = 0, 1, 2, \ldots$.

13.6. Using the method given in Section 13.3.3, evaluate the computational stability of the following two finite difference approximations to the one-dimensional advection equation:

(a) $$\hat{\zeta}_{m,s+1} - \hat{\zeta}_{m,s} = -\sigma(\hat{\zeta}_{m,s} - \hat{\zeta}_{m-1,s})$$

(b) $$\hat{\zeta}_{m,s+1} - \hat{\zeta}_{m,s} = -\sigma(\hat{\zeta}_{m+1,s} - \hat{\zeta}_{m,s})$$

where $\sigma = c \,\delta t/\delta x > 0$. [The schemes labeled (a) and (b) are referred to as *upstream* and *downstream* differencing, respectively.] Show that scheme (a) damps the advected field, and compute the fractional damping rate per time step for $\sigma = 0.25$ and $k \,\delta x = \pi/8$ for a field with the initial form $\zeta = \exp(ikx)$.

13.7. Using a staggered horizontal grid analogous to that shown in Fig. 13.5 (but for an equatorial β-plane geometry) express the linearized shallow-water equations (11.29)–(11.31) in finite difference form using the finite differencing and averaging operator notation introduced in Section 13.6.1.

13.8. Verify the equality

$$\left(\frac{1 - i \tan \theta_p}{1 + i \tan \theta_p}\right) = \exp(-2i\theta_p)$$

given in (13.22).

13.9. Compute the ratio of the numerical phase speed to the true phase speed, c'/c, for the implicit difference scheme of (13.19) for $p = \pi$, $\pi/2$, $\pi/4$, $\pi/8$, and $\pi/16$. Let $\sigma = 0.75$ and $\sigma = 1.25$. Compare your results to those of Table 13.1.

13.10. Starting with the assumed solution (13.61), obtain the normal mode solutions to the linearized versions of (13.58)–(13.60) and hence verify (13.62) and (13.63).

13.11. Show that the projection of the observations onto the Rossby normal mode (13.66) is equivalent to requiring that the quasi-geostrophic potential vorticity of the Rossby mode be proportional to the sum of the observed relative vorticity minus the observed geopotential multiplied by the factor (f/c^2). *Hint*: linearize (4.26) and assume that f is constant.

Suggested References

Haltiner and Williams, *Numerical Prediction and Dynamical Meteorology*, is a textbook that covers both the numerical and physical aspects of weather prediction at the graduate level.

Manabe (ed.), *Issues in Atmospheric and Oceanic Modeling: Part B, Weather Dynamics*, contains a number of excellent articles including overviews of the ECMWF forecasting operation, data assimilation, and predictability.

Burridge and Källén, *Problems and Prospects* in *Long and Medium Range Weather Forecasting*, is another book containing overview articles on a number of topics in numerical prediction.

Appendix

A | Useful Constants and Parameters

Gravitational constant	$G = 6.673 \times 10^{-11}$ N m^2 kg^{-2}
Gravity at sea level	$g_0 = 9.81$ m s^{-2}
Mean radius of the earth	$a = 6.37 \times 10^6$ m
Earth's angular speed of rotation	$\Omega = 7.292 \times 10^{-5}$ rad s^{-1}
Universal gas constant	$R^* = 8.314 \times 10^3$ J K^{-1} kmol^{-1}
Gas constant for dry air	$R = 287$ J K^{-1} kg^{-1}
Specific heat of dry air at constant pressure	$c_p = 1004$ J K^{-1} kg^{-1}
Specific heat of dry air at constant volume	$c_v = 717$ J K^{-1} kg^{-1}
Ratio of specific heats	$\gamma = c_p / c_v = 1.4$
Molecular weight of water	$m_v = 18.016$ kg kmol^{-1}
Latent heat of condensation at 0°C	$L_c = 2.5 \times 10^6$ J kg^{-1}
Mass of the earth	$M = 5.988 \times 10^{24}$ kg
Standard sea-level pressure	$p_0 = 101.325$ kPa
Standard sea-level temperature	$T_0 = 288.15$ K
Standard sea-level density	$\rho_0 = 1.225$ kg m^{-3}

Appendix

B | List of Symbols

Only the principal symbols are listed. Symbols formed by adding primes, overbars, or subscripted indices are not listed separately. Boldface type indicates vector quantities. Where symbols have more than one meaning, the section where the second meaning is first used is indicated in the list.

a Radius of the earth; inner radius of a laboratory annulus (Section 10.5)

b Buoyancy; outer radius of a laboratory annulus (Section 10.5)

c Phase speed of a wave

c_p Specific heat of dry air at constant pressure

c_{pv} Specific heat of water vapor at constant pressure

c_v Specific heat of dry air at constant volume

c_w Specific heat of liquid water

d Grid distance

e Internal energy per unit mass

f Coriolis parameter ($\equiv 2\Omega \sin \phi$)

g Magnitude of gravity

$\mathbf{g}$ Gravity

$\mathbf{g}^*$ Gravitational acceleration

h Depth of fluid layer; moist static energy (Section 9.5)

i Square root of minus one

$\mathbf{i}$ Unit vector along the x axis

$\mathbf{j}$ Unit vector along the y axis

$\mathbf{k}$ Unit vector along the z axis

k Zonal wave number

l Mixing length; meridional wave number (Section 7.5).

m Mass element; vertical wave number (Section 7.4); planetary wave number (Section 13.5)

m_v Molecular weight of water

n Distance in direction normal to a parcel trajectory; meridional index for equatorial waves (Section 11.4)

$\mathbf{n}$ Unit vector normal to a parcel trajectory

p Pressure

p_s Standard constant pressure; surface pressure in σ coordinates (Section 10.3)

q Quasi-geostrophic potential vorticity; water vapor mixing ratio (Section 9.5)

q_s Saturation mixing ratio

r Radial distance in spherical coordinates

$\mathbf{r}$ A position vector

s Generalized vertical coordinate; distance along a parcel trajectory (Section 3.2); entropy (Section 2.7); dry static energy (Section 9.5);

t Time

$\mathbf{t}$ Unit vector parallel to a parcel trajectory

u_* Friction velocity

u x component of velocity (eastward)

v y component of velocity (northward)

w z component of velocity (upward)

w^* Vertical motion in log-pressure system

x, y, z Eastward, northward, and upward distance, respectively

z^* Vertical coordinate in log-pressure system

$\mathbf{A}$ An arbitrary vector

A Area

B Convective available potential energy

C_d Surface drag coefficient

D_e Depth of Ekman layer

E Evaporation rate

E_I Internal energy

$\mathbf{F}$ A force; EP flux (Section 10.2)

$\mathbf{Fr}$ Frictional force

G Universal gravitational constant; zonal force (Section 10.2)

H Scale height

J Diabatic heating rate

K Total horizontal wave number; kinetic energy (Section 9.3)

K_m Eddy viscosity coefficient

L A length scale

L_c Latent heat of condensation

M Mass; mass convergence in Ekman layer (Section 5.4); absolute zonal momentum (Section 9.3); angular momentum (Section 10.3)

N Buoyancy frequency

P Ertel potential vorticity; available potential energy (Section 8.3); precipitation rate (Section 11.3)

Q   **Q** vector

R Gas constant for dry air; distance from the axis of rotation of the earth to a point on the surface of the earth (Section 1.5); diabatic energy generation rate (Section 10.4)

R   Vector in the equatorial plane directed from the axis of rotation to a point on the surface of the earth

R^* Universal gas constant

S $\equiv HN^2/Rm$, stability parameter in log-pressure coordinates

S_p $\equiv -T\,\partial \ln \theta/\partial p$, stability parameter in pressure coordinates

T Temperature

U Horizontal velocity scale

V Speed in natural coordinates

δV Volume increment

U   Three-dimensional velocity vector

V   Horizontal velocity vector

W Vertical motion scale

X Zonal turbulent drag force

Z Geopotential height

α Specific volume

β $\equiv df/dy$, variation of the Coriolis parameter with latitude; angular direction of the wind (Section 3.3)

γ $\equiv c_p/c_v$, the ratio of the specific heats

ε Rate of frictional energy dissipation; thermal expansion coefficient of water (Section 10.7)

ζ Vertical component of relative vorticity

η Vertical component of absolute vorticity; weighting function for heating profile (Section 11.3)

θ Potential temperature

$\dot{\theta}$ $\equiv D\theta/Dt$, vertical motion in isentropic coordinates

θ_e Equivalent potential temperature

κ $\equiv R/c_p$, ratio of gas constant to specific heat at constant pressure; Rayleigh friction coefficient (Section 10.6)

λ Longitude, positive eastward

μ Dynamic viscosity coefficient

ν Angular frequency; kinematic viscosity (Section 1.4)

ρ Density

σ $\equiv -\alpha\, \partial\theta/\partial p$, static stability parameter in isobaric coordinates; $\equiv -p/p_s$, vertical coordinate in σ system (Section 10.3). "Density" in isentropic coordinates (Section 4.6)

τ_d Diffusive time scale

τ_E Eddy stress

ϕ Latitude

χ Geopotential tendency; meridional streamfunction

χ_i Ozone mixing ratio

ψ Horizontal streamfunction

ω Vertical velocity ($\equiv dp/dt$) in isobaric coordinates

$\boldsymbol{\omega}$ Vorticity vector

Γ $\equiv -dT/dz$, lapse rate of temperature

Γ_d Dry adiabatic lapse rate

Φ Geopotential

Π Exner function

Θ Potential temperature deviation

Ω Angular speed of rotation of the earth; angular speed of rotation of laboratory annulus (Section 10.7)

$\boldsymbol{\Omega}$ Angular velocity of the earth

Appendix

C | Vector Analysis

C.1 Vector Identities

The following formulas may be shown to hold where Φ is an arbitrary scalar and $\mathbf{A}$ and $\mathbf{B}$ are arbitrary vectors.

$$\nabla \times \nabla \Phi = 0$$

$$\nabla \cdot (\Phi \mathbf{A}) = \Phi \nabla \cdot (\mathbf{A}) + \mathbf{A} \cdot \nabla \Phi$$

$$\nabla \times (\Phi \mathbf{A}) = \nabla \Phi \times \mathbf{A} + \Phi (\nabla \times \mathbf{A})$$

$$\nabla \cdot (\nabla \times \mathbf{A}) = 0$$

$$(\mathbf{A} \cdot \nabla)\mathbf{A} = \tfrac{1}{2}\nabla(\mathbf{A} \cdot \mathbf{A}) - \mathbf{A} \times (\nabla \times \mathbf{A})$$

$$\nabla \times (\mathbf{A} \times \mathbf{B}) = \mathbf{A}(\nabla \cdot \mathbf{B}) - \mathbf{B}(\nabla \cdot \mathbf{A}) - (\mathbf{A} \cdot \nabla)\mathbf{B} + (\mathbf{B} \cdot \nabla)\mathbf{A}$$

C.2   Integral Theorems

(a) Divergence theorem:

$$\int_A \mathbf{B} \cdot \mathbf{n} \, dA = \int_V \mathbf{\nabla} \cdot \mathbf{B} \, dV$$

where V is a volume enclosed by surface A and $\mathbf{n}$ is a unit normal on A.

(b) Stokes' theorem:

$$\oint \mathbf{B} \cdot d\mathbf{l} = \int_A (\mathbf{\nabla} \times \mathbf{B}) \cdot \mathbf{n} \, dA$$

where A is a surface bounded by the line traced by the position vector $\mathbf{l}$ and $\mathbf{n}$ is a unit normal of A.

C.3   Vector Operations in Various Coordinate Systems

(a) Cartesian coordinates: (x, y, z)

Coordinate	Symbol	Velocity component	Unit vector
Eastward	x	u	$\mathbf{i}$
Northward	y	v	$\mathbf{j}$
Upward	z	w	$\mathbf{k}$

$$\mathbf{\nabla}\Phi = \mathbf{i}\frac{\partial \Phi}{\partial x} + \mathbf{j}\frac{\partial \Phi}{\partial y} + \mathbf{k}\frac{\partial \Phi}{\partial z}$$

$$\mathbf{\nabla} \cdot \mathbf{V} = \frac{\partial u}{\partial x} + \frac{\partial v}{\partial y}$$

$$\mathbf{k} \cdot (\mathbf{\nabla} \times \mathbf{V}) = \frac{\partial v}{\partial x} - \frac{\partial u}{\partial y}$$

$$\nabla_h^2 \Phi = \frac{\partial^2 \Phi}{\partial x^2} + \frac{\partial^2 \Phi}{\partial y^2}$$

(b) Cylindrical coordinates: (r, λ, z)

Coordinate	Symbol	Velocity component	Unit vector
Radial	r	u	$\mathbf{i}$
Azimuthal	λ	v	$\mathbf{j}$
Upward	z	w	$\mathbf{k}$

$$\nabla \Phi = \mathbf{i} \frac{\partial \Phi}{\partial r} + \mathbf{j} \frac{1}{r} \frac{\partial \Phi}{\partial \lambda} + \mathbf{k} \frac{\partial \Phi}{\partial z}$$

$$\nabla \cdot \mathbf{V} = \frac{1}{r} \frac{\partial(ru)}{\partial r} + \frac{1}{r} \frac{\partial v}{\partial \lambda}$$

$$\mathbf{k} \cdot (\nabla \times \mathbf{V}) = \frac{1}{r} \frac{\partial(rv)}{\partial r} - \frac{1}{r} \frac{\partial u}{\partial \lambda}$$

$$\nabla_h^2 \Phi = \frac{1}{r} \frac{\partial}{\partial r}\left(r \frac{\partial \Phi}{\partial r} \right) + \frac{1}{r^2} \frac{\partial^2 \Phi}{\partial \lambda^2}$$

(c) Spherical coordinates: (λ, ϕ, r)

Coordinate	Symbol	Velocity component	Unit vector
Longitude	λ	u	$\mathbf{i}$
Latitude	ϕ	v	$\mathbf{j}$
Radial	r	w	$\mathbf{k}$

$$\nabla \Phi = \frac{\mathbf{i}}{r \cos \phi} \frac{\partial \Phi}{\partial \lambda} + \mathbf{j} \frac{1}{r} \frac{\partial \Phi}{\partial \phi} + \mathbf{k} \frac{\partial \Phi}{\partial r}$$

$$\nabla \cdot \mathbf{V} = \frac{1}{r \cos \phi} \left[\frac{\partial u}{\partial \lambda} + \frac{\partial(v \cos \phi)}{\partial \phi} \right]$$

$$\mathbf{k} \cdot (\nabla \times \mathbf{V}) = \frac{1}{r \cos \phi} \left[\frac{\partial v}{\partial \lambda} - \frac{\partial(u \cos \phi)}{\partial \phi} \right]$$

$$\nabla_h^2 \Phi = \frac{1}{r^2 \cos^2 \phi} \left[\frac{\partial^2 \Phi}{\partial \lambda^2} + \cos \phi \frac{\partial}{\partial \phi}\left(\cos \phi \frac{\partial \Phi}{\partial \phi} \right) \right]$$

D | Equivalent Potential Temperature

A mathematical expression for θ_e can be derived by applying the first law of thermodynamics to a mixture of 1kg of dry air plus q kg of water vapor (q, called the *mixing ratio*, is usually expressed as grams of vapor per kilogram of dry air). If the parcel is not saturated, the dry air satisfies the energy equation

$$c_p \, dT - \frac{d(p-e)}{p-e} RT = 0 \qquad (D.1)$$

and the water vapor satisfies

$$c_{pv} \, dT - \frac{de}{e} \frac{R^*}{m_v} T = 0 \qquad (D.2)$$

where the motion is assumed to be adiabatic. Here, e is the partial pressure of the water vapor, c_{pv} the specific heat at constant pressure for the vapor, R^* the universal gas constant, and m_v the molecular weight of water. If the parcel is saturated, then condensation of $-dq_s$ kg vapor per kilogram dry air will heat the mixture of air and vapor by an amount of heat that goes into the liquid water, and the saturated parcel must satisfy the energy equation

$$c_p \, dT + q_s c_{pv} \, dT - \frac{d(p - e_s)}{p - e_s} RT - q_s \frac{de_s}{e_s} \frac{R^*}{m_v} T = -L_c \, dq_s \qquad \text{(D.3)}$$

where q_s and e_s are the saturation mixing ratio and vapor pressure, respectively. The quantity de_s/e_s may be expressed in terms of temperature using the Clausius–Clapeyron equation[1]

$$\frac{de_s}{dT} = \frac{m_v L_c e_s}{R^* T^2} \qquad \text{(D.4)}$$

Substituting from (D.4) into (D.3) and rearranging terms, we obtain

$$-L_c d\left(\frac{q_s}{T}\right) = c_p \frac{dT}{T} - \frac{Rd(p - e_s)}{p - e_s} + q_s c_{pv} \frac{dT}{T} \qquad \text{(D.5)}$$

If we now define the potential temperature of the dry air θ_d, according to

$$c_p \, d \ln \theta_d = c_p \, d \ln T - R \, d \ln(p - e_s)$$

we can rewrite (D.5) as

$$-L_c \, d\left(\frac{q_s}{T}\right) = c_p \, d \ln \theta_d + q_s c_{pv} \, d \ln T \qquad \text{(D.6)}$$

However, it may be shown that

$$dL_c/dT = c_{pv} - c_w \qquad \text{(D.7)}$$

where c_w is the specific heat of liquid water. Combining (D.7) and (D.6) to eliminate c_{pv} yields

$$-d\left(\frac{L_c q_s}{T}\right) = c_p \, d \ln \theta_d + q_s c_w \, d \ln T \qquad \text{(D.8)}$$

[1] For a derivation, see, for example, Hess (1959, p. 46).

Neglecting the last term in (D.8) we may integrate from the original state $(p, T, q_s, e_s, \theta_d)$ to a state where $q_s \approx 0$. Therefore, the equivalent potential temperature of a saturated parcel is given by

$$\theta_e = \theta_d \exp(L_c q_s / c_p T) \approx \theta \exp(L_c q_s / c_p T) \tag{D.9}$$

Equation (D.9) may also be applied to an unsaturated parcel provided that the temperature used is the temperature that the parcel would have if brought to saturation by an adiabatic expansion.

E Standard Atmosphere Data[1]

Table E.1 *Geopotential Height versus Pressure*

Pressure (kPa)	Pressure (mb)	Z (km)
100	1000	0.111
90	900	0.988
85	850	1.457
70	700	3.012
60	600	4.206
50	500	5.574
40	400	7.185
30	300	9.164
20	200	11.784
10	100	16.180
5	50	20.576
3	30	23.849
1	10	31.055

[1] After *U.S. Standard Atmosphere,* 1976.

Table E.2 *Standard Atmosphere Temperature, Pressure, and Density as a Function of Geopotential Height*

Z (km)	Temperature (K)	Pressure (kPa)	Density (kg m^{-3})
0	288.15	101.325	1.225
1	281.65	89.874	1.112
2	275.15	79.495	1.007
3	268.65	70.108	0.909
4	262.15	61.640	0.819
5	255.65	54.019	0.736
6	249.15	47.181	0.660
7	242.65	41.060	0.590
8	236.15	35.599	0.525
9	229.65	30.742	0.466
10	223.15	26.436	0.412
12	216.65	19.330	0.311
14	216.65	14.101	0.227
16	216.65	10.287	0.165
18	216.65	7.505	0.121
20	216.65	5.475	0.088
24	220.65	2.930	0.046
28	224.65	1.586	0.025
32	228.65	0.868	0.013

Answers to Selected Problems

Chapter 1

1. $\alpha \approx (\Omega^2 a \sin 2\phi)/(2g)$ **2.** 36,000 km   **3.** 0.35 N kg^{-1}
4. 556 m   **5.** 1.46 cm   **6.** 6.22 m s^{-1}
7. Upward force $= 2 \times 10^3$ N, greater for westward travel   **8.** 7.8 m eastward
10. 1 rad s^{-1}, 1.22 rad s^{-1}, 0.17 J   **12.** 5.536 km, 5.070 km   **13.** 3°C
14. 7.987 km   **17.** 5.187 km

Chapter 2

1. -2 mb/3 h   **2.** -2°C/h   **5.** 130 km deflection   **6.** 1.02 cm s^{-1}
9. 135 J kg^{-1}, 16.4 m s^{-1}

Chapter 3

1. -1 m km^{-1} (-0.23 m s^{-1})   **2.** -10^{-3} m s^{-2} **3.** 94 kPa   **5.** $V_{grad}/V_g = 2$
7. North, $R_t = 250$ km; west, $R_t = 500$ km; south, $R_t \rightarrow \infty$; east $R_t = 500$ km

8. North, 10.5 m s^{-1}; west and east, 12.1 m s^{-1}; south, 15 m s^{-1}

10. 25 m s^{-1}, 34°   **11.** -1.5°C h^{-1} **12.** 0.5°C/h in the 90–70-kPa layer

13. 238-km to the left

15. North, 9.4 m s^{-1}; west and east 11.5 m s^{-1}; south, 15 m s^{-1} **17.** 7.66 K

18. -852-km   **20.** 2×10^{-5} s^{-1} **21.** 110%   **23.** 0.96 cm s^{-1}

Chapter 4

1. -2×10^7 m^2 s^{-1}, -2×10^{-5} s^{-1} **2.** -5.5 m s^{-1} (anticyclonic)   **3.** -2.3×10^{-5} s^{-1}

4. 9.4×10^{-5} s^{-1}, -8.4×10^{-6} s^{-1} **5.** 0 (annulus), 5×10^{-5} s^{-1} (inner cylinder)

7. -7.2 m^2 s^{-2} **9.** $h(r) = H + \Omega^2 r^2/(2g)$ **12.** 10^{-5} s^{-1}, -1.29×10^{-5} s^{-1}, -1546 km

14. 0.71° equatorward, 138.6 km upward

Chapter 5

4. 293 s, 1.7 cm   **5.** 3.21×10^3 kg m^{-1} s^{-1} (mixed layer)   **6.** $De = 14.05$ m

8. $w_{max} = 9.6 \times 10^{-3}$ mm s^{-1}, $v_{max} = 0.75$ mm s^{-1} **9.** 155 days

10. $\kappa_s = 0.016$ m^{-1} s

Chapter 6

3. 2.84×10^{-6} s^{-1} **4.** $\omega(p) = -(k^2 c^2 p_0/f_0 \pi) \sin(\pi p/p_0) \cos k(x - ct)$

5. $k^2 = \sigma[(f_0 \pi)/(\sigma \rho_0)]^2$ **8.** $W_0 = -cf_0 U_0 \pi/(\sigma p_0)$

9. $\omega(p) = (V p_0/f_0 \pi)[k^2(c - U) + \beta] \sin(\pi p/p_0) \cos k(x - ct)$

10. $c = U - \beta[k^2 + (f_0^2 \pi^2)/(\sigma p_0^2)]^{-1}$ **11.** 0.047 Pa s^{-1}

Chapter 7

2. $B_r = A \cos kx_0$, $B_i = -A \sin kx_0$, $c_r = \nu/k$, $c_i = \alpha/k$

4. $u' = p'(c - \bar{u})/(\gamma \bar{p})$, $\rho' = p'/(c - \bar{u})^2$ **7.** $u' = h'(c - \bar{u})/H$

8. $u' = -mw'/k$, $p'/\rho_0 = -\nu mw'/k^2$, $\theta'/\bar{\theta} = -iw'(N^2/\nu g)$

9. $\overline{\rho_0 u' w'} = -\rho_0 m|A|^2/(2k)$

12. $u' = (0.66 \ m \ s^{-1}) \sin(kx + mz)$, $w' = -(0.75 \ m \ s^{-1}) \sin(kx + mz)$

15. -24.3 m s^{-1} **16.** -0.633 cm s^{-1}

Chapter 8

1. 4.3×10^4 s   **2.** $\psi_3' = -\psi_1'$, $\omega_2' = \mathrm{Re}\{2i\lambda^2 k\beta\ \delta p\ \psi_1'/[f_0(k^2 + 2\lambda^2)]\}$
3. $\psi_3' = \psi_1'(\sqrt{3} - 1)/(\sqrt{3} + 1)$, $\omega_2' = \mathrm{Re}[ik\beta\ \delta p\ \psi_1'(1 - \sqrt{3})/(3f_0)]$
4. $\psi_3' = \mathrm{Re}[\psi_1'(1 - i\sqrt{3})/2]$, $\omega_2' = \mathrm{Re}[-k^3\ \delta p\ U_T\psi_1'(1 + i\sqrt{3})/(\sqrt{3}f_0)]$
6. ψ_1' lags ψ_3' by $65.5°$ **7.** $B = (4.1 \times 10^6 - 9.1 \times 10^6 i)$ m^2 s^{-1}

Chapter 9

3. 344 K   **4.** 1.42 m s^{-1}

Chapter 10

8. 6.25×10^5 kg s^{-2} **11.** 0.49 cm s^{-1}

Chapter 11

1. 7.59×10^{-8} m s^{-1} **2,** $2u_0 l^2 > \beta$

Chapter 12

1. 6697 m   **2.** 87.6 m s^{-1}, 48.7 m s^{-1}, 28.0 m s^{-1}
3. $|\Phi'| = 200$ m^2 s^{-2}, $|w'| = 1.57$ mm s^{-1}
5. $\hat{w} = \hat{v}(-imy)(\nu^2/N^2)\exp[-\beta|m|y^2(2N^2)]$ **6.** 6.14°

Chapter 13

5. 1.3%, 5%, 19%, 59%   **6.** 1.4% per time step

Bibliography

Andrews, D. G., J. R. Holton, and C. B. Leovy. 1987. *Middle Atmosphere Dynamics*. Academic Press, Orlando, Florida.

Andrews, D. G., and M. E. McIntyre. 1976. Planetary waves in horizontal and vertical shear: The generalized Eliassen-Palm relation and the mean zonal acceleration, *J. Atmos. Sci.*, **33**, 2031-2048.

Arakawa, A., and W. Schubert. 1974. Interaction of cumulus cloud ensemble with the large-scale environment. *J. Atmos. Sci.*, **31**, 674-701.

Arya, S. P. 1988. *Introduction to Micrometeorology*, Academic Press, Orlando, Florida.

Baer, F. 1972. An alternate scale representation of atmospheric energy spectra. *J. Atmos. Sci.*, **29**, 649-664.

Battisti, D. S., A. C. Hirst, E. S. Sarachik, and P. Tian. 1992. A consistent model for the steady surface circulation in the tropics. *J. Atmos. Sci.*, **49** (submitted).

Bengtsson, L. 1985. Medium-range forecasting at the ECMWF. In S. Manabe (ed.), *Issues in Atmospheric and Oceanic Modeling Part B: Weather Dynamics*, Academic Press, New York, 3-54.

Blackburn, M. 1985. Interpretation of ageostrophic winds and implications for jetstream maintenance, *J. Atmos. Sci.*, **42**, 2604-2620.

Blackmon, M. L., J. M. Wallace, N.-C. Lau, and S. L. Mullen. 1977. An observational study of the northern hemisphere wintertime circulation, *J. Atmos. Sci.*, **34**, 1040-1053.

Bourne, D. E., and P. C. Kendall, 1968. *Vector Analysis*. Allyn & Bacon, Boston.

Bowman, K. P., and A. J. Krueger. 1985. A global climatology of total ozone from the Nimbus 7 total ozone mapping spectrometer, *J. Geophys. Res.*, **90**, 7967-7976.

Brasseur, G., and S. Solomon. 1986. *Aeronomy of the Middle Atmosphere: Chemistry and Physics of the Stratosphere and Mesophere*, 2nd ed. Reidel, Dordrecht, Netherlands.

Brown, R. A. 1970. A secondary flow model for the planetary boundary layer. *J. Atmos. Sci.*, **27**, 742–757.

Brown, R. A. 1991. *Fluid Mechanics of the Atmosphere*, Academic Press, San Diego.

Burridge, D. M., and E. Källén. 1984. *Problems and Prospects in Long and Medium Range Weather Forecasting.* Topics in Atmospheric and Oceanographic Sciences, Springer-Verlag, New York.

Chang, C. P. 1970. Westward propagating cloud patterns in the tropical Pacific as seen from time composite satellite photographs, *J. Atmos. Sci.*, **27**, 133–138.

Chapman, S., and R. S. Lindzen. 1970. *Atmospheric Tides. Thermal and Gravitational*, Reidel, Dordrecht, Netherlands.

Charney, J. G. 1947. The dynamics of long waves in a baroclinic westerly current, *J. Meteorol.*, **4**, 135–163.

Charney, J. G. 1948. On the scale of atmospheric motions. *Geofys. Publ.*, **17**(2), 1–17.

Charney, J. G., and J. G. DeVore. 1979. Multiple flow equilibria in the atmosphere and blocking. *J. Atmos. Sci.*, **36**, 1205–1216.

Durran, D. R. 1990. Mountain waves and downslope winds. In W. Blumen (ed.), *Atmospheric Processes Over Complex Terrain*, American Meteorological Society, 59–82.

Durran, D. R., and L. W. Snellman. 1987. The diagnosis of synoptic-scale vertical motion in an operational environment. *Wea. Forecsting*, **1**, 17–31.

Eady, E. T. 1949. Long waves and cyclone waves. *Tellus*, **1**, 33–52.

Emanuel, K. A. 1988. Toward a general theory of hurricanes. *American Scientist*, **76**, 370–379.

Fels, S. B. 1985. Radiative–dynamical interactions in the middle atmosphere. *Adv. Geophy.*, **28A**, 277–300.

Fels, S. B., J. D. Mahlman, M. D. Schwarzkopf, and R. W. Sinclair. 1980. Stratospheric sensitivity to perturbations in ozone and carbon dioxide: radiative and dynamical response. *J. Atmos. Sci.*, **37**, 2265–2297.

Fleming, E. L., S. Chandra, J. J. Barnett, and M. Corey. 1990. Zonal mean temperature, pressure, zonal wind and geopotential height as functions of latitude. *Adv. Space Res.* **10**, No. 12, 11–59.

Gill, A. E. 1982. *Atmosphere–Ocean Dynamics*, Academic Press, New York.

Haltiner, G. J., and R. T. Williams. 1980. *Numerical Prediction and Dynamical Meteorology*, Wiley, New York.

Held, I. M. 1983. Stationary and quasi-stationary eddies in the extratropical troposphere: theory. In B. J. Hoskins and R. Pearce (eds.), *Large-Scale Dynamical Processes in the Atmosphere*, Academic Press, New York, 127–168.

Hess, S. L. 1959. *Introduction to Theoretical Meteorology.* Holt, New York.

Hide, R. 1966. Review article on the dynamics of rotating fluids and related topics in geophysical fluid mechanics. *Bull. Amer. Meteorol. Soc.*, **47**, 873–885.

Hildebrand, F. B. 1976. *Advanced Calculus for Applications*, 2nd ed. Prentice Hall, Englewood Cliffs, New Jersey.

Holton, J. R. 1986. Meridional distribution of stratospheric trace constituents. *J. Atmos. Sci.*, **43**, 1238–1242.

Horel, J. D., and J. M. Wallace. 1981. Planetary scale atmospheric phenomena associated with the interannual variability or sea-surface temperature in the equatorial Pacific. *Mon. Wea. Rev.*, **109**, 813–829.

Hoskins, B. J. 1975. The geostrophic momentum approximation and the semigeostrophic equations. *J. Atmos. Sci.* **32**, 233–242.

Hoskins, B. J. 1982. The mathematical theory of frontogenesis. *Annu. Rev. Fluid Mech.*, **14**, 131-151.

Hoskins, B. J. 1983. Dynamical processes in the atmosphere and the use of models. *Quart. J. Roy. Meteorol. Soc*, **109**, 1-21.

Hoskins, B. J., and F. P. Bretherton. 1972. Atmospheric frontogenesis models: mathematical formulation and solution. *J. Atmos. Sci.*, **29**, 11-37.

Hoskins, B. J., M. E. McIntyre, and A. W. Robertson. 1985. On the use and significance of isentropic potential vorticity maps. *Quart. J. Roy. Meteorol. Soc*, **111**, 877-946.

Houze, R. A., Jr., and P. V. Hobbs. 1982. Organization and structure of precipitating cloud systems. In B. Saltzman (ed.), *Advances in Geophysics*. Academic Press, New York, 225-315.

Iribarne, J. V., and W. L. Godson. 1981. *Atmospheric Thermodynamics* 2nd ed. Reidel, Dordrecht, Netherlands.

Klemp, J. B., 1987: Dynamics of tornadic thunderstorms. *Annu. Rev. Fluid Mech.*, **19**, 369-402.

Lim, G. H., J. R. Holton, and J. M. Wallace. 1991. The structure of the ageostrophic wind field in baroclinic waves, *J. Atmos. Sci.*, **48**, 1733-1745.

Lindzen, R. S., E. S. Batten, and J. W. Kim. 1968. Oscillations in atmospheres with tops. *Mon. Wea. Rev.*, **96**, 133-140.

Lorenz, E. N. 1960. Energy and numerical weather prediction. *Tellus*, **12**, 364-373.

Lorenz, E. N. 1967. *The Nature and Theory of the General Circulation of the Atmosphere*. World Meterological Organization, Geneva.

Lorenz, E. N. 1984. Some aspects of atmospheric predictability. In D. M. Burridge and E. Källén (eds.), *Problems and Prospects in Long and Medium Range Weather Forecasting*, Springer-Verlag, New York, 1-20.

Manabe, S (ed.). 1985. *Issues in Atmospheric and Oceanic Modeling Part B: Weather Dynamics*. Academic Press, New York.

Matsuno, T. 1966. Quasi-geostrophic motions in the equatorial area. *J. Meteorol. Soc. Japan*, **44**, 25-42.

Naujokat, B. 1986. An update of the observed quasi-biennial oscillation of the stratospheric winds over the tropics. *J. Atmos. Sci.*, **43**, 1873-1877.

Oort, A. H. 1983. Global atmospheric circulation statistics, 1958-1973. *NOAA Professional Paper* 14, U.S. Government Printing Office, Washington, D.C.

Oort, A. H., and J. P. Peixoto. 1974. The annual cycle of the energetics of the atmosphere on a planetary scale. *J. Geophys. Res.*, **79**, 2705-2719.

Ooyama, K. 1969. Numerical simulation of the life cycle of tropical cyclones. *J. Atmos. Sci.*, **26**, 3-40.

Palmén, E., and C. W. Newton. 1969. *Atmospheric Circulation Systems*. Academic Press, New York.

Panofsky, H. A., and J. Dutton. 1984. *Atmospheric Turbulence*. Wiley, New York.

Pedlosky, J. 1987. *Geophysical Fluid Dynamics*, 2nd ed. Springer-Verlag, New York.

Philander, G. S. 1990. *El Niño, La Niña, and the Southern Oscillation*. Academic Press, San Diego.

Phillips, N. A. 1956. The general circulation of the atmosphere: A numerical experiment. *Quart. J. Roy. Meterol. Soc*, **82**, 123-164.

Phillips, N. A., 1963. Geostrophic motion. *Rev. Geophys.* 1, 123-176.

Plumb, R. A. 1982. The circulation of the middle atmosphere. *Aust. Meteorol. Mag.*, **30**, 107-121.

Reed, R. J., J. C. Norquist, and E. E. Recker. 1977. The structure and properties of African wave disturbances as observed during phase III of GATE. *Mon. Wea. Rev.*, **105**, 317-333.

Reed, R. J., J. L. Wolfe, and N. Nishimoto. 1963. A spectral analysis of the energetics of the stratospheric sudden warming of early 1957. *J. Atmos. Sci.*, **20**, 256–275.

Richardson, L. F. 1922. *Weather Prediction by Numerical Process.* Cambridge Univ. Press, London (reprt. Dover, New York).

Richtmeyer, R. D., and K. W. Morton. 1967. *Difference Methods for Initial Value Problems,* 2nd ed. Wiley (Interscience), New York.

Riehl, H., and Malkus, J. S. 1958. On the heat balance of the equatorial trough zone. *Geophysica* **6**, 503–538.

Salby, M. L., H. H. Hendon, K. Woodberry, and K. Tanaka. 1991. Analysis of global cloud imagery from multiple satellites. *Bull. Am. Meteorol. Soc.*, **72**, 467–480.

Sanders, F., and B. J. Hoskins. 1990. An easy method for estimation of Q-vectors from weather maps. *Wea. Forecasting*, **5**, 346–353.

Sawyer, J. S. 1956. The vertical circulation at meteorological fronts and its relation to frontogenesis. *Proc. Roy. Soc. A*, **234**, 346–362.

Schubert, S., C.-K. Park, W. Higgins, S. Moorthi, and M. Suarez. 1990. An *Atlas of ECMWF Analyses (1980–1987) Part I—First Moment Quantities.* NASA Technical Memorandum 100747.

Scorer, R. S. 1958. *Natural Aerodynamics.* Pergamon, New York.

Simmons, A. J. 1986. Numerical prediction: some results from operational forecasting at ECMWF. In B. Saltzman (ed.), *Advances in Geophysics.* Academic Press, Orlando, Florida, 305–338.

Simmons, A. J., and L. Bengtsson. 1984. Atmospheric general circulation models: their design and use for climate studies. In J. T. Houghton (ed.), *The Global Climate.* Cambridge Univ. Press, London, 37–62.

Simmons, A. J., D. M. Burridge, M. Jarraud, C. Girard, and W. Wergen. 1989. The ECMWF medium-range prediction models: development of the numerical formulations and the impact of increases resolution. *Meteorol. Atmos. Phys.*, **40**, 28–60.

Smagorinsky, J. 1967. The role of numerical modeling. *Bull. Amer. Meteorol. Soc.*, **46**, 89–93.

Smith, R. B. 1979. The influence of mountains on the atmosphere. In B. Saltzman (ed.), *Advances in Geophysics,* **21**, Academic Press, New York, 87–230.

Sorbjan, Z. 1989. *Structure of the Atmospheric Boundary Layer.* Prentice Hall, Englewood Cliffs, New Jersey.

Stevens, D. E., and R. S. Lindzen. 1978. Tropical wave-CISK with a moisture budget and cumulus friction. *J. Atmos. Sci.* **35**, 940–961.

Stull, R. B. 1988. *An Introduction to Boundary Layer Meteorology.* Kluwer Academic Publishers, Boston.

Trenberth, K. E. 1984. Signal versus noise in the Southern Oscillation. *Mon. Wea. Rev.* **112**, 326–332.

Trenberth, K. E. 1991. General characteristics of El Niño–Southern Oscillation. In M. Glantz, R. Katz, and N. Nichols (eds.), *ENSO Teleconnections Linking Worldwide Climate Anomalies: Scientific Basis and Societal Impact.* Cambridge Univ. Press, London, 13–42.

Turner, J. S. 1973. *Buoyancy Effects in Fluids.* Cambridge Univ. Press, London.

U.S. Government Printing Office. 1976. *U.S. Standard Atmosphere,* 1976. Government Printing Office, Washington, DC.

Wallace, J. M. 1971. Spectral studies of tropospheric wave disturbances in the tropical western Pacific. *Rev. Geophys.*, **9**, 557–612.

Wallace, J. M., and P. V. Hobbs. 1977. *Atmospheric Science: An Introductory Survey.* Academic Press, New York.

Wallace, J. M., and V. E. Kousky. 1968. Observational evidence of Kelvin waves in the tropical stratosphere. *J. Atmos. Sci.*, **25**, 900-907.

Warsh, K. L., K. L. Echternacht, and M. Garstang. 1971. Structure of near-surface currents east of Barbados. *J. Phys. Oceanog.*, **1**, 123-129.

Washington, W. M., and C. L. Parkinson. 1986. *An Introduction to Three-Dimensional Climate Modeling.* University Science Books, Mill Valley, California.

Webster, P. J. 1983. The large-scale structure of the tropical atmosphere. In B. J. Hoskins and R. Pearce (eds.), *Large-Scale Dynamical Processes in the Atmosphere.* Academic Press, New York, 235-276.

Webster, P. J., and H.-R. Chang. 1988. Equatorial energy accumulation and emanation regions: impacts of a zonally varying basic state. *J. Atmos. Sci.*, **45**, 803-829.

Williams, J., and S. A. Elder. 1989. *Fluid Physics for Oceanographers and Physicists.* Pergamon, New York.

Williams, K. T. 1971. A statistical analysis of satellite-observed trade wind cloud clusters in the western North Pacific. *Atmospheric Science Paper No. 161*, Dept. of Atmospheric Science, Colorado State Univ., Fort Collins, Colorado.

Index

International Geophysics Series

EDITED BY

RENATA DMOWSKA

Division of Applied Science
Harvard University

JAMES R. HOLTON

Department of Atmospheric Sciences
University of Washington
Seattle, Washington

*Out of print.